Agricultural and Resource Policy

Principles and Practice

David Godden

Melbourne
OXFORD UNIVERSITY PRESS
Oxford Auckland New york

OXFORD UNIVERSITY PRESS AUSTRALIA

Oxford New York
Athens Auckland Bangkok Bombay
Calcutta Cape Town Dar es Salaam Delhi
Florence Hong Kong Istanbul Karachi
Kuala Lumpur Madras Madrid Melbourne
Mexico City Nairobi Paris Port Moresby
Singapore Taipei Tokyo Toronto

and associated companies in
Berlin Ibadan

OXFORD is a trade mark of Oxford University Press

First published 1997

National Library of Australia
Cataloguing-in-Publication data:

Godden, D. P. (David P.).
Agricultural and resource policy: principles and practice.

Bibliography.
Includes index.
ISBN 0 19 554023 9.

1. Agriculture and state — Australia. 2. Natural resources — Government policy — Australia.
3. Agriculture — Economic aspects — Australia — 1945– .
I. Title.

338.1894

Edited by Jenny Missen
Cover design by Kirstin Lowe
Line drawings by David Godden
Typeset by Syarikat Seng Teik Sdn. Bhd., Malaysia
Printed through Bookpac Production Services, Singapore
Published by Oxford University Press,
253 Normanby Road, South Melbourne, Australia

Preface

Policy analysis is an eclectic discipline. Many of the ideas in this book have been acquired from a variety of mentors over several decades. I have tried wherever possible to give credit to the source of these ideas but it is quite possible, indeed even probable, that I have failed to correctly acknowledge all those to whom I am indebted. To those whose credit I have inadvertently not acknowledged, my apologies—but the real acknowledgment is that I have used your ideas even if I have forgotten where I acquired them. I am particularly grateful to Bruce Standen and Jim Johnston, colleagues and friends, whose continual insistence on intellectual rigour has often prevented me from making lamentable errors. Geoff Edwards, John Kerin, and Deane Crabb also kindly commented on a draft manuscript and rescued me from some follies. I would like to make them responsible for any remaining errors but, being economists, they would probably demand both authorship rights and the royalties.

The first draft of this book was written while I was on sabbatical at the School of Economic and Financial Studies at Macquarie University. I am grateful to David Collins, Head of Discipline, for agreeing to my sabbatical undertaken largely *in absentia* while I wrote in quiet corners, and for providing the necessary office facilities and library access which were so essential to the preparation of this book. And, of course, I am grateful to my colleagues at the Department of Agricultural Economics, University of Sydney, who undertook my teaching during my absence.

This book is dedicated to Judith, Sarah and Nick.

Acknowledgments

The following material has been reproduced with the kind permission of the respective publishers:

K. Anderson 1992, chapter 13 in I. Goldin & L. A. Winters (eds) *Open Economies: structural adjustment and agriculture*, Cambridge University Press: figure 2.10 (his figure 13.1).

R. Bewley & G. White 1990, 'Do high interest rates improve or worsen the current account?', *Economic Papers* vol. 9(4), December: figures 12.4, 12.7, 13.3 (adapted from their unnumbered figure, p. 25).

J. S. Davis, P. A. Oram & J. S. Ryan 1987, *Assessment of Agricultural Research Priorities: an international perspective*, Australian Centre for International Agricultural Research, Canberra: figure 5.6 (their figure 2.1).

A. de Janvry 1978, chapter 11 in H. P. Binswanger, V.W. Ruttan et al., *Induced Innovation: technology, institutions, and development*, Johns Hopkins University Press, Baltimore: figure 5.2 (his figure 11.1).

P. J. Higgs 1986, *Adaptation and Survival in Australian Agriculture*, Oxford University Press, Melbourne: table 12.1 (material drawn from his tables 1.1 and 1.2); table 12.2 (data from his tables 7.1, 7.2, 7.5 and 7.9); table 13.5 (data from his tables 6.1, 6.3, 6.7, 6.10); table 14.2 (data from his tables 8.1 and 8.3).

Human Rights and Equal Opportunity Commission 1988, *Toomelah Report*, Sydney: Boxes 10.1 and 10.2 (quotations from paragraphs 4.1, 4.3, 4.4, 4.5, 4.7, 4.11–12, 4.15, 6.13–14, 6.15, 6.16, 6.17, 6.19).

T. Roberts & R. Pinner 1990, chapter 22 in A. J. Miller, J. L. Smith & G. A. Somkuti (eds), *Foodborne Listeriosis*, Elsevier Science, Amsterdam: figure 8.4 (their figure 1).

C. A. Tisdell 1982, *Microeconomics of Markets*, John Wiley & Sons, Brisbane: figure 2.5 (adaptation of his figure 16.5); figure 2.8 (adaptation of his figure 16.3).

S. J. Turnovsky 1978, chapter 6 in F. G. Adams and S. A. Klein (eds), *Stabilizing World Commodity Markets: analysis, practice, policy*, Jossey-Bass Inc. First published by Lexington Books, Lexington: Box 6.1 (his equations 6.1 through 6.11).

P. G. Warr 1978, 'The case against tariff compensation', *Australian Journal of Agricultural Economics* vol. 22(2): figures 13.5–13.7 (adaptations of his figure 1).

Contents

Tables

Figures

Boxes

Glossary of Acronyms

AAC	Australian Agricultural Council (see also ACANZ, ARMCANZ)
ABARE	Australian Bureau of Agricultural and Resource Economics
ABC	Australian Broadcasting Corporation (previously Commission)
ACANZ	Agricultural Council of Australia and New Zealand (see also AAC, ARMCANZ)
ACTU	Australian Council of Trade Unions
ALP	Australian Labor Party
AMA	Australian Medical Association
APEC	Asia-Pacific Economic Cooperation (forum)
ARMCANZ	Agricultural and Resource Management Council of Australia and New Zealand (see also AAC, ACANZ)
CGE	computable general equilibrium
COAG	Council of Australian Governments
CPI	Consumer Price Index
CSIRO	Commonwealth Scientific and Industrial Research Organisation
DPIE	Department of Primary Industries and Energy (Commonwealth)
EC	European Communities (or Community, see also EEC, EU)
EEC	European Economic Community (see also EC, EU)
EU	European Union (see also EC, EEC)
GATT	General Agreement on Tariffs and Trade
GDP	Gross Domestic Product
GST	Goods and Services Tax (see also VAT)
GWh	gigawatt hour
IAC	Industries Assistance Commission (see also IC)
IC	Industry Commission (see also IAC)
ICAC	Independent Commission Against Corruption (NSW)
MMP	Mixed Member Proportional representation
MRS	marginal rate of substitution
MW	megawatt

MWh	megawatt hours
NFA	National Food Authority
NFF	National Farmers' Federation
NPV	net present value
MEC	marginal external cost
MNP	marginal net profit
MSC	marginal social cost
PAC	political action committee (USA)
R&D	research and development
RAS	Rural Adjustment Scheme (Commonwealth and states)
RDC	research and development corporations (see also RIRF)
RIRF	rural industry research fund (see also RDC)
SCA	Standing Committee on Agriculture (see also SCARM)
SCARM	Standing Committee on Agriculture and Resource Management (see also SCA)
SMA	statutory marketing authority
SWF	social welfare function
VAT	Value Added Tax

Introduction

What is 'agricultural and resource policy'? This question can be answered in two ways. First, it may be seen as applying economic ways of thinking to socio-economic issues in the agricultural and resource sector where 'the government' is somehow involved. Government may be involved in various ways: as the solution to a problem, perhaps as the cause of a problem, or even as an innocent bystander. But, somehow, government is involved by doing—or not doing—something. Analysing policy requires a systematic, theoretical framework for analysing specific problems. This book is concerned with developing just such a framework.

But agricultural and resource policy is more than just theory—it is the application of economics to real world problems. Second, therefore, policy analysis involves asking questions about existing or possible government activities, and using a systematic, theoretical framework to develop answers to those questions. The key to good policy analysis is asking the *right questions* about existing or proposed policies. The kinds of questions that should be asked include:

- *Why* do governments make decisions?
- *How* do governments make decisions?
- What are the *economic consequences* of the decisions that governments make?

The objective of studying agricultural and resource policy is to understand the *principles* underlying the policy-making process, how economists understand and evaluate the policy-making process and, by using examples, to gain practical experience in applying these principles. The approach to analysing agricultural and resource policy in this book does not involve, therefore, an exhaustive survey of Australian policy. Rather, the objective is to develop a framework for analysing natural resource policy issues, and agricultural policy issues in particular.

Natural resources and agriculture

Agricultural policy analysis has often been confined to a rather narrow view of agriculture. This limited view arises from identifying agriculture with farming. A more useful view is that it comprises farms and all their linkages with the rest of the economy, especially the linkages between farm production and natural resources. Agriculture is simply one use of a set of natural resources comprising land, water and air. It competes with other uses for the land base—urban and semi-urban (e.g. hobby-farming), infrastructure (roads, railways), tourism and recreation, mining, water catchment, forestry and wilderness. Agriculture competes with domestic, commercial and industrial sectors for water and environmental flow. Agriculture also affects other users of natural resources. Economists are especially interested in non-marketed spillovers ('externalities') which may occur between agriculture and other industries, or among agricultural firms. Such externalities include natural-resource degradation, such as dryland and irrigation salinity, soil and water contamination (especially from animal wastes of the intensive livestock industries), blue-green algae (together with urban runoff), pests and diseases (of both indigenous and exotic origin), and natural disasters such as bushfires and floods.

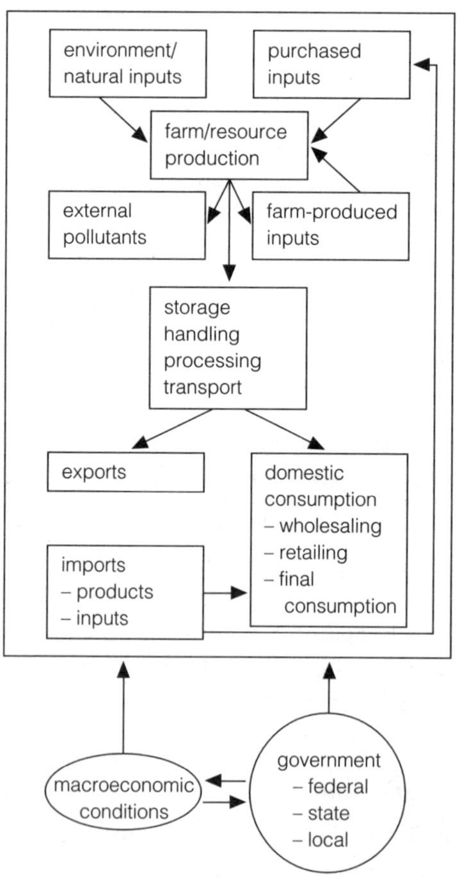

Figure 1 Agricultural and Resources Sector

The scope of agricultural and resource policy addressed in this book is illustrated in figure 1. Agriculture uses both natural resources (land, water and air) and produced inputs (machinery; agricultural chemicals, fertilisers and other farm supplies; services such as machinery repair, accounting and legal, farm technical advice from the private and public sectors; and specialised labour and contracting services such as shearing and pest control) to produce farm output. Some output is consumed within the agricultural sector, such as on-farm conserved feed; grain for seed, and other planting material; and feedstuffs for the intensive livestock industries such as pigs, poultry, dairy, beef feedlots, horses. Some output is non-marketed in the form of externalities. Most farm output is consumed outside the agricultural sector: depending on the industry, most may be consumed domestically (e.g. fruit and vegetables, ornamentals, pigmeat, oilseeds, sheepmeats), predominantly exported (wool, cotton, sugar, wheat, feedlot beef) or shared between the domestic and export markets (extensive beef, rice, barley).

For products primarily consumed domestically, marketing channels have a major impact on the profitability of agricultural production. Marketing channels include all processes from the assembly of agricultural products at farm-gate, through transportation and storage, to processing and wholesale and retail distribution. Marketing channels are especially affected by government intervention, which affects the institutional structure of the marketing chain; such intervention has been a traditional focus of agricultural policy analysis. However, the efficiency of these marketing channels, and their economic impacts on agriculture, are also affected by economic conditions external to agriculture, including general economic conditions and government policy.

For products primarily exported, marketing channels are also crucial to the profitability of agricultural production. These marketing channels too are partly affected by domestic economic conditions. Macroeconomic conditions and national macroeconomic policy, especially relating to exchange rates, have a major impact on export-oriented agriculture. Until the relatively recent past, Australian agriculture has had a major effect on general economic conditions and policy because of agriculture's dominance in Australia's exports. The decline of Australian agriculture's macroeconomic importance, and the implications of this decline for economic and agricultural sector policy, are important issues.

Also important are the effects of international economic conditions and agricultural policies in countries that import Australian agricultural products, or that produce competitive products. These international aspects deserve a special treatment because of the importance of trade theory to their analysis; they are not covered in this book.

Encountering agricultural and resource policy

The reader is likely to have at least three types of encounter with policies for agricultural and resource policy.

An encounter of the first kind is as an *economics practitioner*. Economists are constantly involved in evaluating current economic policies. Those relating to macroeconomic decision-making—setting interest and tax rates and the level of government spending, and the balance between monetary and fiscal policy—are currently the most visible. But of most interest in this book are microeconomic policies. In the agricultural sector, post-Second World War policy was traditionally oriented to the construction of orderly marketing schemes for agricultural products; and economists were extensively involved in evaluating existing policies and proposals for their extension. From the early 1970s these marketing schemes slowly began to fall into disfavour, although it took until the late 1980s, with the deregulation of domestic wheat marketing and the collapse of the wool reserve price scheme, for substantial inroads to be made into marketing regulation. But marketing schemes were not the limit of government intervention in marketing—nor was marketing the only focus of government intervention.

An encounter with agricultural and resource policy of the second kind is as a *participant* in the policy process. Many of the readers of this book will be involved as participants in the governmental process—as politicians; as public service analysts and advisers; as lobbyists, or advisers or public relations 'flaks' for lobbyists; as 'hired-gun' analysts (a major growth industry since the deregulatory fervour of the 1980s); or as economic analysts in the print or electronic media. In these roles, an economist may not necessarily use—or be permitted to use—the full range of economic skills that are available, or even use the selection of economic tools that are presented in subsequent chapters. But, it is hoped, they will be aware of any anorexic version of economics that they use.

An encounter with the policy process of the third—and ultimately the most important—kind is as a *citizen*. If the price of liberty is eternal vigilance, democracy has a similar price. Australian citizens are required to attend polling booths (although *voting* is not compulsory) every three or four years for both federal and state elections (and more frequently for federal elections if House of Representatives and Senate elections are not synchronised). Ratepayers and residents may also vote in local government elections. Referenda for a variety of purposes, including amendments to the Commonwealth Constitution, are held irregularly. If a better-informed citizenry results in a better-performing government, a better understanding of the policy process and a more vigorous scrutiny of its participants will advantage society at large, and this book will have achieved an important objective.

Historical context

The contemporary setting of Australian agriculture is a function of its evolution. Australian agriculture now makes about the same contribution to Gross Domestic Product (GDP) as the Australian mining industry (table 1). Agriculture contributes about as much to the aggregate economy as do the electricity, gas and water industries combined; or the entire transport industry; or the entertainment, hotel and club industries combined (table 1). Agriculture makes about one-quarter of the contribution of manufacturing to GDP; about one-third the contribution of services from dwellings; and about one-half the contribution to GDP of either wholesale or retail trade (table 1). Agriculture's contribution to the Australian economy has, however, changed markedly over the past five decades, and the current institutional framework and policies for agriculture and resources reflect this change.

Table 1 Agriculture Compared to the Australian Economy

	1986–87		*1993–94*	
	GDP ($m)	% of total GDP	GDP ($m)	% of total GDP
Agriculture	13211	4.1	15121	3.8
Forestry, fishing, hunting	1182	0.4	1299	0.3
Mining (excluding services)	12114	3.8	16703	4.2
Services to mining	504	0.2	513	0.1
Manufacturing	52302	16.4	62535	15.9
Electricity	6689	2.1	8600	2.2
Gas	818	0.3	1029	0.3
Water	2817	0.9	3195	0.8
Construction	25151	7.9	26862	6.8
Wholesale trade	24770	7.7	29168	7.4
Retail trade	30540	9.6	37886	9.6
Rail transport	1545	0.5	1880	0.5
Water transport	1568	0.5	1896	0.5
Air transport	2833	0.9	4719	1.2
Road and other transport, and storage	9341	2.9	11681	3.0
Communications	6261	2.0	11525	2.9
Finance, property and business services	38516	12.0	44491	11.3
Public administration and defence	11949	3.7	14681	3.7
Community services	36095	11.3	47431	12.0
Entertainment and recreational services	4811	1.5	6159	1.6
Restaurants, hotels, clubs and personal services	9487	3.0	11253	2.9
Ownership of dwellings	31430	9.8	39041	9.9
Import duties	2783	0.9	4376	1.1
less imputed bank service charges	–7081		–7858	
Total	319636		394186	

Source: Australian Bureau of Statistics (1994a, table 63)

An occasional reaction to the material covered in this book may be that there is too much 'old' material in it. Depending on the reader's age, this might mean that some of the readings, or policies discussed, have dates before 1985—or even 1990. There are several explanations for the 'antiquity' of some of the material included. First, while agents in the policy process are often remarkably inventive, their inventiveness frequently takes the form of recycling old tricks. Policy analysts who do not have a good knowledge of past analysis are condemned to repeat it. Second, some 'old' analysis is remarkably robust and is applicable to contemporary conditions as long as appropriate adjustments are undertaken for the empirical context. Third, what might currently be observed as inappropriate policy might have its origins in policies that were appropriate when first implemented. Policy analysts who neglect evolution in the underlying economic conditions that help determine the appropriateness of policy, or evolution in the policy-generating process itself, are doomed to be constantly surprised by the apparent need to keep adjusting policy. However, if policy is implemented and evaluated in the certain knowledge that it will operate in a world of constant change, there is a better chance of performing relevant policy analyses—and perhaps even ultimately implementing policies that will be relatively robust to the 'real world'.

Agricultural policy since the Second World War can be seen as a process of transition from policies that are sector- and industry-specific to those more general for the economy as a whole. At the beginning of the 1950s, even though agriculture employed a relatively small percentage of the workforce (15.8 per cent in 1947) Australian agriculture played a significant role as the major export earner: 86 per cent of total export produce emanated from the agricultural sector in 1953–54 (Vernon 1965, pp. 515, 1006). With relatively fixed exchange rates, a dominant export sector was crucial for generating the export income to pay for imports (Edwards & Watson 1978, p. 190). Agriculture was central to the development of Australia inasmuch as that development was dependent on imports. Nurturing agriculture, and especially agricultural exports, was necessary for promoting economic development in the face of a balance-of-payments constraint to economic growth. Despite the substantial reorientation of Australia's international relations towards the USA during the war, Australia was still viewed by its own government (and the United Kingdom Government) as an appendage of the United Kingdom's economy. The serious economic difficulties Britain faced as a result of its war financing, and Australia's membership of the 'sterling bloc', meant that Australia's economic policy was also partly seen as an appendage of United Kingdom policy (Box 1). Because of international policy intervention in the form of fixed exchange rates, there existed a (second-best) 'efficiency rationale' for interventions by Australian governments to promote agricultural production and exports. Examples of such interventions included home price schemes (in wheat, dairy, sugar and dried fruits), tax concessions, export promotion, publicly-funded research and development, land development, and the provision of rural infrastructure.

One seminal event, and two accelerating processes, occurred in the 1960s; together these ultimately shattered the pivotal role of agriculture in the Australian economy. The seminal event was the initial attempt by the United Kingdom Government to join the European Economic Community (EEC) in 1961–62. It was clear, even at the time, that although this attempt was abortive Britain remained determined to join the EEC and that once this occurred Australia was unlikely to retain its extensive access to United Kingdom markets for agricultural products (cf. James 1971). Australian farm export industries therefore rapidly developed alternative markets, especially at first in Japan and later throughout East Asia.

Box 1 Agricultural Production Aims

Our task now is to expand agricultural production so as to meet the domestic needs of our growing population and to increase our overseas purchasing power.

The balance of payments problem which confronts Australia underlines the urgent need to step up the rate of expansion and development in our primary industries. These industries have always been the major source of our export income and it is to them that we look for an increase in our overseas earnings.

Grains and sugar are two of the largest items of the United Kingdom dollar expenditure and any increase in the exports of these commodities from Australia to the United Kingdom or to any other country in the sterling area represents a saving of dollars, and thus contributes to the balancing of receipts and expenditures in the dollar area.

The Australian Agricultural Council, in April, 1952, announced a set of production aims for certain agricultural products, the target date being five years hence . . .

Acceptance of these aims by the Australian Agricultural Council means that both the Commonwealth and State Governments have recognised these defined levels of production as bench marks on which to base their policies.

Source: McEwen (1952, pp. 3–5)

The effect of one accelerating process which shaped Australian agriculture in the 1960s was initially muted. American agricultural economists had, from the 1950s, emphasised the relative contraction of the agricultural sector as economic growth proceeded. This analysis was generally conducted, however, in the context of a non-exporting agricultural economy. By contrast, the Australian agricultural sector had a substantial export component. The steady growth in the world economy through the 1950s and 1960s meant that—even in the face of significant trade barriers—there were sufficient outlets for Australian agricultural exports that stagnant or contracting domestic demand had substantially less impact than in other developed economies. A second accelerating process was the post-Second World War Australian mining boom, beginning with the discovery of massive iron ore deposits in Western Australia and subsequently extending to major discoveries and development of other mineral deposits. The relaxation of export controls on iron ore foreshadowed the subsequent enormous development of the Australian minerals industry. This minerals boom provided the conditions for the relative decline of the Australian agricultural sector, analogous to the constraint provided by domestic orientation for US agriculture.

By the early 1970s, political forces had conspired to create an Australian Labor Government which was markedly less sympathetic to the agricultural sector than its Coalition predecessors (e.g. its termination of the superphosphate, dairy and cotton bounties). By commissioning the 'Green Paper' on agriculture, this Government provided a platform for public discussion about the effect on agriculture of the structural changes that were then proceeding in the Australian economy, and from which to reassess the need for special policies for agriculture. By creating the Industries Assistance Commission (IAC), the Government provided a relatively permanent forum for public discussion of economic ideas that both directly affected agricultural policy (cf. Edwards (1980) for early IAC examinations of agricultural policy) and indirectly affected agriculture through examining industry protection in other sectors (Box 2).

Box 2 Green Paper and IAC Objectives

- a improve the efficiency with which the community's productive resources are used;
- b encourage those economic activities in Australia, and the producers of the goods and services concerned, which contribute to improving the efficiency with which the community's productive resources are used;
- c facilitate adjustment to changes in the economic environment by industries and persons affected by those changes;
- d recognise the interests of consumers and consuming industries likely to be affected by the Commission;
- e ensure that any measures for assistance to, and development of, industries are integrated with national economic policy as a whole;
- f ensure that Australia's trade and protection policies are compatible.

Source: Harris et al. (1974, chapter 3, quoting the IAC Act)

International economic disruption associated with US involvement in the Vietnam war and the first oil price shock of 1973 coincided with the Australian Labor Government of the early 1970s. This disruption was followed by instability and diminished economic growth in developed economies (including Australia, and despite its growing trade with Japan and emerging economies of East Asia). The effects of this dislocation were increasing economic criticism of the extent of government intervention in the Australian economy. While progress towards deregulation was not consistent—for example, protection in the motor vehicle industry increased in the second half of the 1970s (Balderstone 1982, table 4.4)—the intellectual debate favouring the demise of industry protection and 'protection all round' was won by the end of the decade. The political debate had intensified with the creation of the National Farmers'

Federation in 1979, and its adoption of 'economic rationalist' policies (National Farmers' Federation 1981). Such policies were adopted, albeit in a very muted form, in the Balderstone Report on agriculture in 1982 (Box 3). Consistent with this theme, the Federal Coalition Government commissioned an inquiry into the financial sector which subsequently provided the arguments supporting deregulation of the financial sector in the 1980s.

Box 3 The End of the 1970s: the Balderstone Report

General aims and objectives of agricultural policy:

- be responsive to national and international economic developments;
- create conditions under which the agricultural sector can handle change and at the same time realise its full potential in contributing to national economic growth;
- aim to encourage the efficient use and conservation of the nation's resources;
- promote the adoption of new technology and increased productivity in the farm sector; and
- be consistent with other national policies and goals. There also should be consistency of policy at the Commonwealth, State and local levels of government.

Source: Balderstone et al. (1982, para. 1.14)

Although there were other contemporaneous pressures for deregulation—e.g. IAC recommendations for deregulating the Australian wheat industry and pressures for tariff reduction—deregulation of the financial sector, beginning in late 1983, was the catalyst for a sequence of as-yet-unended deregulations affecting large parts of the Australian economy. Once the financial sector was deregulated, some forms of government intervention became increasingly difficult to manage (e.g. the Wool Reserve Price Scheme). Other forms of regulation also became increasingly difficult to justify (cf. Box 4); in agriculture, there was deregulation of the domestic marketing of wheat, and substantial deregulation of the storage, handling and transport of grain. But, as deregulation of the financial sector proceeded, the deregulatory fervour it invoked increasingly acquired the air of a religious crusade.

The effects of financial sector deregulation were superimposed on two other processes of the 1980s. The first was an international economic boom. The second, specific to agriculture, was the rapid increase in US agricultural protectionism from the mid-1980s, ostensibly as a bargaining chip to induce West Europeans to reduce agricultural protection. This increased protectionism resulted in greater levels of world agricultural trade, and greater competition for markets, culminating in depressed prices for many agricultural products. Depressed commodity prices were exacerbated by the collapse of farm-gate wool prices with the demise of the Reserve Price Scheme in 1990 and the effects of high interest rates resulting from international economic conditions and domestic macroeconomic policy. The combination financially weakened many agricultural producers just as much of Queensland and northern New South Wales entered a long drought in 1991.

Box 4 Policies for the Deregulatory 1980s

Underpinning the economic policy reforms are the Government's social goals of greater and more equal opportunity, substantial improvements in living standards, an enhanced quality of life for all Australians and care for our environment. While objectives for the primary industries and energy sector are expressed in terms of efficient economic and resource management principles, it is progress towards these social goals that is the ultimate outcome of improved economic performance.

The Government's long-term objectives were:

- enhancing productive capacity through
  - — balanced management and development of natural resources
  - — developing human skills
  - — improving technical capacity and research
- developing a more responsive and productive industry structure, featuring
  - — an economy-wide approach to policy reform
  - — lower protection all round and better-balanced assistance
- influencing and responding to the external environment
  - — better (international) marketing of Australian produce
  - — reducing international trade distortions
- more efficient government business enterprises, and transport and services
- positive assistance for structural adjustment.

Source: Kerin & Cook (1988, pp. 4–8)

It is easy to look back on the past, and characterise it as a process of economic and policy transition. Without a crystal ball, however, it is difficult to see beyond the immediate present. In the large, there was the ostensible determination of the Federal Labor Government of 1983–96, and the subsequent Coalition government, to deregulate the Australian economy, even to the extent of pushing the states to deregulate significant portions of their own domains. Until 1993, the Federal Opposition Coalition had promoted even more extensive structural change in the Australian economy. Most state governments also paid at least lip service to the need for further deregulation of or structural reform in the economy. The evolution of the former Industries Assistance

Commission into the Industry Commission via absorbing the functions of the Inter-State Commission, gave the government a public process for investigating areas of the economy formerly beyond the purview of the IAC. Additionally, the Federal Government commissioned the Hilmer Report to accelerate the progress towards increasing competitiveness.

But it is not the role of this book to divine the future. The objective is to provide a framework that economic analysts might use to undertake their own analysis of policy issues. Those who wish for regular updates on policy specifics might undertake the annual trek to the National Agricultural and Resources Outlook Conferences, wade through the financial, daily and agricultural press, dip lightly into television, and put their name on the mailing lists of a few well-chosen politicians. This book will have achieved its purpose if it helps these future analysts to better organise their approach to policy analysis.

Outline

The art of policy analysis is to combine economic efficiency analysis (Chapter 1) with public choice (Chapter 2) in the context of the relevant institutional framework (Chapter 3). But economic analysis of policy is not the mere mechanical application of theory and techniques. Insight is required to identify relevant problems, to identify the key aspects of these problems, and to combine economic efficiency, distributional and public choice aspects appropriately. Policy analysis itself is partly an economic problem—which combination of economic efficiency, distributional and public choice aspects is optimal to select, given the financial and informational constraints facing the analyst?

A simple framework for integrating efficiency, distributional and public choice aspects of microeconomic policy is utilised in chapters 4 to 11 and outlined below.

Policy issue

Investigating the 'policy issue' means describing the perceived—or, perhaps, unperceived or misconceived—'problem'. Why does the present state of the world require—or appear to require—rectification via government action? The appropriate identification of the policy issue is crucial to good policy analysis. If economists want their analysis to be effective, they must evaluate policy problems that are real as perceived by the policy makers. However, policy makers are not necessarily perfect in identifying real policy issues—they may persist in supporting outmoded policies long after their use-by date; they may be 'captured' by interest groups or simply wish to support the interest groups who support them, irrespective of the effects of resulting policy on social welfare; or they may be ill-advised by their public servants. Policy analysts who do not dance to the policy-maker's tune have, therefore, a vital role. Such independent analysts are unlikely, however, to be employed to work for policy participants—whether for politicians, political parties, lobby groups or in the public service.

Policy intention

Analysing 'policy intention' describes the objective of policy—that is, what policy makers intend to achieve. The intention of policy may be to eliminate or reduce economic inefficiency, to increase incomes of disadvantaged sections of the community, to redistribute income towards members of powerful lobby groups, or to provide sops to less powerful lobby groups. It is tempting to assume that, just because a policy issue has been identified, the intention of any corresponding policy will be consistent with the identified issue. While this may happen, it will not necessarily do so. Participants in the policy process may simply misunderstand the nature of the policy issue and make an inappropriate decision. Or, even if policy makers appropriately match a policy intention with the policy issue they have identified, their intention may be derailed as a consequence of incompetence in implementation; competition within the bureaucracy, which produces an unintended result; or competition among lobby groups, which may derail the original intention. Or government, or one of the other policy participants, may propound one intention while covertly intending to implement another.

Area of sector

The 'area' of the resources sector that is relevant to policy analysis denotes both where the policy issue is perceived to exist, and where the resultant policy is to be implemented. Policy may target the natural resource base, the farm sector, post-farm-gate industries, other sectors of the economy or the economy as a whole. The analyst must be able to identify where policy is intended to have an effect, so that this effect can be estimated or predicted, and evaluated. The analyst must also be aware that, even if a policy carefully targets the intended area of the sector, there will be 'collateral damage'—it is rare that the effects of a policy can be quarantined to the area of intended effect.

Policy orientation

'Policy orientation' refers to the level of government where policy intervention occurs. Both the federal and state governments may intervene in external or domestic policy issues as permitted under the Commonwealth Constitution. The states also legislate to provide the powers of local government. A combined approach to policy involving two, or even all three, levels of government may be necessary or desirable.

Policy mechanism

A policy mechanism is the vehicle by which policy is implemented: Acts or Regulations of Parliament, or executive decision. Government may choose to intervene directly by direct control, ownership of commercial or potentially commercial activities, or ownership of non-commercial undertakings. Government may also intervene indirectly, by seeking to influence individuals' activities or community attitudes. The policy analyst must be sensitive to the wide variety of mechanisms by which governments can intervene, and diligent in discerning their often subtle or hidden forms. The participating policy analyst needs creativity with regard to mechanisms to be able to present policy makers with feasible, perhaps durable, it is hoped unchallengeable and occasionally novel policy instruments.

Institutional structure

The 'institutional structure' of intervention concerns the various organs of government that develop, evaluate, enact and implement policy, and the various interest groups that interact with government to both effect and affect policy. Not only is the structure of these institutions important, but also the form and relative power of the relationships within and between organisations, and the determinants of these structures and relationships, and their forms and power. Without a clear understanding of the processes by which policy is formulated, policy analysts cannot explain why particular policies are enacted (or not enacted), why policies are implemented the way they are, or whether or not possible alternatives are likely to be adopted. Similarly, without such understanding, policy advisers cannot develop policies that are likely to be acceptable to the interest groups for whom they work, nor would their recommended policies have a high chance of gaining approval from other players in the policy process.

Policy performance evaluation

'Policy performance evaluation' concerns why government policy is evaluated, who evaluates policy, when policy is evaluated, how policy is evaluated, and whether or not there is any economic input into policy evaluation. The nature of policy performance evaluation is crucial for policy advisers, who may need to chart a course for evaluation which is feasible in financial terms, bearable in political terms, and satisfies the interests of lobby groups who may wish for, or wish to escape, the review of policies which affect them. Policy performance evaluation is also important for the independent analyst who wishes to influence the policy process; unless the analyst's intervention is carefully targeted, it is likely to be ineffective against the noise of a plethora of competing interests—or even, perhaps, counter-productive. Policy may be evaluated regularly or occasionally, and frequently or rarely. Evaluation may occur publicly or privately, and be either formal or informal.

Economic issues

'Economic issues' refers to the broad economic scope of a policy problem. This scope will include the economic effects of existing policies and alternative policy instruments. Important issues may include efficiency versus distribution, partial equilibrium versus macroeconomic or general equilibrium approaches, theoretical versus empirical analysis, and the desirability of qualitative versus quantitative analysis.

Analytic techniques

Analytic techniques for policy analysis include any reputable economic or econometric modelling. In partial equilibrium analysis, this may include graphical or algebraic analysis of simple partial equilibrium (including analysis of consumer and producer surplus). Alternatively, partial equilibrium relationships may be estimated econometrically as single equations or equation systems, and consumer and producer surplus estimated quantitatively. Partial equilibrium relationships may also be estimated using programming techniques or, if supply and demand curves can be estimated econometrically, they may be combined in mathematical programming models. Ultimately, general equilibrium models may be constructed from estimated supply and demand equations. Where variability (e.g. of weather or prices) is important, models may incorporate stochastic analysis of important relationships. It is salutary, however, to remember Edwards and Watson's (1978, p.192) sage advice that:

> . . . economic policy making often hinges on simple figuring of the key economic magnitudes involved.

The interaction of the agricultural and resources sector with the rest of the economy, and the importance for agriculture of policies external to agriculture, are evaluated in chapters 12 to 14—'intersectoral' issues. As background, simplified models are constructed of the effects on agriculture of changes in exchange rates and interest rates, inflation, and wages policy. It is then shown why, for complete analysis, more comprehensive models such as computable general equilibrium (CGE) models are required to evaluate policy options. Differential sectoral rates of export growth (the 'Gregory' thesis) are examined as an example of the effect on agriculture of changes in the rest of the economy. Both simplified and CGE models are used to examine the effects of differential sectoral rates of protection and tariff compensation. Finally, the effects on agriculture of microeconomic reform outside the agricultural sector—e.g. in banking and indirect taxation, and the Hilmer reforms—are evaluated.

Conclusion

In Isaac Asimov's Foundation trilogy, the psychohistorian Hari Seldon predicted the fall of the Galactic Empire based on Trantor, and the future rise of the planet Terminus. These predictions were based on a general stochastic psychological theory of human emotions and reactions sufficiently good to predict the historical sweep of the future (Asimov 1953). Economists do not, as yet, have economic policy models that enable the prediction of future events with such precision—and they probably never will. Moreover, an inference from the 'Lucas' critique of 1970s macroeconometric modelling is that such policy modelling is actually impossible (Lucas 1976). Economic analysis of government policy—and agricultural and resources policy in particular—is a judicious blend of science and art.

Economic analysis of government policy has two key requirements. First, good policy analysis demands an ability to ask the right questions—those that identify the central issues of the problem. Good analysis partly requires a knowledge of economic theory, which can illuminate the problem; but such analysis needs to be creative, to be able to see beyond a conventional formulation of the problem. Good policy analysis also requires an adequate empirical understanding of both the economic and institutional conditions of the relevant industry or industries, as well as an understanding of how the present has evolved from the past, and how the present is evolving into the future. Second, good policy analysis requires the deployment of the right tools to obtain appropriate answers to the questions that have been asked. But the ways of answering the relevant questions need not necessarily be complex: Edwards and Watson (1978, p.192) argued that 'economic policy making often hinges on simple figuring of the key economic magnitudes involved'. The objective of this book is to assist readers in developing their creative and analytical faculties and therefore their skills in policy analysis.

If students merely rote learn 'facts' from this book, the author has failed in his intention. The author has presented *a* method of doing agricultural and resource policy, not *the* definitive approach. Readers should critically appraise the arguments presented in this book. Is the approach to conducting agricultural and resource policy appropriate? Have the arguments been adequately justified on theoretical and empirical grounds? Has economic theory or the empirical facts changed so that the conclusions reported are no longer valid or relevant? Even where it is adopted as appropriate, the book's argument must be continually reassessed—and refined and developed—for future evaluation of policy issues.

Part I

Policy Analysis Framework

Why do governments do what they do? Economists have typically provided two explanations: one is that governments identify inefficient economic performance, and act to correct these deficiencies; and the other that governments are primarily interested in redistributing income. The first might be termed a 'benign' explanation of government intervention—that governments intervene to do good. Analysis of what constitutes economic efficiency, and how government might act to secure this objective, is presented in Chapter 1.

A note of caution is desirable. Boulding (1970) argued that market transactions, on which Chapter 1 is primarily based, reflect voluntary relationships among individuals. However, these are only one of three possible relationships affecting resource allocation within market economies—threat and 'integrative' relationships are also important. Government dominates threat relationships since it is the only source of legitimate threat in society. Legitimised threat authorises taxation and the use of these revenues in delivering services such as external and internal security, health and education, transport infrastructure, and welfare transfers. Since governments in developed economies typically acquire 30 to 40 per cent of GDP, the 'threat' economy is clearly of major importance. Boulding's 'integrative' relationships are characterised by the mutual acceptance of higher and lower status. The classic integrative relationship is the family. There also are important integrative relationships within market participants such as firms, or within non-market organisations such as not-for-profit or philanthropic organisations. Until recently, organised religion dominated integrative relationships, acquiring great wealth through 'tithing' and similar forms of taxation, and disposing of this wealth especially in constructing vast religious monuments. Although conventional religion has declined in Western European culture, its integrative role has been partly replaced by charismatic and mystical religions, and secular movements such as science and environmentalism. The capacity of integrative relationships to affect resource allocation is still substantial because of the extensive wealth commanded by the family, church and secular movements, and the capacity for resource allocation decisions by government and private firms to be modified by integrative behaviour. Recognising these other forms of non-voluntary behaviour is important because efficient resource allocation is not just affected solely by voluntary market transactions, and the canons of efficient resource allocation derived from voluntary behaviour will not necessarily be automatically relevant to resource allocation where other relationships dominate.

The second explanation of government action might be judged either 'benign' or 'malign'. It might be benign if there are groups in society who are so disadvantaged that intervention on their behalf improves social welfare. However, it might be termed malign if it worsens income distribution so that social welfare is reduced. In the latter case, redistribution might occur essentially in favour of those groups most influential in ensuring that politicians achieve their objectives. These objectives may be, for the government to retain power or, for political parties out of power, to regain it. Economic aspects of the policy process—i.e. the ways by which political groupings seek to retain or obtain power—and the symbiotic groups that are involved in this quest (political parties, interest groups and the bureaucracy) are discussed in Chapter 2.

Analysis of neither benign nor malign cases is exclusively restricted to democratic government. Non-democratic governments may be interested in improving some aspects of economic performance or social welfare through distributional measures. Non-democratic governments also have to secure continuing support—especially from the armed forces—to retain power.

Neoclassical economics is sometimes assumed, especially by its more ideological technicians and political acolytes, to operate in a social vacuum. However, economies are systems of institutions, and formal theoretical representations of these systems are simply useful abstractions from empirical reality. To understand how economic policy analysis proceeds, and how it informs the policy process, requires an understanding of the policy process itself. Economic policy analysis—whether of agriculture and resources, or any other sector—cannot occur in an institutional vacuum. It is essential, therefore, to understand the institutional structure in which Australian agricultural and resource policy occurs (Chapter 3).

It is impossible, however, to write a definitive study of the institutions of agricultural and resource policy in Australia. Politicians and some bureaucrats—and many voters—confuse 'change' and 'progress'. Institutions frequently change their names, sometimes their functions, and often their managers—and especially the last when their political masters are embarrassed by institutional shortcomings. Some of the contemporary institutions considered in Chapter 3 may well have changed, or even disappeared, by the time this book is published. Emphasis in this chapter is placed on the enduring institutions and the relationships among them, but the reader must be prepared to discover differences between the personnel, name and functions of the institutions as described, and the institutions that the reader may actually observe in practice.

The following discussions of economic efficiency, economic aspects of the policy process, and the institutional structure of agricultural and resource policy in Australia are essential for economic policy analysts to understand the policy process, and to be equipped to analyse it.

1 Governing the farm: economic efficiency foundations

If governments intervene in the economy to improve economic efficiency, then they—and economic analysts—must be able to identify economic *efficiency* and, more importantly, economic *in*efficiency. But this, surely, is easy: microeconomists have a set of rules that determine whether or not markets are efficient. All that is required is to implement those rules and, where the market fails the efficiency test(s), government should intervene. A major objective of this chapter is to demonstrate that this is a naive view of economic policy analysis. The chapter commences with a discussion of economic efficiency in its simplest, partial equilibrium context and then identifies the limitations of this approach with reference to its underlying assumptions. The issues of property rights and associated transactions costs are emphasised. The discussion is then broadened to consider the nature of efficiency in a general equilibrium context, and the limitations of simple models of this genre. The discussion is developed by linking market failure and market imperfections to the previous analysis, and the chapter concludes with a discussion of government failure.

1.1 STANDARD PARTIAL EQUILIBRIUM ANALYSIS

A policy economist must be able to use the basic tools of the applied neoclassical economist's trade confidently. The tools of partial equilibrium economics are briefly reviewed in this section.

Standard model

The fundamental concept is the demand/supply diagram for a perfectly competitive market (figure 1.1). The supply curve S1 represents the quantities that a large number of small producers are prepared to supply, parameterised on price. Other variables may also shift the supply curve—important supply shifters in agriculture are weather conditions, the prices of other agricultural products, input prices and technology. The demand curve D is the *farmgate* demand curve for the product, again assuming a large number of small buyers. The price c is the efficient and equilibrium price for the commodity, at which price quantity q will be produced and sold. In this book linear supply and demand curves are normally assumed for simplicity, but there is no particular reason for assuming linearity. It will often be necessary to estimate supply and demand curves by some quantitative technique; normally this will be econometric, but other methods may also be used.

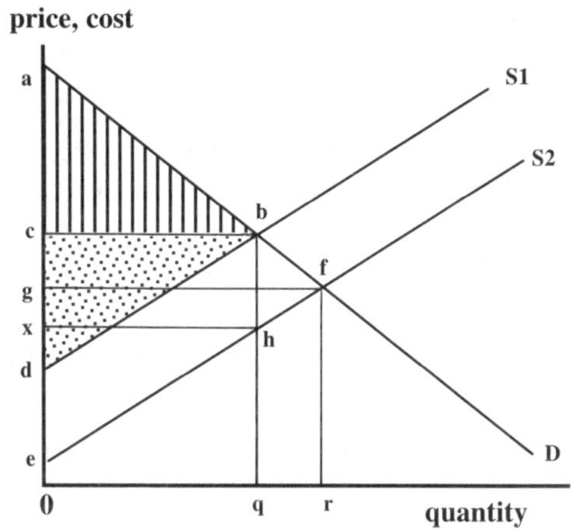

Figure 1.1 Supply, Demand and Marshallian Surplus

In figure 1.1, the vertically hatched area abc is the Marshallian consumer surplus, representing the value all consumers obtain from consuming quantity q additional to the expenditure required to purchase q (represented by area cbqO). When quantity q is sold at price c, the dotted area cbd is the Marshallian producer surplus, representing the difference between producers' aggregate revenue from selling quantity q (area cbqO), and the opportunity cost of the inputs to produce q (area dbqO). The total Marshallian surplus is the sum of consumer and producer surplus (i.e. the area abd). Should the supply curve shift to S2, the change in the total Marshallian surplus is the area dbfe—the difference between original total surplus abd, and the total Marshallian surplus with supply curve S2 which is area afe. The change in consumer surplus if the supply curve shifts from S1 to S2 is area cbfg; thus the change in producer surplus is area dbfe minus area cbfg.

Underlying assumptions

Policy economists become so adept at drawing supply and demand diagrams that they sometimes forget the stringent assumptions that underlie these models. A common assumption is perfect competition or, at least, atomistic competition with no collusion. In this model there is assumed to be a sufficiently large number of buyers and sellers, all of whom are sufficiently small so that they cannot influence the market price. It is not difficult to make simple modifications to this model to consider three additional cases: monopoly seller facing atomistic consumers; monopoly buyer ('monopsonist') facing many sellers; and bilateral monopoly of monopolist facing monopsonist (e.g. Varian 1990, pp. 430–3). More difficult problems arise with a small number of sellers relative to consumers (oligopoly), or a small number of buyers facing producers (oligopsony), and game theoretic solutions may be required (e.g. Varian 1990, chapter 26). For many agricultural products, at least if intermediate marketing stages are ignored, large numbers of producers and consumers enable the atomistic competition case represented by figure 1.1 to be deployed without too great a violation of reality. Where, as with many Australian agricultural products, single seller marketing boards have been implemented, the monopoly seller model may be conveniently used to represent the sale of agricultural commodities. In this last case, however, the 'monopolist' is not necessarily a profit-maximising firm.

Some technical assumptions are also worth noting. The model depicted in figure 1.1 represents a *homogeneous* commodity. Applied economists often happily aggregate across different classes or qualities of a commodity, giving very little thought to the implied aggregation conditions. It may, however, be necessary to disaggregate from a generic product (e.g. meat) to its constituents (e.g. beef, sheepmeats, pigmeats, poultry), to individual types (e.g. mutton and lamb) or even to individual cuts (e.g. forequarter, leg). It may also be desirable to analyse the problem with several linked markets —e.g. studies of meat demand typically disaggregate into the main meat types. A policy study to examine the effects of some intervention such as price stabilisation for one meat type would need to be sufficiently disaggregated to determine the effects of this stabilisation on supply of and demand for other types of meat.

Another important technical assumption is that the supplies and demands for product depicted in figure 1.1 occur at the same *place*. If there are multiple supply regions and a single consuming region, then differential transport costs from each region must be added to each region's supply curve to obtain a single supply curve in the consuming region. With m supply regions and n consuming regions, market equilibrium is not easily obtained from diagrammatic representations because demand for one region's output is a function of the quantities supplied by other regions. Programming models such as spatial equilibrium models are required to obtain solutions to such problems (cf. Chapter 11). Economic analysis of government policy may require complex modelling where there are multiple consuming and producing regions.

The analysis represented by figure 1.1 implies that the decision problem is *timeless* because supply and demand curves, and the resultant equilibrium, occur at an unspecified time, and this time period is disconnected from previous and subsequent time periods. Except in the very short run, with highly perishable commodities, this assumption is a gross departure from the observed economy. Firms that invest in produced or human capital, or research and development, are clearly making decisions over multiple time periods. Where storage is available, output is not necessarily closely tied to consumption at a single point in time—if price is deemed too low, the product will be stored for later sale. With consumer durables, or where consumer storage is available, consumption decisions are also not disconnected from later time periods. An additional dimension of the assumed 'timelessness' in simple economic models is that adjustment from one level of output and consumption to another is both instantaneous and costless. For example, in considering the supply curve shift from S1 to S2 in figure 1.1 it is commonly assumed that the supply curve shift itself, and the consequent expansion of optimal output from q to r, occurs within the time period and incurs no adjustment cost. Empirical economic models of markets generally relax these assumptions by either introducing arbitrary lags, or formally modelling physical adjustment processes (e.g. herd or flock dynamics), arbitrary structural change processes (e.g. adaptive expectations), or the economic costs of adjustment. Thus, in a market model, the *length of run* of the problem must be clearly specified.

Perfect competition not only assumes atomistic markets but also *perfect knowledge* of the present and, if the timeless assumption is relaxed, the future. However, even in an atomistic, timeless market, firms will not necessarily know their existing supply curves perfectly; they are also highly unlikely to know their competitors' supply curves or the aggregate demand curve. They are unlikely to know the current market price with certainty, and will not know the optimal level of their own output in the current time period. A simplistic way of handling imperfect knowledge is to assume that part of the supply and demand curves are known with certainty, but there is an additional random component. Since the 1970s, economic modelling has often assumed the existence of 'rational expectations'—i.e. that economic agents know the process by which equilibria emerge (i.e. know the relevant supply and demand curves in the microeconomic context, but have different sets of knowledge at different points in time).

The market represented by figure 1.1 also implies that producers and consumers *interact directly*. This assumption is inappropriate for most markets, including agricultural markets. This model ignores the roles of assemblers (who gather product from producing firms), transporters, storers, processors, wholesalers, exporters and retailers. Since the demand for agricultural products is a derived demand—i.e. most products are transformed before end-use consumption—it is desirable to examine the effect of marketing stages on the derived demand for farm products.

The model represented by figure 1.1 is only directly relevant for *marketed goods*—equilibrium is achieved through notional adjustments of prices and quantities until the price–quantity pair (c,q) occurs. The method may also be used to evaluate non-marketed commodities such as the provision of irrigation storage. But, although there may be markets to allocate the impounded water, there are no markets to determine the optimal quantity of water to impound. Governments often decide on the investment in water storage without reference to water markets. Should water storage be provided which delivers a non-stochastic amount q of water per period (figure 1.1), then the gross benefit of its provision is abqO. Should additional irrigation investment deliver a total quantity of water r per period, the per period increase in gross benefit is bfrq. Economic evaluation of investment in additional water storage requires comparing a stream of future benefits of bfrq per period, and the costs of augmenting the irrigation infrastructure. With some non-marketed commodities, such as externalities (e.g. 'pollution'), analysis must be extended to allow for the evaluation of the non-marketed commodity (cf. following section).

1.2 PROPERTY RIGHTS AND TRANSACTIONS COSTS

Existence of markets

The market models discussed above assume that market equilibrium occurs at a non-zero quantity and a positive price. There may, however, be 'markets' that do not satisfy this condition. Consider figure 1.2 where S represents supply conditions for a high cost good to potential consumers whose demand curve is D. The supply and demand curves do not intersect; market equilibrium is represented by zero sales with a divergence between the price consumers are willing to pay, and the price at which producers are willing to sell. A 'potential' market exists in the sense that, if the demand curve shifted outwards (e.g. with income growth), or if the supply curve shifted down to the right (e.g. with technical change or a reduction in the cost of inputs), or both, then intersection of S and D in the quadrant depicted in figure 1.2 would lead to a non-zero level of sales at a positive price.

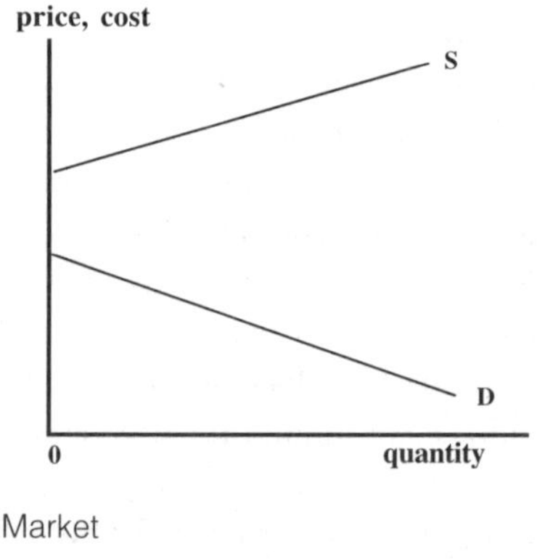

Figure 1.2 'Non-existent' Market

Does the reverse hold? If supply and demand curves intersect, will production and consumption necessarily occur? The answer is also 'no'. To see why, first consider the components of a supply curve. In a market model where the demand curve is that of final consumers, the supply curve incorporates not only the raw material costs (e.g. from agricultural products) but also the transport, processing, storage, wholesaling and retailing activities that are required to transform 'raw material at farmgate' to 'final consumer product at retail'. So, for example, in figure 1.1 supply curve S2 might represent the supply curve of raw material at farmgate, and supply curve S1 incorporates both S2 and the cost of activities required to transform the raw material to the final consumer product at the point of consumer purchase. A quantity q of the consumer product would be sold at price c; farmers would be paid price x for raw material contained in each unit of the final consumer product. Area Oqbd represents the opportunity cost of materials used to produce the final product (Oqhe farm costs + ehbd post-farm costs).

However, activities that firms are observed to undertake are not the only ones required for a market to exist. Firms may undertake 'invisible' activities, while many other activities are undertaken on their behalf. Without these invisible activities, markets would not exist even if supply and demand curves intersected in the appropriate quadrant. The key explanation of this paradox is 'transactions costs' in which property rights play a central role.

Property rights

Although supply curve S1 (in figure 1.1) appears to incorporate all the costs of producing and marketing the final product, there is a vast array of productive activities whose costs are not included in firms' visible activities. Consider the activities that are required for trade in goods and services to occur. In a market economy, trade is assumed to occur voluntarily—two agents agree to exchange a commodity (good or service) for a consideration (currency, or some other form of payment). To guarantee that the seller is entitled to have the commodity in the first place, societies have created the institution of *property rights*, which define the legitimate owner of a commodity who is thus entitled to enjoy services that flow from its possession. The property rights institution defines, for broad classes of commodities, characteristics that the right bequeaths—these characteristics include the *extent* to which the owner may enjoy use of the commodity, and the conditions under which it may be *transferred* to another owner. Perfectly non-attenuated property rights have no restrictions on extent of enjoyment or transfer. In general, property rights are attenuated in some way: there may be limits on how the commodity may be used or transformed; or there may be limits as to how, when or to whom ownership of a commodity may be transferred.

In day-to-day transactions of commodities, little or no thought is required about the vast institutional structure underpinning the property rights institution. When consumers purchase at a supermarket, they assume the store has the right to sell the item and they have the right to buy. However, there are many common transactions where it would be risky to make such assumptions. Consider the private purchase of a second-hand car advertised in a newspaper: just because the car is advertised does not mean that the seller is either the legitimate owner (the car may have been stolen) or the beneficial owner (the car may be owned by a financial institution). Unless purchasers satisfy themselves as to the legal status of the vehicle's seller, they may subsequently find the car returned to the rightful owner or repossessed without compensation. The institutional structure supporting property rights thus includes an information system that defines, and in some cases records, legitimate ownership; and an enforcement system that provides redress to holders of property rights deprived of the legitimate exercise of those rights. This institutional structure is largely, but not exclusively, public. There are public registers of real property and motor vehicles (and their encumbrances), and a police force to provide protection against theft and trespass; there are also private security firms that offer similar forms of protection and detection. Disputes about the legitimate ownership of commodities may be resolved in criminal or civil courts and, increasingly, through public or private mediation services.

For many commodity transactions considered below, social convention and publicly funded enforcement via police and the criminal courts provide all the necessary enforcement for everyday transactions. In these cases, the benefits of trade represented by total Marshallian surplus in figure 1.1 are sufficient to permit this trade to exist. But consider those cases where resort to private enforcement is required: suppose public enforcement institutions could not guarantee the security of the market represented by figure 1.1. Suppose sellers considered employing a firm to 'protect' the market during the transaction period, but the cost of this security exceeded the producer surplus (area bcd). Unless sellers could persuade consumers to contribute to the cost of protecting the market, it would not be economic for sellers to produce and sell; the market would not exist. If the cost of protecting the market exceeded the *total* Marshallian surplus (area abd), then under no circumstances would the market exist. Thus, even if there appear to be profitable opportunities for trade, the conditions may not be met if the costs of establishing and protecting a market exceed the net benefits from trade. Since the benefits from establishing effective property rights may endure over time, the appropriate comparison is between the costs of creating, maintaining and protecting the property rights, and the resulting *stream* of Marshallian surpluses over time.

Market establishment costs—e.g. generically defining property rights and creating an appropriate enforcement system—are termed *transactions* costs. Transactions costs should not be confused with *marketing* costs such as sales force costs: the latter are (broadly) related to the quantity of commodity being traded. Transactions costs, by contrast, are the costs of investing in the existence of the market, and are largely invariant to the quantity traded.

Other transactions costs

In addition to property rights, there are *information* costs, which may also be included as transactions costs. Consider, from a seller's perspective, the costs of establishing the conditions of demand for a commodity that would lead to the definition of a demand curve. For most commodities, this presents no problems—consumers regularly visit supermarkets, department and speciality stores and car-yards, and their purchasing behaviour can be (roughly) determined. In other cases, the definition of demand conditions may be much more difficult, and therefore more costly. For new products, not only is research and development required to produce a workable commodity or service, but extensive market testing is required to determine whether and at what price consumers will purchase the commodity. Similar problems exist for firms considering new markets, especially where these markets are overseas and require knowledge of languages and cultures with which the firm is unfamiliar. If the costs of determining demand conditions exceed the potential benefits from trade, the firm is unlikely to venture into these markets even if potential total Marshallian surplus exceeds zero. Once market conditions have been defined, these costs may be largely invariant to the number of time periods in which trade occurs. Thus, the firm's (or society's) decision is not simply to compare information transactions costs with Marshallian surpluses in a single period—the appropriate comparison is between investment in these set-up costs and the 'present value' of future Marshallian surpluses.

Information obtained by a firm about a potential new market may be useful not only to that firm but also to market competitors. Ignoring potential monopoly advantages to the firm initially generating the data, consider the potential problems involved if the first firm attempts to sell information about a new market to another firm. The potential purchaser is unlikely to pay a price for the information about the new market unless it can be convinced that the information is of value to it. But it can only be ultimately convinced of the value of the information when it is in complete possession of it. However, once the second firm possesses the information, it has no incentive to pay the purchase price to the first firm. Quite complex contracts might arise to govern the transfer of valuable information between firms; such contracts are an example of an additional form of transactions costs, that of *negotiation* transactions costs. Especially for single-event transactions like the transfer of market or technological information, negotiation transactions costs may be high.

There are many other negotiation transactions costs, many of which are borne by society. Even behind everyday transactions, like purchases at a supermarket, there are very substantial transactions costs that have taken centuries to develop. Two such key transactions costs are *systems of measurement*, in which characteristics such as mass, volume, length and area are defined and their use monitored, and the use of *currency*. The central importance of these *revenue collection* transactions costs being administered by the state can be appreciated by considering their absence. Suppose every transaction required definition of the physical unit of transaction, and an exchange mechanism. Without these key public services, a market economy would promptly collapse. There may be particular occasions, however, when measurement or the unit of exchange are defined for individual transactions. For example, during reconstruction of the economy of the former USSR in the late 1980s and early 1990s, firms often resorted to barter because there was not a sufficiently stable internal currency. Again, in these cases, the costs of establishing single-use transaction conditions must be compared with the potential gains from trade in the form of Marshallian surpluses.

Excludability and rivalry

Considering the problems of trading in information leads naturally to a discussion of two generic characteristics of property rights. These characteristics are *excludability* (or 'exclusiveness') and *rivalry*. A commodity is (*price*) *excludable* if an agent other than the owner can only gain access to the commodity by paying the owner for this access. For example, a producer's apple is price excludable because a consumer can only legitimately enjoy the apple by purchasing it from the owner. By contrast, even if an agent laid claim to fish in the sea, they are generally not price-excludable (i.e. they are 'non-excludable' or 'non-exclusive')—property rights cannot be sufficiently well-defined that anglers can only gain access to the fish by buying access from the owner. Of course, once a fish is actually caught, it becomes price excludable like the apple. A commodity is *rival* if its consumption by one agent automatically precludes its consumption by another. Thus, for example, an apple is a rival commodity in that (unless shared), it can only be consumed by a single individual. By contrast, 'free-to-air' radio and television broadcasts are *non-rival* because their reception by one listener does not preclude their reception by others. Examples of each of four archetypal categories of excludability and rivalry may be defined (figure 1.3). The non-excludable/non-rival category is of considerable relevance because of the interaction of this characteristic with the structure of agriculture (cf. chapter 5 on research and development).

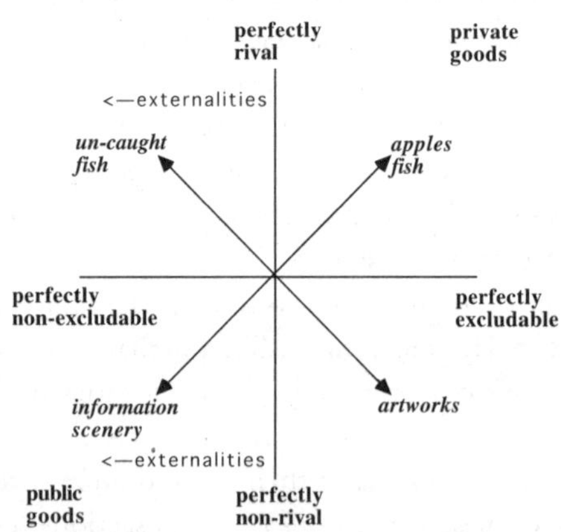

Figure 1.3 Combinations of Price-excludability and Rivalry

Whether or not a commodity is price-excludable may be a matter of choice, or a consequence of technology. For example, Australian beaches could be made excludable (as in other countries), but Australian authorities generally choose not to do this (although beaches are sometimes closed for private activities like filming). Conversely, ocean fish are (currently) non-excludable because technology is not currently available to prevent fish 'rustling'; however, *access* to ocean fisheries is increasingly restricted by quotas or similar mechanisms to conserve fish stocks. Traditionally, exclusion from road facilities has not been widely practised, because of the cost of doing so; although particular road facilities such as freeways, bridges and tunnels often attract tolls because limited entrances to these facilities enable revenue from toll collection to exceed the costs of collection. However, changes to the technology of recognising, monitoring and recording road use by individual motor vehicles is making much more widespread charging for road use increasingly feasible. Radio and television free-to-air broadcasts are non-excludable unless they are 'scrambled' so that they can only be decoded by purchasing technology from the broadcaster.

Whether or not a commodity is *rival* in use is largely a technological matter. Radio and television broadcasts do not diminish in intensity simply because a particular receiver is turned on. Scenic views and works of art do not disappear because they have been observed. There may, however, be technical ways to circumvent non-rivalry. For example, renting films on video cassette or CD-ROM converts the non-rivalry of free-to-air or cable broadcasts into a rivalrous viewing of the film on home electronics (as long as the consumer cannot re-broadcast or copy). The item becomes rivalrous for a particular viewing, although the item is non-rivalrous through time as there will be other renters.

Externalities

Externalities are an additional aspect of property rights relevant to agricultural and resource policy. Markets may be lacking for particular commodities because property rights have not or cannot be defined for them; such commodities may be either rival or non-rival. These commodities have the additional characteristic that they are produced jointly with other marketed commodities. The archetypal case of such commodities is pollutants. Pollutants are waste products for which there is no market, and which are produced jointly with some marketed commodity. 'Externalities' is the generic term covering pollutants (negative, or detrimental, externalities) and a rather smaller class of positive externalities.

The effect of an externality on a market is illustrated in figure 1.4 for a perfectly competitive firm. The marginal private cost curve (MPC) is the firm's conventional marginal cost curve; given the price the firm receives for its product (P), its profit-maximising output is q^*. If the firm produced no wastes, or disposed of all its wastes and the costs of this disposal were included in MPC, then the firm's optimal output would be q^*. However, the firm also produces a pollutant (W) at a rate proportional to the level of output ($W = \Omega.q$); thus the privately optimal level of pollution is $W^* = \Omega.q^*$. The 'marginal external cost' curve could be understood in one of two ways: either it is the marginal cost imposed on others because of the pollutant's detrimental effect; or, alternatively, it is the cost that the firm would bear to prevent discharge of the pollutant (i.e. if the cost of the pollutant were 'internalised'). The marginal social cost of producing commodity q is the firm's marginal private cost, plus the marginal external cost of the wastes. The socially optimal level of commodity output is therefore q** (marginal social cost equals marginal revenue) and the optimal level of pollution is W** = Ω.q**. Australian agriculture, and other natural-resource-using activities such as mining, potentially produce a range of externalities, especially saline water.

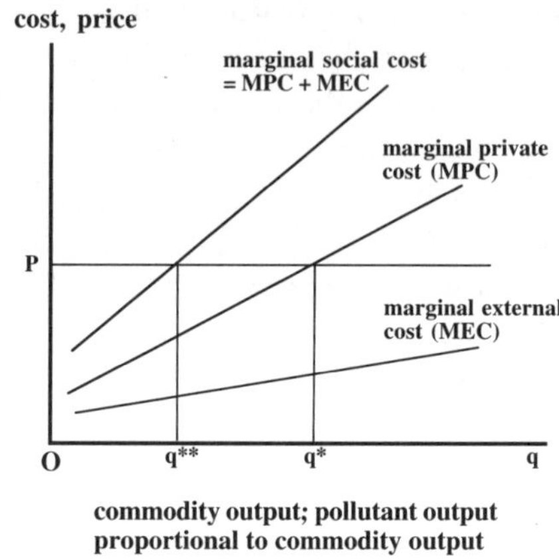

Figure 1.4 Production with Externality

1.3 STANDARD GENERAL EQUILIBRIUM ANALYSIS[1]

In the preceding sections, models were presented relating to a single firm, a single commodity or a single industry. These models allowed for some interaction among different enterprises, or for competition in consumption between different grades of a commodity, or different types of a generic commodity such as meat. It is sometimes desirable to examine the interactions of all firms, industries and consumers—i.e. to examine effects on all the components of the agricultural sector of such things as changes in exchange rates, or changes in the efficiency of non-agricultural industries like banking. Two possible methods of examining such effects are possible: the first is to aggregate all agricultural industries into a single composite agricultural industry; and the second is to examine disaggregated agriculture along with disaggregated non-agricultural industries (e.g. Part IV below).

The necessary conditions for economic efficiency in a general equilibrium context can be simply interpreted in price terms. The first efficiency condition is that the relative prices of inputs will be the same, no matter what commodity they are used to produce. The second condition is that efficient consumption means all consumers will pay the same relative prices for commodities that are homogeneous with respect to time, place and form. The third condition is that the relative price all firms charge for their output will be the same for commodities homogeneous with respect to time, place and form. The fourth condition is that the relative prices charged by producers will be identical to the relative prices paid by consumers.

General equilibrium efficiency conditions are developed by considering voluntary markets in inputs and outputs. While the above four conditions are necessary for economic efficiency, this does not mean that the existence of voluntary markets (i.e. a free market or capitalist economy) is a necessary condition for attaining economic efficiency. It is conceivable that a benevolent dictator could—by knowing all factor endowments, all production technologies and all individual preferences—unilaterally set all the relative prices consistent with the above conditions, and direct the disposition of inputs and outputs to firms and consumers determined by these relative prices and technologies and preferences. The economy would thus be efficient without having free markets. This outcome is, however, highly unlikely, because of the enormous costs required to aggregate information about endowments, technologies and preferences to some central analyst; for this analyst to compute the appropriate prices and consequent physical flows; and to ensure that the flows of factors and products to firms and consumers were as computed. Since information is itself an economic commodity, an inherent virtue of a free market economy is that it economises on the use of information—information is only transmitted to those agents who require it for decision-making.

The standard general equilibrium model has a considerable number of assumptions, not all of which are obvious (e.g. Rowley & Peacock 1975, pp. 11–12). Since this book is primarily concerned with policy analysis, the approach taken is to note the restrictions of the model and be prepared to relax those assumptions which are manifestly inappropriate to a particular problem being examined.

1.4 MAXIMISING WELFARE

In a simple two-person economy, a utility possibility frontier may be constructed which represents the set of (Pareto) efficient points for the economy (figure 1.5). A Pareto efficient frontier comprises that set of points where it is not possible to improve the welfare of one individual without harming the welfare of another. Under Pareto efficiency there is an infinite number of possible combinations of utility of the two individuals—and, therefore, an infinite number of allocations of inputs among different goods and firms, and an infinite number of allocations of the commodities among individual consumers.

If there was a 'rule' for determining how well off society was with different combinations of utility of its citizens, then it would be possible to determine the corresponding combinations of the utility of individuals that gave rise to this maximum welfare. One such rule might be to define the level of social welfare (SWF) as a function of the utilities of individual consumers (Uj):

$$SWF = SWF(Uj, \text{ for all consumers } j)$$

Assuming the existence of such a rule, the curve SWF1 in figure 1.5 represents different combinations of welfare of the individuals M and F that give rise to a constant level of social welfare. Point Z represents the maximum level of social welfare. The combination of allocations of inputs to production, production among firms, and commodities to consumers that gave rise to this maximum level of social welfare could conceptually be determined.

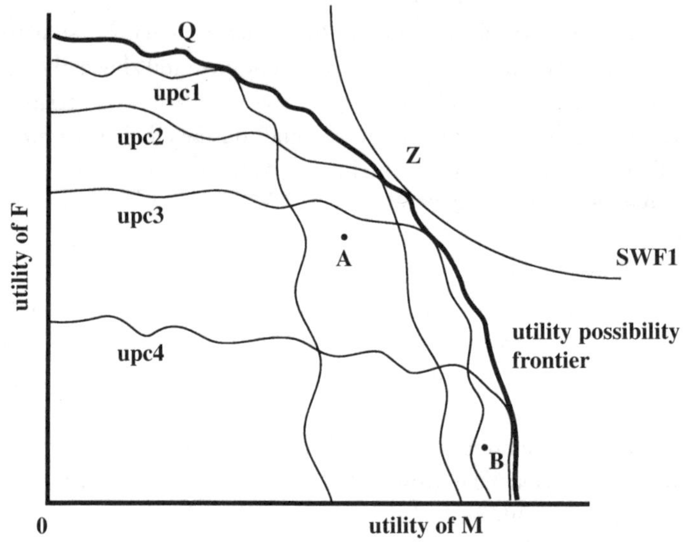

Source: Randall (1987, figures 5.9 and 7.2)

Figure 1.5 General Equilibrium—Utility Possibility Curves and Frontier

However, it is unlikely that an economy would be on its Pareto efficiency frontier. Because of many inefficiencies in the economy, and government intervention, it is more likely that the economy is somewhere inside the frontier. If there were many inefficiencies in the economy, it is tempting to think that each correction of an inefficiency would point the economy towards an improved outcome. However, Lipsey and Lancaster's 'General Theory of the Second Best' indicates that elimination of a particular inefficiency will not necessarily improve aggregate welfare (e.g. Stiglitz 1988, p. 479). This result may be illustrated using figure 1.5—suppose inefficiencies result in the economy being at point A. Suppose one of the inefficiencies at A is eliminated, and the utility of individual consumers changes. Efficiency may improve, and the welfare levels of M and F move towards the Pareto efficient frontier; the welfare position might move towards Z, or it might move to B. Depending on the shape of the social welfare function, a move like that to B might result in a lower level of welfare even though efficiency has improved. It is therefore a matter of empirical analysis whether or not any individual improvement, or set of improvements, in efficiency will increase economic welfare.

An additional point can be illustrated using figure 1.5. Consider any point Q on the utility frontier other than Z. Since any point on this frontier represents an efficient point, shifting around the frontier cannot improve efficiency. Such movements, however, result in welfare changes—as long as a rule can be defined for deriving social welfare from individual utility (i.e. a social welfare function exists). Despite Arrow's theorem ostensibly proving the impossibility of formulating a democratic constitution to derive such a welfare function (e.g. Mueller 1989, pp. 384–99), *applied* economists quite happily work with implicit social welfare functions. For individual commodities, changes in economic surpluses (either Marshallian or Hicksian) are used to measure changes in the level of welfare; applied economists use Gross Domestic Product as the indicator of national social welfare.

In applied policy, the importance of moves *around* the utility frontier cannot be overstated: social welfare may be improved even if efficiency is not increased. Indeed, turning 'Second Best' on its head, it is possible for social welfare to be improved *even if the economy changes from being efficient to being inefficient*. Thus the search for better social welfare outcomes need not necessarily be confined to identifying efficiency improvements, although economists traditionally focus on these. Government policy action that is not purely efficiency-oriented cannot necessarily be identified as inappropriate on economic grounds alone. Since government policy action for efficiency improvement or welfare redistribution uses resources, welfare redistribution may be rationally preferred as a policy option by government if the net welfare benefits of redistribution (the gross welfare benefits less the cost of welfare improvement) are greater than the welfare benefits of efficiency improvement (the gross welfare benefits less the cost of efficiency improvement). Which measure is to be preferred is a matter of empirical analysis.

1.5 MARKET FAILURE, MARKET IMPERFECTIONS AND GOVERNMENT FAILURE

Traditional ways of testing the desirability of government intervention on *efficiency* grounds were to identify particular types of problems, and to select the best means of overcoming these problems. Two problems identified were 'public goods' and 'externalities'. The contemporary approach is to disaggregate these problems into the characteristics that give rise to them, and to attempt to find solutions to the problematical characteristics. Thus, for example, 'public goods' fall into the bottom left-hand quadrant of figure 1.3 where commodities are 'non-price-excludable' and 'non-rival'. 'Externalities' are commodities with property rights problems (i.e. are on the left-hand side of figure 1.3) and are jointly produced with marketed goods. Thus, for example, modifying property rights might eliminate conditions creating non-excludability. However, not all efficiency problems can be associated with property rights problems, and a broader characterisation of the causes of inefficiency is required. This broader characterisation leads to a useful dichotomy between *market failure* and *market imperfections.*

Market failure and market imperfections

Market failure is characterised by those situations where *transactions costs* are so high that no market exists or, at best, there is only a vestigial market. For example, acute property rights problems in basic research result in such research being seriously under-funded, and therefore under-provided, in the absence of public funding, private philanthropic funding or commercial 'not-for-profit' funding (cf. Chapter 5). Property rights and informational transactions costs also justify public funding of information collection such as that undertaken by the Australian Bureau of Statistics. Negotiation and revenue collection transactions costs justify the public provision of a currency[2], the public definition of weights and measures, provision of internal and external security forces, and provision of monitoring and enforcement procedures for identifying and punishing breaches of laws relating to personal security and private property (cf. private detection and civil dispute resolution procedures). Such activities have been so long provided by the public sector that they are frequently taken for granted and not recognised as examples of market failure.

In the (agricultural and resource) policy context, there are two contemporary challenges. The first is to enquire of existing government activities, which are supposedly justified on efficiency grounds, whether or not there are market failure grounds that justify continued government intervention. For example, have social institutions of property rights changed sufficiently that previously non-price-excludable commodities are now price excludable and thus could be provided by private firms; or has technology sufficiently changed that previous problems of non-rivalry can be circumvented using new technologies? The disappearance of market failure reasons justifying public provision of services may not be sufficient reason for government withdrawing from service provision. For example, even though the provision of 'high culture' such as classical music, ballet, opera and theatre is not characterised by exclusion or non-rivalry problems, there *may* be social reasons justifying their continued subsidy from the public purse—although the subsidies may primarily be provided to the quite well-off. The second contemporary policy challenge is to enquire of proposed policy whether there are market failure grounds justifying the implementation of such policy—with caveats similar to those above.

Market 'imperfections' characterise those situations where there is an absence of *technical conditions* for perfect competition other than those related to transactions costs (cf. Jones 1994b). Market imperfections therefore include:

i natural market concentration (monopoly, oligopoly or monopolistic competition), where efficiencies of scale, size or scope preclude the emergence of a sufficiently large number of firms for atomistic competition;

ii other barriers to entry (e.g. monopolising of key technologies via intellectual property rights or trade secrets leading to prohibitive costs of entry to the industry), which give rise to market concentration;
iii cartels and anti-competitive behaviour; excessive competition (short-run behaviour dominates); factor immobility.

Additional factors particularly relevant to the agricultural and resources sector include risk (especially production risk), which is largely ignored in many microeconomic models (cf. Chapter 7 on marketing stabilisation), and divergent rates of social and private time preference.[3]

Moral hazard and adverse selection

Creation of a market or a policy framework establishes a set of incentives for economic agents. The incentives that are established, however, need not necessarily lead to an efficient outcome. Where there is risk or uncertainty about future events, and economic decisions are affected by the distribution of outcomes, incentives may be created that have perverse outcomes. This is the 'moral hazard' problem. Consider an economic decision which has a distribution of possible outcomes (e.g. rainfall) given by figure 1.6a. With an initial incentives set, the pay-off distribution is given by the curve PP in figure 1.6b—this pay-off distribution is the consequence of an economic agent's decision taking into account the distribution of outcomes. Suppose there is intervention in this activity: for example, an insurance company offers insurance which will provide a pay-out if the pay-off falls below zero (the outcome is less than Q)—the pay-off below Q becomes the curve P1. The economic agent might then choose not just to be compensated below Q, but entirely change its decision-making. The agent might change the selection of enterprises, for example, so that the pay-off is better if the outcome exceeds Q, but the pay-off is worse if the outcome is less than Q—curve P′P′ in figure 1.6b—although the agent continues to be compensated to P1 below outcome Q. Typical examples of moral hazard occur with private insurance—e.g. insurees may become less rigorous with the security of their property if they know an insurance company will replace losses; insurance companies may respond by requiring insurees to undertake specific security activities such as the installation of alarm systems. Other examples include that, where individuals are compensated by government or private philanthropy for natural disasters such as flood or fire, they make take fewer precautions in locating their activities in higher risk areas for these events. Similar activities may characterise agricultural production if government programs compensate for low pay-offs: individuals might be induced to undertake higher-risk strategies because low pay-offs from their optimal risk strategy are subsidised. A classic case is where fodder subsidies during low-rainfall periods defined as 'drought' induce graziers to retain animals that would otherwise be sold because of lack of feed.

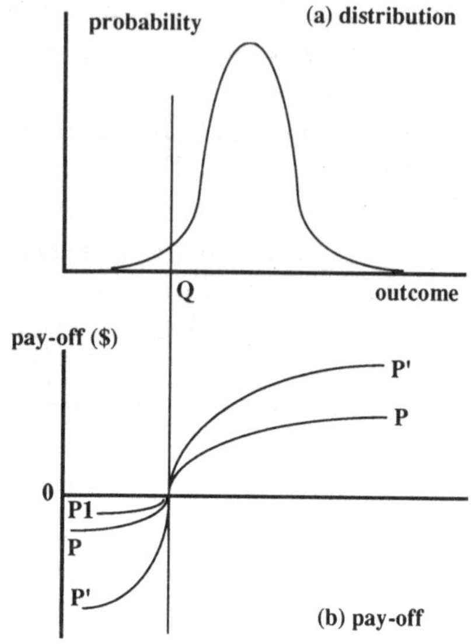

Figure 1.6 Moral Hazard

Imperfect information may give rise to adverse selection. Suppose the distribution of individuals' pay-offs is as shown in figure 1.7 and suppose an insurance scheme is offered to compensate all individuals when their income falls below Y. Individuals whose pay-off always exceeds Y have no incentive to join the scheme even if they were not required to contribute, since they would obtain no benefit from doing so. Only individuals with pay-offs sometimes less than Y could be induced to join. If the scheme required contributions from potential beneficiaries, individuals whose mean benefit from the scheme was less than the contribution rate would be unlikely to join. Individuals whose mean pay-off was below Y would only have the incentive to join if the mean contribution was less than the mean pay-out from the scheme. Suppose the financial feasibility of the scheme requires that all individuals represented by figure 1.7 contribute, but the agency running the scheme cannot distinguish between those whose mean pay-offs are above or below Y. Since only potential beneficiaries would join, the scheme could not be self-financing. Some statutory marketing schemes may have some of the characteristics of adverse selection. A marketing scheme may need to market a high proportion of the crop to achieve competitive marketing costs, but many growers with high quality produce may be able to provide these services themselves or buy these services from private firms at low cost; thus only firms with low quality produce will use the marketing scheme. The scheme would either only receive low prices for the products it markets, or incur high costs in undertaking some form of processing (e.g. grading out contaminants) to achieve higher prices. Since adverse selection would result in only firms with low quality product using the scheme, it would either fail or require (extensive) government funding to keep it solvent.

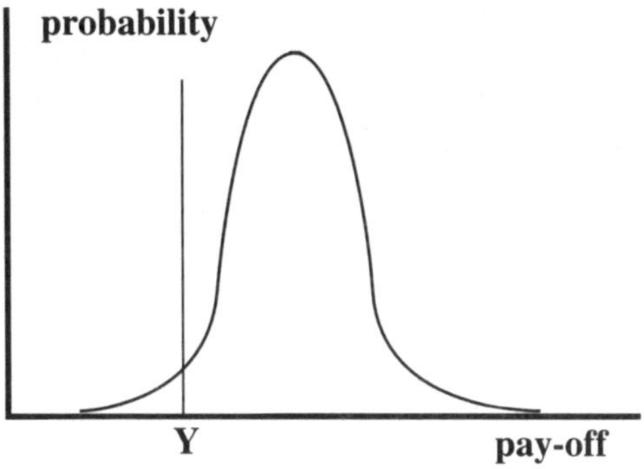

Figure 1.7 Adverse Selection

Potential solutions

Market failure or market imperfections might be overcome by many different means. Invention of institutions might sometimes be required to permit private markets to emerge (e.g. creation of property rights), or additional monitoring and enforcement be provided to prevent the emergence of undesirable forms of uncompetitive behaviour (e.g. the *Trade Practices Act* and the Australian Competition and Consumer Council). Past intervention in Australia has often been seen to require the creation of public production or distribution authorities, for example:

- agricultural statutory marketing authorities;
- banks (both general banking and agricultural sector specific);
- processors (e.g. the former Homebush abattoir and State Brickworks in New South Wales);
- service providers (electricity generation and distribution; water and sewerage; irrigation water; ports, airports and airlines; state agriculture departments' research, extension and veterinary services; CSIRO); and
- transport and storage providers (railways, state grain-handling authorities).

The philosophical and political tide has turned against such public provision, so that many of these once extensively publicly-provided commodities have been corporatised or (partially) privatised. Other forms of intervention include:

- price control or monitoring (e.g. the former Prices Surveillance Authority, gas and milk price control);
- command and control (prohibition or regulation of the production, distribution, use and disposal of dangerous or noxious substances, including pharmaceutical

drugs and agricultural chemicals, controls on fire lighting, controls on land development and building);
- taxation (to reduce use of 'bads' such as tobacco and alcohol, and subsidies to increase use of 'goods' such as investment, especially in R&D and environmentally desirable practices such as tree planting and erosion control); and
- moral suasion to promote 'desirable' behaviours (such as a healthy lifestyle) and to condemn 'undesirable' behaviours (such as risky driving or littering).

A nice example of information transactions costs and policy reaction in agriculture relates to seasonal labour requirements for manual crop picking. In March 1995, a Commonwealth Employment Service (CES) officer, interviewed on Sydney radio when providing information on fruit picking employment in southern New South Wales, commented that the CES relied heavily on the media to disseminate information about employment prospects. Farmers were clearly relying heavily on the CES and the media to recruit casual employees. The CES officer was quite explicit that, except for experienced pickers, the employment was not very well paid.

Government failure

Even if market failure or market imperfections appear to justify some form of government intervention, it does not necessarily follow that such intervention is appropriate. Market failure or market imperfections may be *necessary* grounds for government intervention to improve efficiency but they are not, by themselves, *sufficient.* First, it may not be appropriate to intervene on efficiency grounds because the benefits of intervention may never exceed the costs of doing so. Second, despite its best intentions, government may not be able to improve efficiency. Government failure is said to occur where the costs of intervention exceed the benefits of such intervention.

Conversely, however, simply because the efficiency benefits of government intervention are less than the costs of intervening, this intervention is not necessarily inappropriate. If government intervenes on 'benign' distributional grounds—the net welfare benefits of intervention are estimated to exceed the net welfare effects of achieving the same distributional effects by other purely distributional or efficiency policies—such intervention is economically justifiable. If the welfare gains were unequivocal, and the social welfare function on which the distributional benefits were estimated were unassailable, then the economist as an analyst would be forced to accept that such a distributional policy was superior to an efficiency one—even if the economist as a private citizen happened to disagree with the social welfare function on which the decision was based.

It has become fashionable among economists to label cases of government intervention where intervention costs clearly exceed efficiency benefits of intervention as 'government failure'. There are four possible interpretations of empirical examples of supposed government failure:

i There is government failure. Government has chosen to intervene even though the efficiency benefits of that particular intervention or, indeed, of any possible alternative, could never exceed the costs of doing so even if government intervention were costless.

ii Government is incapable of intervening at a sufficiently low cost that the potential benefits exceed the attainable costs. For example, government might be so hamstrung by the social need for accountability that the costs of this accountability exceed the potential benefits of intervention; or the management or coordination difficulties of operating in a not-for-profit environment may be so large that cost-effective programs are not possible.

iii The benefits of intervention have been inappropriately estimated. For example, if government's objectives are distributional as well as efficiency, it is not sufficient merely to estimate efficiency benefits to compare with the resource costs of programs.

iv Government may be intervening not to promote social welfare, but to enhance its own private objectives, principally the retention of power (cf. Chapter 2).

Policy economists who are neither ideologues nor apologists for particular interests are well advised to be cautious in drawing conclusions about the advisability or inadvisability of particular government policies from efficiency analysis alone.

ISSUES FOR DISCUSSION

1 Evaluate the following argument:
'Markets based on secure property rights provide the best means of efficiently allocating goods and services between different, often conflicting needs.

For market systems to be fully effective in environmental matters, a more comprehensively defined and more securely vested system of property rights is required. But even where collective ownership of property rights is unavoidable, the incentive systems embodied within markets can still have a valuable role in ensuring efficient arbitration between different goals' (*Tasman Institute Annual Review*, December 1992, p. 16).

2 Evaluate the following argument:
'Market failure is a necessary condition for government action, but government intervention in the economy is characterised by government failure. Therefore there are no good grounds justifying government intervention in, or for, Australian agriculture.'

3 'Unless we understand what is meant by "efficiency" we cannot explain the rationale for government intervention in the economy.' Discuss.

4 Discuss whether economic analysis of agricultural and resources policy should be more concerned with efficiency or distributional issues.

5 The following letter appeared in the *Sydney Morning Herald.*

As a new Landcare member, I am only just beginning to understand how to apply for Government funding. Correct me if I am wrong, but it seems to go something along these lines.

Year 1: Application for funding of $500 to aid in the purchase of fencing materials and trees to control gully erosion.

Result: Application rejected—however, please find enclosed a cheque for $100 to assist in Incorporation and Establishment Fees of a new Landcare Group.

Year 2: We have just established the Washaway Gully Landcare Group. Can we now get $500 to aid in gully erosion control?

Result: Application rejected—however, find enclosed a cheque for $1000/member for a Whole Farm Plan and $800 for a Total Catchment Management (TCM) Policy. This will enable your group to define the boundaries and assist in detecting other potential degradation problems which may impinge on the future sustainability of your catchment area.

Year 3: We have just completed our Whole Farm Plan and TCM policy. Is it now possible to get $500 for gully control?

Result: Application rejected—however, we have agreed to appoint a Landcare coordinator to help you and six other groups fill in your application forms correctly [approx cost $40,000 per annum].

Year 4: We thank you for the Landcare coordinator and you will note that all our forms are now correctly filled in. Can we please have $500 for the gully?

Result: Application rejected—but we enclose $1000 for a Gully Control Expert to assess your gully and to recommend a series of strategic control measures to be implemented.

Year 5: We have just had our gully checked by J. Blow (Diploma of Gullies). He says we have a C3 gully with a 4 score erosion component and he suggests we fence it off and plant trees. Can we now *please* have $500 for the gully?

Result: Application rejected—however, we have agreed to give you $5000 to make it a Demonstration Gully Control Site so other Landcare coordinators, soil conservationists, agronomists, CALM experts, etc. can see how to control a gully.

John Hindmarsh, April 4, Nangus

Endnotes

1 General equilibrium economics at the level necessary for this course is well covered in a variety of texts including Randall (1987).

2 Private currencies have existed, such as the legal note issues of the Scottish banks, and still do in the form of 'barter' currencies such as LETS or Local Economic Trading Schemes.

3 Generally given a more complete treatment in natural resource economics—e.g. Randall (1987), Pearce and Turner (1990), Tietenberg (1992).

2 Farming the government?: economic analyses of the policy process

In Chapter 1 it was argued that there are two possible 'benign' explanations of government intervention in the economy. Government might seek to improve economic efficiency or, if the distribution of income is (manifestly) unjust, government might seek to redistribute income to improve social welfare. These explanations of government intervention imply that either all individuals in the policy process are altruistic and so the process itself is altruistic or, despite the non-altruism of individuals, the process itself is altruistic. Neither of these propositions seems likely. The convictions of former government ministers in New South Wales, Queensland and Western Australia; the 'Metherell' affair in New South Wales; the 'Sports Rorts' affair at the federal level; and the Lockheed bribery scandal in Japan[1] strongly suggest that not all individuals participating in the policy process are altruistic. Conversely, it seems improbable that, even if the individual participants are not altruistic, the policy process is by chance altruistic.

An alternative possibility is to explain the policy process by analogy with neoclassical economics. In economics, consumers are assumed to maximise utility, and producers are assumed to maximise profits. These ideas are sometimes expanded so that, for example, households undertaking production and consumption activities are assumed to maximise utility. It seems reasonable to suppose that policy process participants are also utility maximisers, and to investigate the consequences of this supposition. With a loosely economic framework for analysing policy decision-making, economic methods are useful for evaluating the actions of politicians, political parties, governments, bureaucrats and interest groups. As Mueller (1989, p. 1) noted 'Public choice can be defined as the economic study of non-market decision-making, or simply the application of economics to political science.'[2]

In this chapter, various ways are examined of using economics to examine public policy decision-making. While the ideas are useful, too much should not be made of them. Economists can sometimes become social science imperialists, attempting to explain everything via economic logic. There may, however, be many situations in which economic analysis can be pushed too far. There is an analogy here with microeconomics: ideas of optimising behaviour, supply and demand curves, and market equilibrium do not necessarily explain every decision made by every decision-maker. However, microeconomics is useful in explaining the broad sweep of market behaviour. A similar courtesy should be accorded public choice theory—it might be used to explain the broad tendencies of policy decision-making, but not expected to predict every individual decision or account for every nuance of political behaviour.

2.1 BACKGROUND

Analogously to private markets, in political 'markets' participants may be assumed to come armed with utility functions and an optimising objective. The participants of interest are the voting public, political parties, bureaucracies and interest groups; and the individual voters, politicians, bureaucrats and lobbyists who comprise these groups. The political markets are parliaments, the voting processes by which parliamentary representatives are selected, and the back rooms where deals are made. Where policy-making does not involve parliamentary decisions, the debates and consequent decisions similarly occur behind closed doors, and are *sometimes* recorded in parliamentary debates, sometimes in government announcements, but frequently never publicly recorded at all.

Several ideas about the working of the political process are schematically represented in figure 2.1. On its left is a simple model of consumer choice with utility maximisation. On the right are three models of how the political process might work: the first of these, labelled 'altruistic' ('benign') depicts participants in the political process as implicitly analysed in Chapter 1—i.e. these participants are assumed to seek the socially optimal welfare level. 'Altruistic' models of the political process have sometimes been labelled 'public interest' theories of politics. To the right of the 'altruistic' model is a model in which politicians, bureaucrats and interest groups are assumed to maximise utility. This model forms the underlying theme of the present chapter. Finally, on the far right of figure 2.1, the notion of political participants being utility maximisers is welded to the existence of a dominant interest group. This model of political behaviour is often called the 'capture' theory of politics, because the political process is captured by the dominant interest group (Posner 1974).

2.2 DIRECT DEMOCRACY

In democratic Athens of the fifth century BC, public decision-making occurred through direct democracy. Those entitled to vote—adult free males, but not women or slaves—met in the market place to debate and decide. When a decision was required, a vote was taken of all present. This form of 'direct' democracy is rarely used in modern public decision-making, although some forms still exist. Conversely, voting to elect members of parliament, or directors of companies or the committee members of clubs, is called 'representative' democracy (see section 2.3). Direct democracy occurs where the costs of individuals' participation in government—especially the costs of voting—are low relative to the benefits from having all eligible voters participate. It occurs principally with major decisions, such as to change the constitution of some states (e.g. Australia), where this vote is called a referendum, or to change the articles of association of organisations like companies, or the constitution of clubs or societies.

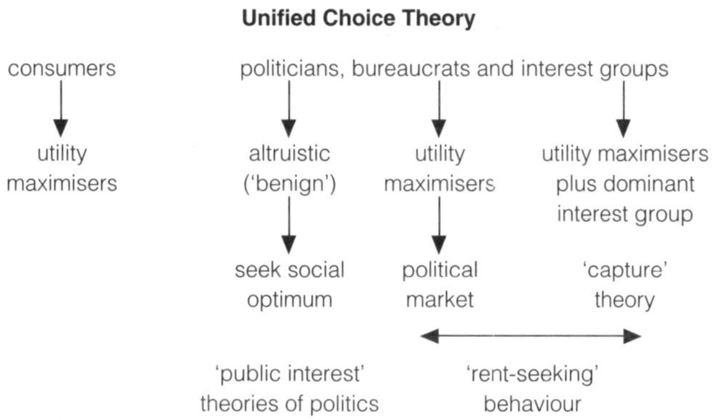

Figure 2.1 Unified Choice Theory

Optimal majority

Where voting is used to decide an issue, there must be a rule to decide the outcome of the vote. The most familiar of these rules is majority voting—the winner is that proposition which achieves 50 per cent plus one of eligible votes. However, the majority rule (or 'simple' or 'bare' majority) is just one of a myriad of possible rules. If a group of voters is called a 'coalition', then potentially the winning coalition in a vote could range from just over 0 per cent to 100 per cent (unanimity) (figure 2.2). However, coalitions of less than a simple majority are not decisive. Suppose votes were taken with a 30 per cent rule and proposition A gained a 30 per cent coalition. A completely contrary proposition B might easily be passed by constructing a 30 per cent coalition from among the 70 per cent of voters who opposed proposition A. The smallest winning coalition where this could not happen is a simple majority of 50 per cent plus one. Thus a simple majority appeals to a commonsense requirement of democratic decision-making, that of preventing contradictory decisions.

But avoidance of contradictory decisions is the most that can be said for a simple majority. If a proposition only achieves a bare majority, this means that just over half of the population could oppress just under half of the population. This leads to a consideration of the notion of an 'optimal' majority. Figure 2.2 shows three total cost curves relating to the size of the winning coalition (S) on some particular issue.

Consider the upward sloping curve, the 'total cost of obtaining a winning coalition' (TCW(S)). Real resources must be expended to convince voters of the merits of a proposition sufficiently that they can be persuaded to support it. These resources include either time and travelling or personal communication costs to contact individual voters, or some form of mass communication. (The possible 'transfer' costs of bribery are ignored here.) The larger the size of the winning coalition required, the larger the cost of constructing it. Not only does the total cost of obtaining a winning coalition rise as the size of the required coalition rises, but it also rises more steeply as the size of the winning coalition increases—voters who are easy to reach and persuade are cheap to add, but voters who are difficult to reach or persuade are expensive (i.e. marginal costs increase).

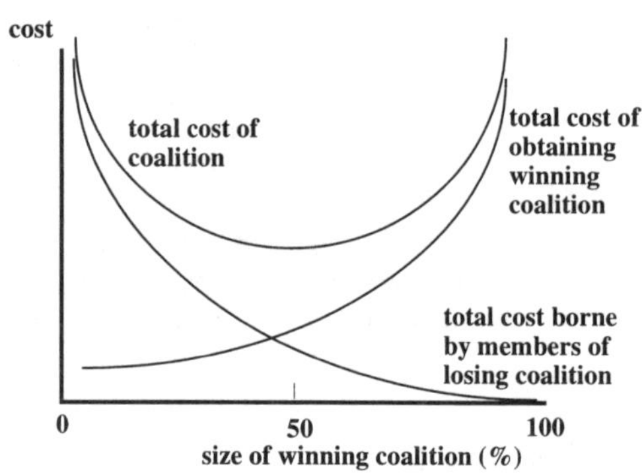

Figure 2.2 Costs of Voting Rules

Conversely, the downward sloping curve in figure 2.2 represents the costs of being in the *losing* coalition. Where the size of the winning coalition is small, many voters are in the losing coalition and there is a high total cost of having a proposition passed with which these losing voters disagree. As the size of the winning coalition rises, the number of voters in the losing coalition falls, and so the total cost borne by members of the losing coalition or TCL(S)—also falls (figure 2.2). This curve is drawn so that, as the total cost of losing falls, so too does the marginal cost of being on the losing side. The total cost of a winning coalition (TCWC(S)) of any given size is therefore the sum of the two total cost curves in figure 2.2:

$$\text{TCWC(S)} = \text{TCW(S)} + \text{TCL(S)} \quad (2.i)$$

An optimal voting rule is one which minimises the costs of constructing a winning majority. Minimising equation (2.i) with respect to S (i.e. taking the first derivative of the right-hand side and setting equal to zero, and assuming the second-order conditions hold) yields the following rule for an optimal majority:

$$\text{MCW(S)} = -\,\text{MCL(S)} \tag{2.ii}$$

where MCW(S) is the first derivative of TCW(S), and MCL(S) is the first derivative of TCL(S). This result is depicted in figure 2.3: the optimal majority occurs where the marginal curves intersect. The minimum total cost can also be seen in figure 2.2: this minimum generally does *not* occur where the two total cost curves cross (unless the absolute values of their slopes happen by chance to be equal where they intersect).

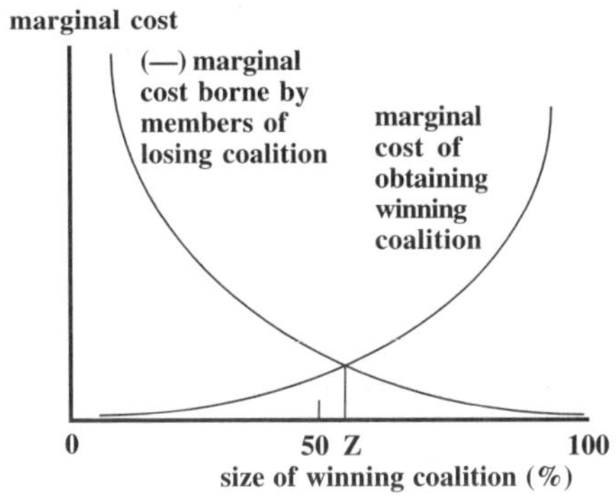

Figure 2.3 Optimal Size of Winning Coalition

These ideas may be interpreted intuitively as follows:

1 The more rapidly the costs of obtaining a winning coalition rise, the higher the corresponding marginal cost curve, and the lower the optimal majority *ceteris paribus*. If voting becomes very expensive in resource terms, the lower is the optimal level of resources and the smaller the optimal majority.
2 The higher the marginal cost curve of being in the losing coalition for any given size of winning coalition, the larger is the optimal majority *ceteris paribus*. For example, the more likely it is that the winning coalition would impose high costs on (i.e. 'oppress') losing voters, the larger the optimal majority.

Referendums

The effect of the size of a majority on the potential costs borne by losers can be used to explain rules often used in direct voting. As previously noted, direct voting is often used when constitutions (of states or clubs) or articles of association (of companies) are being changed and the decision rule may require a majority exceeding 50 per cent. Part of the rationale for such a requirement is that change to a constitution not only determines what will happen to a group now, but will change how that group operates in the future. The potential costs of change may be higher, and more uncertain, than the immediately apparent costs. To help minimise the oppression of large majorities, a large proportion of eligible voters is required in order for the proposed constitutional change to pass.

In figures 2.2 and 2.3, the cost of being in a losing coalition was assumed to be characterised by falling marginal cost. However, the larger the size of the winning coalition, the larger the powers that might be exercised. If oppression occurs then, as the size of the winning coalition rises, the average cost of being in the losing coalition also rises. In figure 2.4, for example, the fall in the number of losing voters is exactly off-set by the rising average cost of being a loser, so that the total cost for losing voters remains constant up to unanimity. The marginal *total* cost of the losing coalition is zero up to unanimity, but the *individual's* marginal cost of being in the losing coalition rises to unanimity—if average cost is rising, so too is marginal cost.[3]

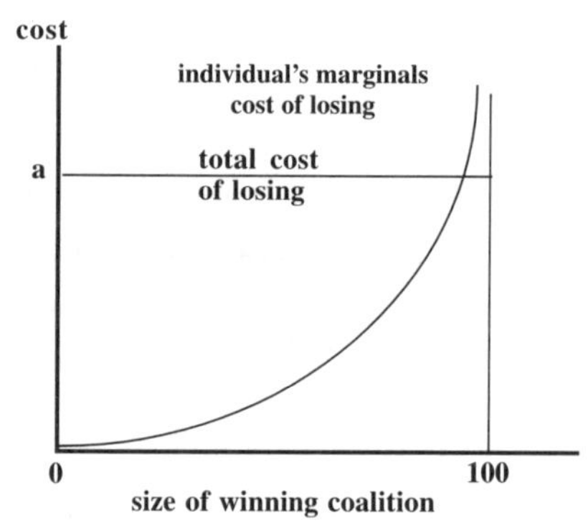

Figure 2.4 Oppression Offsets Rise in Majority Size

Other rules to change constitutions may be more complex than the mere construction of a majority. In the case of the Australian Constitution, the constituency is not simply all Australian electors but is divided into six separate constituencies comprising each of the states. A successful proposition to change this Constitution requires that a majority of all voters *and* a majority of voters in a majority of states approve of the measure. This prevents a small number of states from being oppressed—e.g. a majority of voters in either the two smallest states (Tasmania and South Australia) or the two largest states (Victoria and New South Wales) would only have to convince a majority of voters in one other state to join them to defeat a referendum proposal they found oppressive.

Optimality of majority voting

Once the optimal majority is chosen, it is tempting to assume that the economic problem of voting has been solved. Unfortunately this is not so, because there is a difference between the efficiency of the voting rule, and the efficiency of decisions taken using this rule. Suppose the optimal majority is used to make a decision at a referendum about whether or not to develop a given area of land—e.g. whether or not to log an old growth forest for woodchips. Tisdell (1982, figure 16.5) showed that, for such a decision, majority voting is not Pareto efficient unless a 100 per cent majority rule (unanimity) is used (cf. figure 2.5). Two groups are assumed: developers who would prefer to log all the trees, and development opponents who would prefer that no trees were logged. The marginal costs imposed by development on its opponents is shown as the rising curve; the more that development occurs, the higher the marginal impact. The socially optimal level of development occurs at point R—as, for example, clear-felling R per cent of the area, or logging R per cent of the trees over the whole area. However, majority voting is 'winner-take-all': if developers won the referendum, then they would prefer to log all the trees; if development opponents won the vote, they would prefer no logging. A 'winner-takes-all' vote would not achieve the socially optimal level R. Hence, even though the voting *rule* is efficient, individual *decisions* taken using the rule may be inefficient.

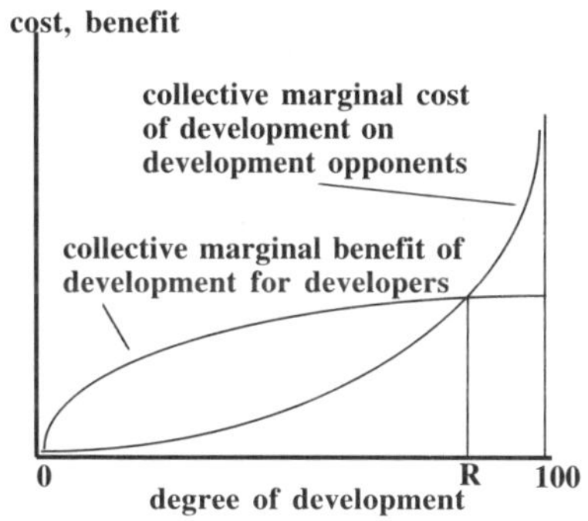

Source: after Tisdell (1982, figure 16.5)

Figure 2.5 Majority Voting Pareto-inefficient unless Unanimous

A unanimity rule, while likely to be inefficient as a *rule*, is Pareto efficient as to *outcomes*. This can most easily be seen by considering a simple exchange model of an Edgeworth–Bowley box (cf. Varian 1990, pp. 276ff.). A bargain by individuals to achieve a point on a contract curve is like a vote with a unanimity voting rule—both parties have to 'vote' in favour of accepting the bargain or the deal is off. That the two individuals would vote to achieve a point on the contract curve where, relative to the starting point, at least one was better off while neither was worse off, demonstrates that the unanimity rule is Pareto efficient.

The key to the inefficiency of majority voting is that, in conventional voting, voters cannot express their preferences for particular actions. In referendum-type voting it is difficult for the large numbers of voters to get together to express preferences and 'trade' votes; this is more a characteristic of representative democracy and is therefore discussed in section 2.3.

Cycling

Sometimes a unique decision may not be obtainable via voting for, as shown above, if the majority rule is less than '50 per cent plus 1', another coalition may be found to obtain a majority which is inconsistent with, or even contradicts, the first decision. However, even with a simple majority rule, inconsistent decisions are possible—for example, where an infinite number of coalitions are possible on such a simple decision as the splitting of $100 among three people (cf. Mueller 1989, table 5.1).

The possibility of such 'cycling' explains why there may be rules governing the procedure by which decisions are made. For example, in New South Wales local government, a 'recision' motion may be lodged to attempt to overturn a majority decision; however, once a recision motion has been put, there is a statutory limitation on subsequent votes (Stuckey 1994, pp. 116–17).

Median voter theorem

Where direct voting is used to make a decision, an important issue concerns how the proposition on which voters will vote is selected. If it is pre-selected, and there is competition to propose alternatives, then the interaction between the proposing of alternatives and the (likely) distribution of voter support is of great interest. Consider voters choosing to support a proposition about a single continuous variable (e.g. tax rates), with two groups vying to propose a tax rate, and the probability distribution of voter support for various values of the tax rate as given in figure 2.6. Voters preferring a lower tax rate than T0 would prefer that rate to the higher T1 (these voters are represented by the dotted area in figure 2.6); voters preferring a higher tax rate than T1 would prefer that rate to T0 (oblique hatched area). The voter who prefers a tax rate exactly halfway between T0 and T1 (at M) is indifferent to T0 and T1, and is called the 'median voter'. Voters preferring tax rates between T0 and M would prefer T0 to T1 ('bricked' area); conversely, voters between M and T1 prefer T1 to T0 (vertically hatched area). Thus, in terms of votes, proposition T0 attracts a proportion of the voting population represented by the dotted and bricked areas; proposition T1 attracts a proportion of the voting population represented by the vertically and obliquely hatched areas. The latter area is larger than the former and so, using the simple majority rule, proposition T1 is chosen over T0.

If, however, the group proposing the losing T0 can modify its proposal—either before the vote is taken, or if there is a series of votes (e.g. each year)—then the group proposing T0 might decide to increase its chances of winning by choosing a higher tax rate—say, T0* in figure 2.7. The voter who prefers a tax rate at the midpoint of T0* and T1 (at N) is the new 'median' voter who is just indifferent between the two rates: voters represented by the dotted area now prefer the low tax proposition T0*, and voters represented by the obliquely hatched area prefer T1. If the group proposing the 'low' tax rate actually wanted to win the vote, they would choose a tax rate just to the right of that supported by the 'median' voter of the entire voter distribution. Of course, the group proposing the 'high' tax rate T1 is unlikely to remain impassive when the 'low' tax rate group is seeking to increase its support by changing the rate it supports. The 'high' rate group will reduce its preferred rate in order to reduce the erosion of its support. Ultimately both groups would offer rates very close to that preferred by the 'median' voter, and would have converged to indistinguishable policies. This result is therefore called the 'Median Voter Theorem' (Mueller 1989, chapter 10).

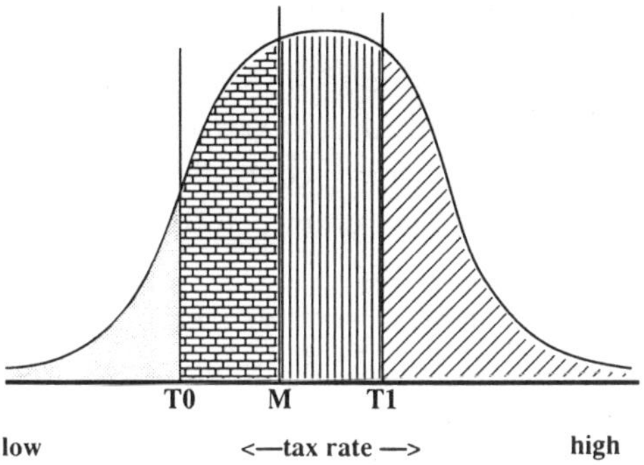

Figure 2.6 Median Voter Theorem—1

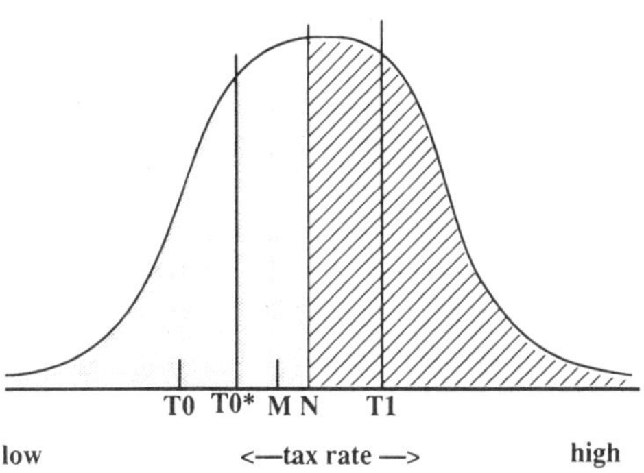

Figure 2.7 Median Voter Theorem—2

The Median Voter Theorem is critically dependent upon underlying assumptions, the principal one being that the proposition being voted on is a continuous variable. Thus, it is possible for proposers of a policy to choose from an infinite number of values of this variable. A second major assumption is that the probability distribution representing how the population regards the policy variable is a continuous distribution. A third major assumption is that this probability is symmetric and uni-modal. These assumptions are often not relevant to policy choices confronting voters in direct democracy. In referendums voters are generally faced with voting for or against a single proposition, rather than choosing between two competing propositions drawn from a continuous policy dimension. There are some proposals where only binomial choice is possible—for example, a voter can only be for or against abortion or the death penalty. Even if the policy variable is continuous, and the probability distribution representing voter support is also continuous, there is no particular reason why this distribution should be symmetric or uni-modal.

2.3 REPRESENTATIVE DEMOCRACY

In section 2.2 it was assumed that, at least for some decisions, direct democracy was both optimal and feasible. If, however, the costs of obtaining a majority exceeding 50 per cent of voters (necessary to avoid passing contradictory propositions) is so high that a simple majority is not feasible because of its high cost, some other form of voting is required. The form to which most modern democracies have turned is 'representative' democracy: instead of being required to vote for every public decision, voters choose representatives, from time to time, to make decisions on the their behalf.[4] The interval between choosing representatives varies considerably: strictly two years for members of the US House of Representatives, a maximum of three years for the Australian House of Representatives, strictly four years for the US President, four years for the New South Wales Legislative Assembly (following a referendum in 1995) and so on. Longer election cycles confer benefits of reduced average annual costs of elections (because there are fewer of them) and more opportunities for governments to take unpopular decisions. The costs of longer election cycles are that governments may become less accountable to the electorate, and thus less democratic.

Choosing representatives

In all lower houses of the Australian federal and state parliaments except Tasmania, parliamentarians represent 'single-member' constituencies: that is, one member of parliament is chosen to represent each electorate. Australian Senators, however, represent states (i.e. a multi-member electorate), where each state comprises a single electorate. The lower house of the Tasmanian Parliament uses the State's five federal seats as five state electorates, but elects seven members for each electorate. Until 1996, New Zealand had single-member electorates for its single chamber of parliament; from 1996 it will use mixed-member proportional representation (MMP), in which half the members are elected in single-member electorates, and the other half from a single national electorate.

The 'parliaments' to which representatives are elected are generally bicameral—that is, the parliament comprises two separate 'houses' or 'chambers'. The Australian Federal Parliament and all Australian state parliaments except Queensland are bicameral (Queensland abolished its upper house in 1922). The New Zealand Parliament is unicameral (it only has one chamber). The United Kingdom's Parliament is bicameral, but representatives are only *elected* to one chamber, the House of Commons; the House of

Lords comprises hereditary and life peers and some ex officio judges and bishops (Rose 1989, p. 119).

Parliaments generally adopt a simple majority rule for their own decision-making. Conceivably, these majorities could form and re-form for each individual decision. However, for a variety of reasons—not the least of which is the cost of decision-making even by quite small groups—representatives tend to coalesce into relatively stable groups of relatively like-minded individuals, which are called political 'parties'. In the twentieth century, particularly in parliaments of English-speaking countries, there have tended to be broad parties of the Left and the Right. In all these countries, however, there have also been minor or regional parties (e.g. Liberals, Social Democrats, Scottish Nationalists and Ulster Unionists in the United Kingdom; Democrats in Australia; and the New Democrats and French Québécois parties in Canada).

In Westminster-style democracies, a party or coalition of parties with more than half the members in the lower house forms the government; the remaining members form the opposition. The leader of the government becomes the Head of Government (generally Prime Minister or Premier), and some other members of the government become ministers, who share executive power (i.e. the power to make decisions). In some cases, it may not be possible to create a formal coalition, in which case a minority government forms, where representatives not in the formal government coalition explicitly or implicitly agree to support the minority government, at least on key legislation.

In monarchies like Australia, New Zealand and the United Kingdom, the monarch is the Head of State; in the first two, a Governor-General is the day-to-day Head of State except when the foreign monarch comes visiting. Democracies may also be republics whose presidents may have largely ceremonial powers (e.g. Ireland, Germany) or executive powers (e.g. France, USA). In democracies with executive presidents, the president is generally elected independently of the house(s) of parliament and is both Head of Government and Head of State. In general these houses of parliament have legislative power (to make laws) but not executive power (to implement laws), which is reserved for the president.

While forming a majority government in a Westminster-style democracy requires at least half the representatives, this does not mean that the government must win at least half of all the votes cast in an election. Theoretically, all that is required is that a party wins '50 per cent plus 1' of the votes in '50 per cent plus 1' of the electorates: i.e. just over one-quarter of the total votes are actually required although such results are unusual.

The method of voting for representatives also varies widely. The United Kingdom's House of Commons uses the simplest system: the winner of an election is the candidate who scores the most votes irrespective of whether or not that candidate obtains a simple majority ('first-past-the-post'). Where there are only two major political parties, the result is the same as majority voting. However, where there are more than two major parties, the winner may obtain substantially less than 50 per cent of the votes.

An alternative to first-past-the-post is often used: in this variant, the two (or perhaps more) candidates scoring the highest number of votes in a first round of voting, contest a run-off election and the winner is declared elected. The French President is elected by such a process. A disadvantage of this system is that, where there are more than two candidates in the first round, a second election may be necessary; thus the cost of holding a single election is doubled.

Preferential voting (or 'exhaustive' voting, Mueller 1979, p. 60; or 'plurality' voting, Mueller 1989, p. 112) is a variant of multi-stage voting used for federal and state elections in Australia. In preferential voting, voters are asked in a *single* ballot to number all the candidates in order of preference. In counting ballot papers, if a candidate fails to obtain a simple majority, the candidate with the *least* number of votes is excluded, and their second preferences (the other candidates' names which have a '2' against them) are distributed as if they actually had a '1' against these names. If still no candidate obtains a simple majority, the candidate with the least number of votes is again excluded and their second (and perhaps third) preferences are distributed to remaining candidates. This procedure is repeated until one candidate obtains a simple majority comprising their own first preference votes and the preferences of excluded candidates. The outcome is similar to a multi-stage voting process in which the candidate with the least number of votes in each stage is excluded from the subsequent stage. In multi-stage voting, however, voters have information at each stage about who has been excluded; in preferential voting, voters have to guess who will be excluded at each stage to exactly replicate the outcome of multi-stage voting. 'Optional' preferential voting occurs where not all candidates need to be numbered in order of preference for the vote to be valid; in 'list' preferential voting, voters only vote for the party they prefer, and preferences are distributed according to the party's (pre-announced) preferences.

More complicated systems of vote counting are required when preferential voting is used in multi-member constituencies. The Droop quota system is used in Australian Senate elections and for the Legislative Councils of Western Australia, South Australia and New South Wales, while in the multi-member electorates of the Tasmanian Lower House, the Hare-Clark system is used (Jaensch 1995, pp. 30–1). There are also indirect forms of election: in elections for the President of the United States, voters select delegates—members of the 'Electoral College' in their state—and members of the 538-member Electoral College formally elect the President.

With bicameral parliaments, it makes sense to have different forms of voting for each house, otherwise each chamber would tend to mirror the other. Thus in Australia for example, single-member constituencies in lower houses may be contrasted to state-wide or large provincial electorates in upper houses with proportional representation. The term of upper house members is also frequently different, often being twice the term of a member of a lower house. These differences may simply reflect the historical process of change from highly restricted franchises in lower houses and frequently the appointment rather than election of members to upper houses.

Political parties as firms

The policy process uses resources, and the key resource is information. The range of issues on which a government is required to make decisions is enormous: economic (including macroeconomic management, taxation and expenditure, industry policy), defence, foreign affairs, legal, health, education, and income support. Part of the response to this information problem is to form a bureaucracy to administer the policy that government (or parliament) chooses, and to advise government upon which policies to enact, and how. In theory, at least, the bureaucracy only solves the technical side of the information problem, and only for the political party forming the government. Government still has to solve the problem as to which policies electors and interest groups would be more likely to support. Thus, as well as a bureaucracy, the representatives also require a mechanism to advise them of the (likely) popularity of measures the government might enact.

Because of the wide range of information required by parliamentary representatives, and because—as discussed in Chapter 1—information has characteristics of price excludability and non-rivalry, there are potential problems in the efficient production of political information. Partly for this reason, politicians have banded together to form political parties to economise on the resources required to produce information. Such parties began as loose groupings of like-minded representatives, and broadened to create a *party* bureaucracy. Because of the resource efficiencies of party organisation, even if political parties had not developed with the historical development of democracy, it would be necessary to invent them. As well as economies in the production and use of information, political parties also provide the opportunity for specialisation of labour in the myriad issues on which parliamentarians require specific information.

In the context of the Median Voter Theorem (cf. section 2.2), the groups offering competitive policies from which voters choose are generally (although not exclusively) political parties. The Median Voter Theorem may be used to explain both the activities of parties, and also of individual candidates. Consider first the party: suppose that the characteristics of parties can be broadly categorised on a Left–Right political spectrum, and that this spectrum is a continuous variable with an infinite gradation of political stance from 'Extreme Left' to 'Extreme Right' (figure 2.8). As long as the distribution of voter support along this Left–Right spectrum satisfies the assumptions required for the Median Voter Theorem—that the voter distribution is continuous, uni-modal and symmetric—then the previous logic can be used to show the consequences of parties choosing particular positions on the Left–Right spectrum. If the broad characteristics of political behaviour are roughly consistent with the Median Voter Theorem, a political system with two major parties would be expected to converge around the mean. For example, after the Federal Labor Cabinet decided to introduce resources security legislation on 7 March 1991, Seccombe (1991) reported that 'All the major Green groups now say that they find Labor and the conservative parties virtually indistinguishable on environmental issues'.

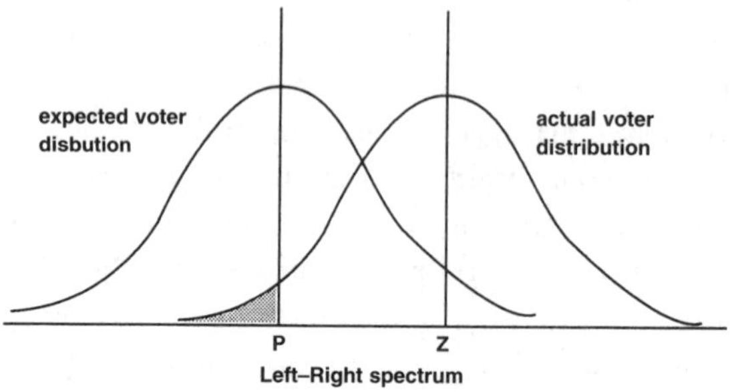

Source: Tisdell (1982, figure 16.3)

Figure 2.8 Expected and Actual Voter Distributions

One inference of using the Median Voter Theorem to explain political party behaviour is that, in attempting to construct a winning position on the Left–Right spectrum, parties must appeal not just to voters of like minds, but to coalitions of voters with different views. Thus, parties' platforms are likely to be formulated with a degree of imprecision so that voters who do not agree with the precise form of the final decision can be induced to vote for the party's general form of the policy prior to the decision actually being taken. Downs noted that 'each party traditionally produces a large output of sanctimonious platitudes praising the flag, motherhood, and the home' (quoted in Mueller 1979, p. 117).

Similarly, political parties themselves are coalitions of like-, but not necessarily identically-, minded individuals. The Australian Labor Party, for example, is made up of three formal factions: the Left, the Centre-Left and the Right (i.e. conservative). J. Howard, leader of the Federal Liberal Party, described that party as a 'broad church' and a 'coalition of liberals and conservatives' (AM Program, ABC Radio 8 March 1994). However, the spectrum of political views within political parties is not likely to represent the entire Left–Right political spectrum, or even that part of it from which the party gains its adherents. For example, voters around the mode of the Left–Right spectrum in figure 2.8 are likely to have more in common with those voting for the other party, than for people who vote the same way as they do but are on the extremes of the spectrum. Individuals around the median, therefore, are less likely to join a political party, and membership of a political party is likely to be more weighted towards the centre of the coalition from which it draws its support.

This difference between the distribution of political attitude among party members and among the population as a whole has implications both for the party and its candidates. For parties that hold internal elections (pre-selection), aspiring candidates for public office first go through an election with a voting population that represents party members, but which may be unrepresentative of the electorate as a whole, and even unrepresentative of the voters likely to support that party. Thus a candidate may have one 'platform' (set of policies) for winning pre-selection, and then develop another platform, or adopt the party's platform, when facing an election. This problem of different platforms for candidacy and election is especially complicated where a candidate stands for an electorate which is atypical of the electorates in which their party is usually successful. Thus, for example, rural candidates and parliamentary representatives of the Australian Labor Party have typically been much more conservative than their city counterparts.

The analysis of political party behaviour is not much changed if a uni-modal distribution of voter preferences is asymmetric, but may be substantially upset if the distribution is not uni-modal. Further, the preceding analysis has assumed that voters' voting intentions are known with certainty, whereas political parties can only *predict* voters' reactions to party policy platforms and thus their voting intentions. The implications of wrong predictions—and thus the value of information about voting behaviour—is illustrated in figure 2.8. If both political parties selected P on the basis of the expected voter distribution, but the actual voter distribution was symmetric around point Z, then the 'Left' party would only capture the proportion of voters shown by the dotted area and would be annihilated in the election. The potential consequences of incorrectly predicting voters' preference explains the considerable resources political parties devote to public opinion polling.

Barriers to entry

A consequence of democratic systems with two major political parties staking out their claims towards the median of a symmetric uni-modal distribution of voter intentions is the effect on new parties. Consider figure 2.7 if the tax rate were replaced by a Left–Right spectrum as the variable over which voters have preferences. Two parties stake out their position on this Left–Right spectrum as T0* (Left) and T1 (Right), and establish a tradition of support among voters on either side of their policies. The two parties can afford to a large extent—although, perhaps, not completely—to ignore the extreme of the spectrum on their side of the median. Thus, for example, the Australian Labor Party has tolerated some of its members flirting with extreme groups, but has generally fought not to have its policies dragged too far from the centre. Similarly, the Liberal Party has tolerated groups and policies towards the extreme Right. Major parties can afford to be tolerant of extreme parties because, by definition, a party established with extreme policies is unlikely to be able to appeal to sufficient voters to have its policies accepted, and preferential voting means the parties closer to the centre can capture their votes anyway. Similarly, because the major parties tend towards the centre of the distribution, there is little room for a centrist party to establish, create an identity separate from the existing parties, and create a band of faithful supporters. Establishing a presence is more difficult against incumbent duopolists, who generally have good access to resource funds for advertising their messages, while parliament itself provides an excellent form of free advertising to which new parties, by definition, do not have access.

Barriers to entry in the political market are reinforced by the behaviour of the established duopolists. Especially in a first-past-the-post voting system (as in the United

Kingdom), but even with preferential voting (as in Australia), a new party has to overcome a substantial threshold in order to compete with existing parties. Thus, for example, neither UK Conservative nor Labour governments have been interested in moving from first-past-the-post voting or to multi-member constituencies which would have given a larger representation to the long-established Liberal Party or to the more recent Liberal–SDP Alliance. For the 1987 British election, Rose (1989, table X.1) estimated that, compared to an outcome from 'pure proportional representation', the Conservatives won 100 more seats (in the 650 seat House of Commons), Labour won 29 more, and the Liberal–SDP Alliance won 125 seats fewer. Similarly, most Australian governments have shown little interest in moving towards multi-member constituencies, which would more closely align votes cast with the percentage of seats won. Only in Australian parliamentary chambers which have multi-member constituencies have minor parties established a permanent toe-hold. Thus the Greens in Tasmania have won up to five seats in the lower house, the Democrats in the Federal Senate have won up to seven seats, and the Democrats and the religious-conservative Call to Australia Party have each won two seats in the New South Wales Legislative Council. Further, Australian incumbent political parties have reinforced their strategic advantage against new parties by establishing public funding of political parties, where the value of the public funding is related to the proportion of total votes the party received at the previous election (Jaensch 1995).

An implicit assumption of the previous argument is that either there is a single electorate (and, for example, proportional representation) or individual electorates are homogeneous (and thus the distributions of their voters all mirror the national distribution). Clearly, however, electorates differ. In Australia, rural towns tend to be relatively small, and thus their interests have tended to be those of rural Australia, particularly agriculture and industries servicing agriculture: electorates in these areas tend to vote National (formerly Country) Party in New South Wales and Queensland, or Liberal in the other states. The exceptions to this generalisation are coastal Queensland, which tends to have large cities (Townsville, Cairns, Rockhampton), and areas where mining is important (e.g. Broken Hill, Kalgoorlie, Mt Isa), which often elect Labor members. Similarly, within major urban areas, wealthier suburbs tend to vote Liberal, and poorer suburbs tend to vote Labor. Thus, each of the Australian political parties tends to have a large number of seats that it is unusual for that party to lose and the real contest is over the 'marginal' seats, those which change parties relatively frequently. Political parties are increasingly learning to target the seats more likely to change—in particular, to concentrate their attention and resources on the 'swinging' voters in these marginal electorates. Opinion polls tend to be taken frequently in marginal seats and policies are especially directed to their interests. The economic aspect of marginal seat targeting is that it makes more efficient use of a party's scarce resources. Rather than averaging its campaign resources over all electorates, and thus 'wasting' them in electorates it would have won even without effort or which it could never realistically hope to win, a party targets its resources on those seats which either it could win from its opponents, or which it might lose without special effort. A party cannot afford to ignore its traditional support, but it will appeal to them as cheaply as possible (e.g. via generic advertising on television and radio and in the printed media).

Political parties not only erect barriers to entry against potential rivals, but also for existing rivals. There are two classic forms of such barriers to entry, both commonly referred to as 'gerrymander' (Johnson 1991, pp. 188–90). In the first form, the boundaries of electorates are arranged so that there is a manageable minority of voters who traditionally support other candidates, but never so many as to permit the opposing party to win the seat. The second form of gerrymander, more practised in Australia, is more accurately called 'malapportionment'; it occurs where a party can arrange to have seats with relatively small numbers of voters where its support is strong, and seats with relatively large numbers of voters where its support is weak. In Australian federal electorates until the 1970s there was a 20 per cent 'tolerance' in the number of voters in a electorate: rural seats generally had fewer electors than urban seats. Large differences in the size of electorates for state governments persisted in Queensland and Western Australia well into the 1980s (Jaensch 1995, chapter 4). The ostensible rationale for rural electorates having smaller number of voters is because low population densities make communication between local member and constituents more difficult and thus more costly than in more densely populated urban seats.

Duopolistic political competition

A consequence of the barriers to entry that political parties manage to erect is highly concentrated political markets. For example, Westminster-style parliamentary democracies such as in Australia are traditionally dominated by two parties. One argument proposed to support such domination is the 'instability' of government where there is a larger number of parties. In Western Europe, France before the Fifth Republic (1958) and post-War Italy, which has had over 50 governments (SBS 1994), are typically used to demonstrate the undesirability of highly competitive—and therefore highly unstable—political markets. Other countries have used mechanisms to constrain the burgeoning of political parties; Germany, for example, requires a party to obtain a minimum of 5 per cent of votes to be eligible to be represented in parliament.

Similar arguments have been proposed in Australia where, particularly in New South Wales in the period 1991–94, non-party politicians (independents) held the balance of power over a minority Coalition government, and contributed to the downfall of a premier. Some have even argued against the right of independents to be elected to parliament (e.g. Baxter 1994). Arguments about the 'excessive' power wielded by independent parliamentarians are somewhat disingenuous. Even if a political party has a clear parliamentary majority, that party is a coalition of individuals. Just as independents or small parties may wield decisive power over a minority government, so individuals or small factions *within* a party may wield similarly decisive power over a *majority* government. Small factions within governments may therefore dominate parliament even more effectively than minor parties or independents. Further, curtailing the power of independents would reinforce the barriers to entry enjoyed by existing parties.

Political parties, being coalitions, suffer the risk of defections on parliamentary votes. Labor parties, in Australia and elsewhere, pioneered the concept of 'caucus', in which the decision made by the parliamentary party was binding on the individual parliamentarian. A member who breaks the caucus can expect expulsion from the party. The ALP's concept of caucus is often implicitly followed by other parties—although usually more discreetly—who may threaten a parliamentarian's pre-selection if they do not toe the party line, or do not toe this line sufficiently often. Party rigidity under duopoly reinforces the barriers to entry enjoyed by existing parties, which are conferred by the historical accident of early entry. In recognition of the 'coalition' nature of political parties, some parliamentary votes are 'conscience' votes, when the nature of the decision is so controversial that requiring caucus could split the party. In Australia, social issues like abortion, the death penalty and euthanasia are typically matters for a conscience vote.

The previous analysis has considerably simplified the political 'market' by assuming that voter intentions can be represented by a uni-dimensional Left–Right spectrum. Clearly, the real world is more complex than this, even though this simple model does seem remarkably robust in explaining political behaviour.

2.4 PARTICIPANTS

To examine what politicians do once they are elected it is necessary to examine their motives in becoming politicians. Breton argued that a politician has in mind 'personal pecuniary gains, personal power, his own image in history, the pursuit of lofty personal ideals, his personal view of the common good' (quoted in Mueller 1979, pp. 148–9).[5]

For the individual politician, all these aspects determining utility are contingent on being elected, or re-elected, to public office. Thus the politician's problem may often usefully be collapsed into the 'reduced form' of the politician's problem—the problem of election. Public choice theory generalises the notion of the politician's optimisation problem and assumes that all participants in the political process are utilitarians, and have objective functions that they attempt to maximise subject to any constraints that they face (cf. section 2.1). Participants in the political process whose actions may be considered in this way include not only individual politicians and political parties (and their incarnations in parliament as 'government' or 'opposition') as 'firms' of politicians, but also voters, the bureaucracy, interest groups, professional lobbyists, and the media.

Bureaucrats' objectives are not necessarily the same as politicians', even though their nominal role is as the independent implementers of government policy. Niskanen argued that bureaucrats' objectives include 'salary, perquisites of office, public reputation, power, patronage, output of the bureau, ease of making changes, and ease in managing the bureau' (quoted in Mueller, 1979 p. 158).

While the 'bureaucracy' is often thought of as a monolith, it may also be usefully considered as a set of firms whose general function is to support the government but whose constituent parts are often competitors. For example, finance departments—in Australia the macroeconomic policy department of Treasury and the department of Finance, which monitors the expenditure of other government departments—are often in conflict with 'spending' departments (especially in the areas of health, education and social security), which see their role as maximising spending for the 'public good'. Developmentally oriented departments—especially those concerned with mining, forestry, water and agriculture, and which see their roles in terms of the intensified use of natural resources—are often in conflict with departments such as environment and national parks, which see their role as conserving or preserving natural resources.[6] The 'bureaucracy' includes not only government departments but also a multitude of quangos, which carry out specific functions (cf. Chapter 3).

Interest groups

Interest groups are those who seek to persuade government to undertake specific actions such as legislating or modifying existing legislation. There is a wide range of interest groups: associations representing broad or specific industry groups; professional associations; trades unions; and cultural, sports, ethnic, religious or welfare groups. Additionally there are ideological 'think tanks', more common in the USA but of growing importance in Australia; professional lobbyists; and consultants (cf. Chapter 3).

The public electronic and print media form a separate interest group. They are a lobby group in the sense that they have specific interests to protect and enhance. For example, the Australian Federal Government has rules on the degree of foreign ownership of domestic media, and there are also cross-media ownership rules to limit the degree of concentration in the combined print and electronic media. However, the media are also the politicians' (and other groups') principal conduit for disseminating messages to other interest groups and the public/voters. The media is also the conduit for paid advertising by political parties and interest groups at the strategically important election periods.

'Private' media, which are attached to specific lobby groups such as farm organisations, have newspapers and newsletters to communicate with their constituents. The rapid reduction in printing costs since 1980 with the personal computer revolution, and accompanying reduction in costs of electronic communication, has greatly increased opportunities for communication between organisations and their constituents. This cost reduction has especially benefited smaller lobby groups, and has dramatically increased their influence over political parties.

The further development of electronic communication, particularly facilitated by national and international computer networks (e.g. the Internet and its commercial rivals) carrying print, audiovisual and interactive facilities, will offer both the public and private communication media extensive opportunities for developing communication with their clients.

2.5 POLITICAL DECISION-MAKING IN AN ECONOMIC FRAMEWORK

Standard neoclassical microeconomic theory is made easy by the assumption that consumers consume, and producers produce, homogeneous goods. The problem is also often simplified by assuming relatively simple functional relationships defining consumption and production behaviour. Consumers' and producers' optimisation problems are easily represented algebraically, and these individual decisions are easily aggregated into demand and supply curves to examine optimal allocation of resources and equilibrium. However, in public choice theory, there are no homogeneous goods which regulators produce, and interest groups and voters consume. The 'market' for regulation is a messy one of discrete choice.

Voter behaviour

In public choice theory, the benefits (B) that a voter obtains from the public sector are a function of the baskets of publicly provided goods and services they consume (Mueller 1979, p. 97). Voters potentially have some control over these benefits by allocating their vote at periodic elections. However, because in general the individual voter is one of many, the return that they obtain from their individual vote is stochastic, and depends on the probability (P) that their individual vote will make a difference during the election. Thus, the utility an individual obtains from voting is the product of B and P (equation 2.iii) where R is the net return from voting. The voter must incur some expenditure (C) in voting, even if this is only the opportunity cost of attending a polling booth. The more informed the voter wishes to be, the greater the costs that must be incurred in obtaining information on which to base a choice. As well as the pecuniary benefits of voting, the voter may obtain some vicarious pleasure (D) from participating in the election process (e.g. the entertainment value of seeing politicians plead for votes). Thus equation (2.iii) represents the voter's decision problem: R constitutes the benefits from voting; only if R>0 will the voter bother to vote:

$$R = B.P - C + D \tag{2.iii}$$

But the likelihood that an individual voter's vote will be 'decisive'—i.e. have any impact on the result—tends to zero (P—>0). For example, in Australian federal electorates with approximately 80 000 voters, P = 0.0000125; in New South Wales state electorates with about 20 000 voters, P = 0.00005. The expected benefit from voting in terms of the impact this will have on the voter's ability to obtain public services is vanishingly small. Whether or not R>0—and, thus, whether or not the voter would rationally choose to vote—depends on the relationship between the costs of voting (C) and the vicarious pleasures of voting (D). This result is so bizarre—i.e. that whether or not a voter will vote depends largely on the vicarious pleasures of voting—that it has been termed the 'voter paradox'.

The voter paradox might be used to explain several features of voting. If there is no incentive for the individual to vote, some external pressure might be necessary to ensure that all voters at least get to the polling booth. In Australian federal and state elections—but not local government elections—attendance at a polling booth is compulsory on polling day. It is *not* compulsory to *vote*: what *is* compulsory is to be given voting papers; the voter could then mark the voting paper according to the rules (a 'formal' vote) or in any other way ('informal'). There is a continuing debate in Australia as to whether or not compulsory voting should continue; this debate is complicated by the belief that compulsory voting results in more Labor Party voters voting than would occur under voluntary voting (Jaensch 1995, p. 21). There is thus a strong incentive for the conservative parties to introduce non-compulsory voting. Conversely, in the United Kingdom, where voting is not compulsory, political parties expend considerable efforts at the local level to discover their potential voters and, if necessary, arrange transport for them to attend a polling booth (thus reducing C for the individual voter). These practices acknowledge the partial validity of the voter paradox—without some form of subsidy (in the form of transport), some voters rationally would not vote.

There are other, less satisfactory aspects of the voter paradox. If the voter paradox is valid, then no voter would vote—but each voter would deduce that no other rational voter would vote. If this supposition were true, then each voter would expect that they would be the only voter—in which case the probability of an individual voter's vote being decisive would tend to one (P—>1). If this were true, then every voter would vote. But, after repeated elections, voters would learn that (many) other voters do in fact vote. Thus it becomes an interesting game theoretic problem as to whether or not a voter would vote.

The above analysis is couched, like neoclassical economics, in a utilitarian framework. However, extension of the right to vote to all adults in Australia (and elsewhere) required an extended political campaign through the nineteenth and twentieth centuries. Many people, knowing how difficult and bitter this political struggle was, still regard the right to vote as a solemn obligation (Mueller 1989, pp. 361–2). The utilitarian framework of equation (iii) is not sufficiently rich to capture an individual's feeling of obligation to vote. However, this obligation is clearly not felt by all individuals, otherwise compulsory voting would be unnecessary, as would special efforts by political parties to ensure their supporters vote where voting is not compulsory. Translating these ideas into the utilitarian framework of equation (iii), if B includes the benefit of fulfilling an obligation to vote then high values of C may be justified and borne.

Demand for economic regulation

Why is there a demand for economic regulation? Why would firms in the private sector, who could be expected to believe that private sector activity is more efficient than the public sector, wish to involve the government in their activities? Several answers can be found to these questions. Consider a private sector, profit-maximising firm whose profits are defined in the standard way:

$$\text{(2.iv)} \quad \Pi = \Sigma i\,(pi.qi) - \Sigma j\,(pj.xj) \qquad i = 1, \ldots N \text{ products}, j=1, \ldots, M \text{ inputs}$$

where (pi,qi) are the price and output level of the firm's i-th product, and (pj,xj) are the price and usage of the firm's j-th input (and prices are exogenous). The firm can only maximise profits by (a) selecting the efficient combinations of inputs and outputs; (b) for given qualities of inputs and outputs, finding the lowest priced markets for its inputs and the highest priced markets for its outputs; and (c) seeking new technologies which improve its profitability. But the firm may also improve profitability through *collective* action—e.g. by inducing government to intervene to increase its output prices, reduce its input costs, improve its production efficiency, or arrange beneficial transfer payments. Such interventions may be 'benign' (e.g. overcoming market failures or diminishing existing government failure) or 'malign' (e.g. government permitting the firm to undertake anti-competitive behaviour). Not-for-profit interest groups may similarly wish government to intervene on their behalf.

For all individuals or interest groups, the key question is: can the individual or interest group obtain its objectives *more cheaply* via government intervention rather than by private market action or *cooperative* collective action? For each potential intervention on behalf of the individual or group there are benefits to be won, and there are also real resource costs such as collecting and processing information, lobbying government, or persuading the public and other affected groups of the appropriateness of the action. The individual or group therefore has a 'benefit–cost' analysis to undertake: would it be better off seeking its own private solutions, or convincing government to solve the problem for it? Individuals or groups, whether or not interested in seeking potential government intervention, also have to be constantly alert to the potential effects on themselves of regulations being sought by other groups. Hence there are both 'offensive' and 'defensive' aspects of regulation seeking.

A schematic representation of interest groups lobbying government is presented in figure 2.9. Interest groups lobby parliamentarians, executive government and the bureaucracy to implement legislation, regulations or administrative decisions favourable to their interests. However, these lobbying activities are not costless. Organising interest groups costs money and lobbying the various arms of government is also costly. Implementing interest group lobbying is also costly: government is costly to run, and the enactments of legislation, regulations and administrative decisions sought by interest groups require government resources. It is, however, difficult to distinguish the costs of running a 'benign' government from the costs of implementing those decisions that result from interest group activity.

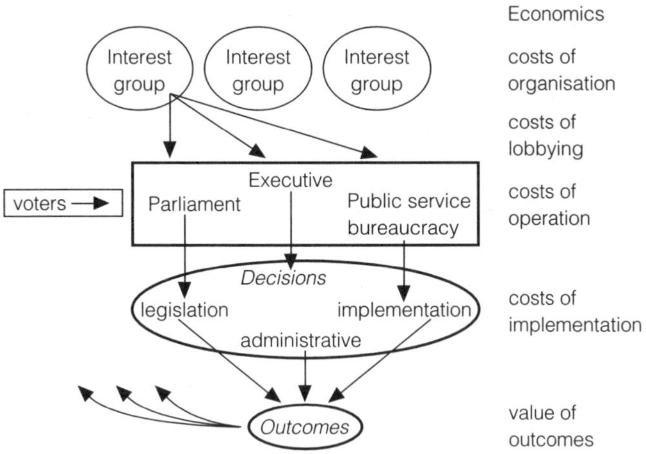

Figure 2.9 Interest Groups and their Economics

The kinds of resources potentially involved in interest group activity are shown in table 2.1 which gives illustrative data for *donations* to political parties in a number of democracies. Brooks and Carter (1994, chapter 5) illustrated the importance of interest group funding of candidates, especially via 'Political Action Committees', in the USA; for example, agricultural interests gave $17.8m in 1989–90. Hutcheon (1995) reported estimates by the Taiwanese Central Bank that 398 candidates for the island's 1995 legislative elections would spend $500m on vote buying in the last three days of the election campaign.

Table 2.1 International Interest Groups and their Resources

		p.a.
USA	Political Action Committees	65.0 m
	Emily's List (women to Congress)	6.2 m
($USm)	National Association of Realtors	3.0 m
	American Medical Association	2.9 m
	International Brotherhood of Teamsters	2.4 m
	Association of Trial Lawyers	2.4 m
	National Education Association	2.3 m
	United Parcel Service	1.5 m
	AT&T	1.3 m
Japan	Keidanren (Assoc Employers)—to LDP *alone* *(allocated to federations and cos.)*	162.5 m
($Am)	Oil Companies Assoc.	1.1 m
	Japan Steel Assoc.	1.1 m
	7 banks (*each*)	1.1 m
	Nippon Steel	0.9 m
	Sohyo Centre (Socialist Dem.)	1.3 m

Table 2.1 Cont'd

		p.a.
Germany	Daimler-Benz	1.5 m
	Deutsche Bank (conservatives)	0.5 m
($Am)	State Metal Industry Assoc.	
	conservatives	0.6 m
	socialists	0.1 m
	East German govt (communists)	65.0 m
France	large construction cos.	no data
UK	United Biscuits (Cons.)	0.1 m
	Taylor Woodrow (Cons.)	0.1 m
(£m)	P&O (Cons.)	0.1 m
	Hanson (Cons.)	0.1 m
	Trades Union (Labour)	4.3 m
Italy	Olivetti (since 1978)	21.0 m
	Saipem (eng. co., 1987–92)	121.0 m
($Am)	Nuovo Pignone (eng. co., 1987–92)	26.6 m
Canada	Canadian Pacific	0.1 m
	Ernst & Young	0.1 m
($Cm)	Imasco	0.1 m
	RBC Dominion Securities	0.1 m
(Cons.+Libs)	Royal Bank of Canada	0.1 m

Source: *Financial Times*, 26–27 June 1993, p. 4

The economics of interest group activity has dimensions similar to those of the economics of other commodities.

1 *Industry size*. The value of regulation to a small industry is likely to be less than the value of regulation to a large one. However, aggregate size alone does not account for the presence or absence of regulation. A second issue is the number of economic 'agents' (firms, consumers etc.) who benefit from the regulation. If the number of producers is small, the average value of regulation per economic agent may be higher than if the number of producers were large. Where industries are small and average returns from regulation high, the resources each agent might commit to interest group activity is likely to be relatively larger, and the costs of collecting those resources commensurately smaller.
2 *Distribution*. The extent to which the value of regulation can be quarantined to small numbers of beneficiaries is also important. If the value of the regulation is shared with a large workforce, or with input suppliers, the residual value of regulation is smaller than if it could be retained just by the owners or managers of the benefited companies, or controllers of other interest groups such as trade unions. The extent of quarantining benefits is also important for the acquisition of resources enabling the regulation to be sought in the first place. If the benefits of

regulation are non-rival and/or non-excludable, some potential beneficiaries will have no incentive to provide the resources to undertake lobbying.

3 *Monitoring and enforcement.* If maintaining the value of regulation requires the monitoring of beneficially affected economic agents (to ensure that they are not 'cheating' on the regulations, for example), then the smaller the number of beneficially affected agents, the lower the private per-unit cost of monitoring and enforcing the regulation.

4 *Specialisation.* The degree of specialisation of the interest group is also likely to be important: the more narrowly it can focus its lobbying, the more likely it is to be successful. Additionally, the more the commodity which the interest group supplies is seen as a merit good, the better its chances of securing favourable regulation—when milk was seen as an important component of public health, its suppliers had good prospects of securing favourable regulation, as does the medical profession.

Supply of government output—government as monopolist

Although the government is a monopolist with respect to the legal use of coercive power, it does not follow that government unilaterally implements the policies it has chosen. Rather, there is a two-way relationship between political parties (current and potential governments) and interest groups that involves both proposing potential policies and evaluating their likely benefits. The benefits to the political parties are the aggregate effects of policies on the parties' chances of election or re-election to power. The effects on the interest groups are the efficiency or distributional consequences of the suite of policies that a political party might choose to implement.

The integration of the 'supply' of and 'demand' for policies is more complex than corresponding supply and demand relationships in neoclassical microeconomics because there is no continuous 'regulation' variable from which simple functional relationships might be constructed. Rather, the supply of and demand for regulation might be illustrated as in table 2.2. Each 'program' in the first column of the table represents a policy that might be proposed by an interest group or a political party. The second and third columns in the table represent the effect of each individual program on 'social' welfare (assuming that a social welfare function can be defined) and 'economic' welfare (for example, using Gross Domestic Product as a proxy for economic welfare), respectively. In both columns, the operator 'Δ' indicates emphasis on the *change* in social or economic welfare. In the remaining columns of the table, Bjk represents the benefits acquired by group j from program k, and Cjk represents the costs to group j of having program k implemented. Thus, for example, the fourth and fifth columns represent the benefits and costs of each program to two arbitrary political parties, government and opposition; the sixth column represents the impact of a program on the bureaucracy; the seventh and eighth columns represent effects on two arbitrary interest groups; and the final column represents the effect on the media. There may be differential impacts for sub-groups of politicians, bureaucrats and media.

Table 2.2 Offering Choices and Making Decisions

Program	*Social welfare*	*Economic welfare*	*Participant*					
			Government	*Opposition*	*Bureaucracy*	*Interest group 1*	*Interest group n*	*Media*
1	ΔSW1	ΔGDP1	Bg1 Cg1	Bo1 Co1	Bb1 Cb1	B11 C11	B21 C21	Bm1 Cm1
2	ΔSW2	ΔGDP2	Bg2 Cg2	Bo2 Co2	Bb2 Cb2	B12 C12	B22 C22	Bm2 Cm2
3	ΔSW3	ΔGDP3	Bg3 Cg3	Bo3 Co3	Bb3 Cb3	B13 C13	B23 C23	Bm3 Cm3
4	ΔSW4	ΔGDP4	Bg4 Cg4	Bo4 Co4	Bb4 Cb4	B14 C14	B24 C24	Bm4 Cm4
5	ΔSW5	ΔGDP5	Bg5 Cg5	Bo5 Co5	Bb5 Cb5	B15 C15	B25 C25	Bm5 Cm5
6	ΔSW6	ΔGDP6	Bg6 Cg6	Bo6 Co6	Bb6 Cb6	B16 C16	B26 C26	Bm6 Cm6
. . .								
N	ΔSWN	ΔGDPN	BgN CgN	BoN CoN	BbN CbN	B1N C1N	B2N C2N	BmN CmN

Facing choices such as posed by the simple set illustrated in table 2.2, how might a political party construct a choice of programs to constitute its platform? The costs in table 2.2 have dollar dimensions. The benefits to interest groups are likely to have dollar dimensions, being the value of regulation to these groups. However, for politicians, the benefit of enacting regulation is the increased *probability* of attracting votes. But this support could be obtained in other ways such as by advertising, which does have dollar dimensions. Denominating the benefits of regulation in *opportunity cost* terms—i.e. the least cost alternative means of obtaining the same benefit—enables both benefits and costs in table 2.2 to be notionally defined in similar terms. Thus, rules might be defined for the political parties to choose among the programs they themselves develop, or which are offered by interest groups or the bureaucracy.

However, the rules political parties might use as government or opposition to choose among programs will partly depend upon the form of government in which these choices are made. The key dimensions of this problem are:

- how many parliamentary chambers there are;
- if there is more than one chamber, whether or not the government has control over all chambers; and
- whether party discipline is tight (as tends to happen in Westminster-derived democracies) or loose (as in the USA) (cf. table 2.3). In the examples below, a majority government in a uni-cameral legislature, or bicameral legislature with government majorities in both houses, is assumed.

Table 2.3 Dimensions of Complexity in Government Decision-making

Number of house of government	*Majority/minority government*	*Control of Upper House*	*Party discipline*	*Degree of difficulty of political decision-making*
Uni-cameral—e.g. Queensland, New Zealand	Majority	Not applicable	Strong Loose	Relatively easy Possibly manageable
	Minority	Not applicable	Strong Loose	Possibly manageable Difficult
Bicameral—Australia and states (except Queensland), United Kingdom	Majority	Yes	Strong Loose	Relatively easy Possibly manageable
		No	Strong Loose	Sometimes manageable Difficult
	Minority	Yes	Strong Loose	Manageable Difficult
		No	Strong Loose	Sometimes manageable Impossible
Tri-cameral—e.g. USA (considering the President as a third 'House')	President same as both Houses	Not applicable	Loose	Manageable
	President different from both Houses	Not applicable	Loose	Probably difficult
	President same as only one House	Not applicable	Loose	Probably difficult

The government might have a rule of the form 'choose a policy which maximises the difference between the net benefits of government and opposition, subject to lower limits (which could be negative) on the effects on social welfare and economic welfare and the net benefits to all other groups'. In algebraic terms, this rule could be represented as:

(2.v) choose j: $\max \{(Bgj - Cgj) - (Boj - Coj)\}$

such that $(Bbj - Cbj) > b0$ and $(B1j - C1j) > I1$ and $(B2j - C2j) > I2$ and $(Bmj - Cmj) > m0$ and $\Delta SWj > W0$ and $\Delta GDPj > G0$

for given $b0, I1, I2, m0, W0, G0$

A corresponding rule for an opposition might be to 'choose a policy which maximises the difference between the net benefits of opposition and government, subject to lower limits (which could be negative) on the effects on social welfare and economic welfare and the net benefits to all other groups'. In algebraic terms, this rule could be represented as:

(2.vi) choose j: max {(Boj – Coj) – (Bgj – Cgj)}

such that (Bbj – Cbj) > b0 and (B1j – C1j) > I1 and (B2j – C2j) > I2 and (Bmj – Cmj) > m0 and ΔSWj > W0 and ΔGDPj > G0

for given b0, I1, I2, m0, W0, G0

Politicians might implement more complex decision rules. For example, J. McEwen, when Leader of the Australian Country Party, commented, 'It has always been the policy of my Party that those who produce, own and sell a product are the best judges of the way in which their own property should be treated' (quoted in Warhurst 1990, p. 113). This decision rule could be represented as 'allow the relevant industry group to choose its preferred strategy'. Given this opportunity, a rational industry group might select a decision rule such as: 'choose a policy which maximises the net benefits of our members, subject to lower limits (which could be negative) on the effects on social welfare and economic welfare and the net benefits to all other groups, including political parties'. So, for example, for industry interest group 1, this rule could be represented in algebraic terms as:

(2.vii) choose j: max (B1j – C1j)

such that (Bgj – Cgj) > g0 and (Boj – Coj) > o0 and (Bbj – Cbj) > b0 and (B2j – C2j) > I2 and (Bmj – Cmj) > m0 and ΔSWj > W0 and ΔGDPj > G0

for given g0, o0, b0, I2, m0, W0, G0

None of the above three representations of decision problems is trivial, even with the relatively small number of interest groups and potential programs that are illustrated. The decision rule requires that decision makers estimate the net returns obtainable for each interest group and define some minimum level of net returns acceptable for each. Additional complexity could be modelled if the benefits a group obtained from a program were conditional on benefits it obtained in another; or if the minimum level of benefits a group was prepared to accept for one program were conditional on the benefits it actually obtained from another. In this case, 'log-rolling' may be practised by political participants as they trade support on one program for support in another.

Log-rolling describes a process where A agrees to support B's pet project in return for B supporting A's preferred project (Mueller 1989, pp. 82–6). Log-rolling thus allows expression of intensity of preference in a voting system where each vote nominally has the same value. It is more a characteristic of *representative* democracy than *direct* democracy, because of the high costs of log-rolling among large number of participants. Log-rolling appears more prevalent in the USA, possibly because its nominal two-party system has less rigid party discipline than contemporary Westminster-style parliamentary democracies like Australia. However changes to voting procedures, particularly in Australian upper houses, have resulted in the balance of power often being held by minor parties who might be expected to log-roll with the major parties. There are also clearly log-rolling opportunities *within* political parties.

There are many policy areas—external security, internal security, macroeconomic management, industry policy, health, education, social security, environment, sport, science—each with many sub-components, and many potential policies and programs within these policy areas. There is, correspondingly, a huge range of potential interest groups. Policy decisions are taken not only at a point in time but also sequentially through time. Formal solution of these problems in a game theoretic framework is probably infeasible. In the absence of formal decision rules for solving the huge complexity of problems that they face, it is amazing that politicians seem to be able to comprehend the choice sets of such a daunting decision task, let alone make a reasonably adequate set of choices.

Continuous analogue

Anderson (1992) argued that differential rates of agricultural protection between developed and developing countries could be understood in terms of different marginal costs and benefits of agricultural protection in such countries. Anderson's model postulated that the costs and benefits of political support are continuous functions of a single variable—the effective rate of protection offered by government policies (figure 2.10). Although the costs of protection in developing countries are low, the benefits in terms of political support to the government are also low and thus the average rate of protection is low. Conversely, in developed countries, although the costs of protection are high, the benefits provided by the protected group to the government are high, and thus the average level of agricultural protection is also high.

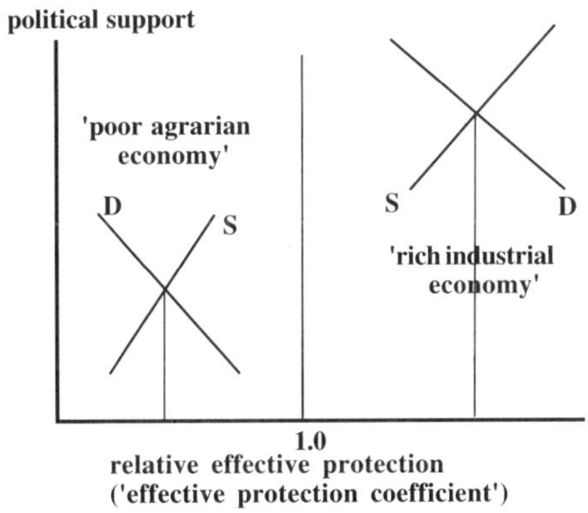

Source: Anderson (1992)

Figure 2.10 Differential Agricultural Protection

Budget-maximising bureaucrat

Most public comment on the government process concentrates on the role of politicians and major interest groups. The bureaucracy has a much lower profile, although bureaucrats play a major role in the choice, design and carriage of policy through the parliamentary process, and its implementation. Each of these aspects of policy bequeaths considerable power, including economic power, on the bureaucracy. Influence over policy choice and design may help determine whether or not policy contributes to increasing economic efficiency, improves or diminishes distributional justice, or particularly favours particular groups and especially interest groups. The role of senior bureaucrats in helping the government guide legislation through the legislative process may influence the degree to which non-government objectives are met, including rewards to interest groups not generally favoured by the government. Since most government funding or output is channelled through the bureaucracy, implementation of legislation—especially where discretion is available in its application—confers considerable economic power on bureaucrats through their being able to decide who merits government subsidies or penalties.

In a theory of public choice, it is the objectives of decision makers that determine how individual agents act. Recent comments during a review of the Australian public service (Johns 1994), indicate that a problem which bedevils politicians is ensuring that their own objectives are implemented rather than those of the bureaucrats. How such a problem may arise is illustrated by the problem of the budget-maximising bureaucrat, which is a conventional principal-agent problem (Mueller 1989, pp. 250–9).

A 'bureau' is a discrete part of a bureaucracy. Suppose the output of the bureau is non-marketable (e.g. as with most transfer payments, defence and external affairs, policy for agriculture). The bureau does not supply an identifiable number of units of output, but only levels of activities from which output levels may be inferred. For example, the level of external security actually provided by the armed forces may only be inferred from numbers of military personnel and the quantity and quality of hardware; the level of public education may only be inferred from the numbers of teachers, buildings and students. The lack of identifiable output and a means of valuing this output create monitoring problems for government as the bureaucracy's nominal masters, 'Just as the separation of ownership and control between stockholders and managers creates incentive and monitoring problems for the firm' (Mueller 1979, p. 158).

If government resources are constrained, the politician's optimum is to equate the marginal benefits of resources used in a bureau's programs with the marginal cost of the bureau's operations. But, since output cannot be easily observed, the bureau may have control of identifying both benefit and cost functions. Hence, the politician's optimum, if they were fully informed, might differ greatly from the bureau's definition of its optimum production under the politician's objective of constrained optimisation. Further, however, the bureau may not be interested in meeting the politician's objective of constrained optimisation, but may actually seek to maximise its size (if bureau size is a determinant of bureaucrats' utility). If the bureau could circumvent the politician's optimisation objective, the bureau might be substantially larger than that required to deliver services optimally.

2.6 DYNAMICS—INDUCED INSTITUTIONAL INNOVATION

In the preceding models, the politicians' choices were implicitly once-for-all decisions made at an arbitrary point in time. Few political decisions are, however, immutable—a week is a long time in politics. Any particular decision is likely to depend on the previous evolution of policy, and also on related contemporaneous decisions being made by other decision makers, or even 'unrelated' decisions that might affect the relevant interest group(s). Thus, while the modelling of preceding sections concentrated on how political decisions might be *made*, an equally important study is how decisions are *changed*.

Ruttan (1978, pp. 340–1) proposed that institutional change could be modelled as the outcome of the interaction between the demand for and supply of institutional change. As previously, demand and supply are not continuous relationships between price and consumption or production respectively, but are relationships among the benefits and costs of regulation and the amount and type of regulation. The concept of institutional innovation involves aspects of the economics of regulation and public choice theory (Ruttan 1978, pp. 342–56).

An explanation as to why regulation changes take place requires definition of the benefits of regulation (cf. discussion surrounding equation (2.iv)). Non-profit oriented firms may be similarly affected by regulation via its effects on prices, technologies or transfer payments; non-producing economic agents (e.g. consumers, welfare agencies) may be affected by regulation via the prices they pay for commodities or the net transfers they receive from government. The net benefit of regulation is also determined by the cost of obtaining regulation; anything that reduces the costs of lobbying—e.g. changes in communications technology—will influence the relationship between the costs of lobbying and its return. Thus, changes in a firm's net profitability function or another group's net benefit or utility function will cause an economic agent to reassess its attitude to the current regulatory framework.

However, while beneficiaries' benefit functions help determine the state of regulation, so too do the political parties' supply functions. If the benefit a political party receives from a given piece of regulation changes—e.g. the interest group support or votes a party obtains from a particular form of intervention changes because of a change in the beneficiaries' attitude to the regulation—or the cost of providing regulation changes, so too will the net benefits of any particular regulation to a political party.

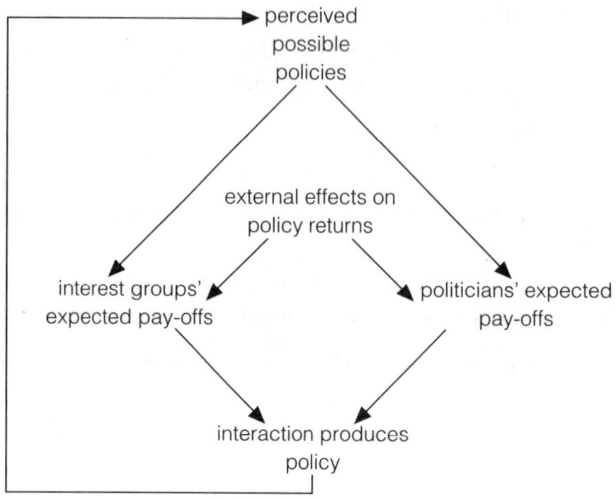

Figure 2.11 Policy process

This model is summarised in figure 2.11. Suppose there is a set of policies which a government could implement, or which a political party seeking government could promise to implement ('perceived possible policies' in figure 2.11). These policies give rise to expected benefits to the interest groups or their members ('interest groups' expected pay-offs'), with commensurate benefits to the politicians who can deliver these policies ('politicians' expected pay-offs'). The pay-offs to interest groups and politicians are also affected by events beyond their control, such as exogenous economic conditions or changes in technology ('external effects on policy returns'). As in preceding analyses, the interaction between interest groups' demands and politicians' supply of regulation produces particular policies ('interaction produces policy'). This policy structure then modifies the set of 'perceived possible policies' because some of the previously perceived policies have now been implemented.

Maintaining a given level and form of regulation is therefore conditional upon the *constancy* of the government's net benefit from maintaining it; the *constancy* of its the net value to the successful interest group; and *constancy* of its net cost to the groups who unsuccessfully opposed its implementation. The level or form of regulation may change if the value of regulation to affected groups changes. Changes in the value of regulation—and consequent reassessment of its value—may occur because:

- imperfect knowledge of the effects of a regulation before it is implemented leads to 'unintended' consequences for those affected by it;
- prices of commodities affected by regulation change through exogenous changes in market conditions, or through the development of new technologies (i.e. changes in 'external effects on policy returns');

- relative factor or product prices change, or new technologies are developed that economise on either the use of relatively scarce factors or the production of relatively expensive commodities (i.e. changes in 'external effects on policy returns');
- other regulations are subsequently developed that affect the profitability of production for a regulated commodity, or there is unexpected synergy between existing regulation and new regulation; or
- new technologies change the market for regulation—particularly information technologies, which change the cost of government finding out about interest groups' demands, or change the cost of interest groups' lobbying.

Any of the above will change the (potential) distribution of income among factor owners. Such (potential) income redistribution changes the previously optimal allocation of resources devoted to maintaining the institutional status quo, and these changes in the allocation of resources in the market for institutions may cause institutional change.

2.7 CONCLUSION

Public choice theory is useful in explaining *ex post* the decisions that are taken in the political market. However, to be useful, public choice theory should also illuminate the future. Again, there is a useful analogue with price theory. It is not expected that microeconomic theory should be able exactly to predict the future. But, for public choice theory to be a useful intellectual endeavour, it should be broadly useful in predicting how the political market will react to change. The challenge for practitioners of public choice theory is to use it in predicting the broad sweep of future political change as a consequence of changes in fundamental economic variables.

ISSUES FOR DISCUSSION

1 The following letter appeared in the *Sydney Morning Herald* in April 1991 as a comment on the Royal Commission into 'Western Australia Inc.':

> Sir, I am amazed at the cynics who believe that large payments to political parties from firms and entrepreneurs are simply a downpayment for future favours.
>
> I, for one, do not believe that political parties can be bribed in this way. However, I do believe that there are fairies in the bottom of my garden.
>
> R.F. Smith
> Nelson Bay

Comment on this letter in the context of public choice theory.

2 In the wake of the Liberal Party's Federal Election loss in 1993, a former Federal President of the party said that 'politics is all about the middle ground'. The

Executive Director of The Sydney Institute, a conservative think tank, advised the Liberal Party to ditch the 'Land of Hope and Glory set' because they would vote for that party anyway. Discuss these comments in the context of economic theories of politics.

3 The following passage appears in Frederick Forsyth's novel *The Devil's Alternative*:

> The President of the United States read the memorandum with an expression of increasing horror. 'This is appalling,' he said. 'Whichever option I choose, men are going to die.'

Discuss whether or not this statement is consistent with politicians' thoughts, feelings and actions as evaluated in public choice theory.

4 Small parties or Independents hold or have held the balance of power in the Australian Senate and both houses of the New South Wales Parliament. Discuss this phenomenon in the context of Hotelling's median voter theorem and the 'representativeness' of government decision-making.

5 Peter Collins, New South Wales Treasurer and Minister for the Arts, interviewed on ABC Radio's 2BL on 18 June 1993 about being treasurer commented that, having served in a number of portfolios after the Coalition came to power in 1988, he knew 'where to look for reasonable savings that won't cause political mayhem.' Comment.

6 Find an item of local, state or national interest—or, preferably, all three. Write to appropriate local, state and federal politicians (both government and opposition) and/or interest groups questioning their stance on this issue. Check their replies against your original letter, and any public statements they may have made. Did they answer your questions? Whom do you think wrote their letters? Do you think your communications did or would affect their attitude to this issue? Why or why not?

7 After the 1995 New South Wales state election, the two most common refrains were, from the losing Liberal–National Coalition Government 'we wuz robbed' because they won a majority of the two-party preferred vote but not a majority of seats and, from ordinary voters, 'why did it take so long to count the votes?' Comment on:

i should the votes in a close election be counted in less time than it takes to play a Test cricket match?

ii under what conditions would the party or coalition which gained the majority of (two-party preferred) votes *always* obtain a majority of seats?

8 A former New South Wales Liberal State Member of Parliament, who lost his party's preselection put it like this:

'When they wanted me to go, when Fahey pushed me, it could have been handled better. If they'd come to me, if John Fahey had come to me, and we'd gone out together, him and his wife, me and my wife, and he'd said to me, "Barry you've been in the party for 45 years, and I've got to ask you not to run, but after the election I'll put you on a board of one of those privatised bodies or something . . ."

If he'd said that, well I'd have said yes. That's what Labor would have done. They would have managed the situation. But our blokes, they're all worried about the ICAC [Independent Commission Against Corruption]. Oo-er, corruption! Bloody bullshit, bloody bullshit. If you've been in power then there's got to be a few lollies along the way. Labor understand that. Our lot don't' (Simons 1995, pp. 19–20).

Comment.

Endnotes

1 The last-mentioned led to sixteen people being convicted, including a former Prime Minister who received $6.7m, and took nineteen years to conclude (Hill 1995a).

2 Models of democratic processes with private utility maximisation have a variety of descriptions including 'public choice' theory (Mueller 1979, Martin 1990, Johnson 1991), 'neoclassical economic theory of politics' (Tisdell 1982, chapter 16), 'economic theory of regulation' (Stigler 1971, Posner 1974), and Krueger's 'theory of rent-seeking behaviour' (Johnson 1991, p. 12).

3 It is actually easier to oppress minorities simply by not extending the vote to those whom decision makers wish to oppress. The franchise was only slowly extended to the 'lower' classes in Britain during the nineteenth century, and only to women in the twentieth. In parts of Australia, voting rights were only extended to Aborigines after 1962 (Chapter 3). Unless there are unamendable parts of a constitution that absolutely forbid genocide, it is theoretically possible in a democracy for one group to sufficiently oppress others as to permit genocide.

4 Changes in electronic communications make it feasible for every voter to vote on every issue without having to attend the modern equivalent of the Athenian market place. In this case, however, only part of the cost of public decision-making has been reduced; but the cost of acquiring sufficiently good information to make a good decision may still work in favour of representative democracy, because only the representatives need to expend resources in acquiring the necessary information.

5 With apologies to mesdames Bandaranaike, Bhutto, Brundtland, Ghandi, Meir and Thatcher.

6 Competition between bureaus was exemplified by the 1994–95 brawl over woodchips between federal ministers J. Faulkner (environment) and D. Beddall (resources). Such conflicts may be good for eliciting information but bad for the efficient implementation of policy.

3 Cultivating power: institutions of Australian agricultural and resource policy

Applying economic efficiency and public choice models to the analysis of government policy is impossible in an institutional vacuum. From an *efficiency* perspective, the institutions of government are the means by which policies to enhance economic efficiency are implemented. From a *public choice* perspective, institutions include the interests that attempt to sway government decision-making in their favour; political parties and the bureaucracy (attempt to) deliver these favours. The institutional framework also comprises rules or conventions that constrain the behaviour of policy makers. At the top level, these rules may include the relatively permanent constraints of written or unwritten constitutions defining permissible behaviour. Conventions and legislation create somewhat looser—because more easily modified—constraints on policy makers' actions.

Because the institutional structures of different countries vary considerably, policy must be analysed in a specific setting. This setting must be firmly located in both space (i.e. a particular country) and time (because institutional structures change over time). The institutional setting used in this book is primarily that of late twentieth century Australia, and this requires a functional description of the relevant Australian policy institutions. The emphasis is on the performance of the institutions—in particular, how they act and relate to other institutions, and how they evolve.

3.1 GOVERNING AUSTRALIA

Countries may be broadly classified as unitary states, where all power devolves from a single authority, and federations, where there are competing levels of power. The United Kingdom and New Zealand are typical unitary states while Australia is a federation. Prior to 1 January 1901, the government of each of the Australian colonies was ultimately responsible to the UK government. Upon their federation into the Australian nation, the former colonies conferred some of their powers on the new national government, but retained all those powers that were not explicitly bequeathed.

The new states also retained direct links to the UK government through the state governors, and through opportunities for legal appeals to the Privy Council independently of the nation's High Court. Thus, Australian states are, notionally at least, partners of the Commonwealth rather than its servitors (cf. lower levels of government in unitary states). The states retain considerable autonomy—whatever has not subsequently been allowed to fall into decay or bequeathed to the national government, been usurped by the national government or obtained by the national government via referendum. Local government in Australia has the same dependent relationship with the states as local government has with the national government in unitary states.

Under the doctrine of the 'separation of powers' deriving from the Westminster system, Australian governments have three 'arms', which independently exercise functions. *Parliament* enacts laws, that provide the framework of government. The *executive*—the monarch or the monarch's representative plus the ministers, who comprise the Executive Council—make decisions; the bureaucracy may also make decisions under authority delegated by the ministers or parliament itself. The *judiciary* arbitrates on disputes arising from legislation, or disputes where there is no statute law (the common law). The structure of the federal government shown in Table 3.1 illustrates Australian government structures generally.

Table 3.1 The Organs of the Federal Government and its Interacting Groups

Parliament (bicameral)	*Executive*	*Judiciary*
Members	Cabinet/Ministry	High Court
• government	Governor-General	
• opposition	Executive Council	
Processes:	Bureaucracy	Constitutional arbiter
• legislation		
• formal inquiries—House, Senate (including Standing,	Inner government	Final appeal court
Joint House, and backbench Committees)	Outer government	
• general debate	quangos (e.g SMAs)	
<— Political parties —>		
<— Interest groups —>		
	Ombudsman, Royal Commissions <— Commissions of Inquiry —>	

Head of state

Australia's head of state, and the head of each of the states, is the British monarch. In the absence of the monarch, the acting head of state is the Governor-General

(Commonwealth) or Governor (states). One of the state governors may act as Administrator of the Commonwealth, and states have lieutenant-governors for the same function. The roles of the Governor-General and the governors are largely ceremonial—the public role ranging from opening parliament (in the monarch's absence) to opening fetes. The private ceremonial role ranges from chairing the Executive Council, signing legislation and promulgating various other instruments of government to entertaining foreign and domestic dignitaries. Occasionally a Vice-Regal representative becomes embroiled in controversy—e.g. New South Wales Governor P. Game's dismissal of Labor Premier J. Lang's government in 1932, and Governor-General J. Kerr's dismissal of Labor Prime Minister G. Whitlam's government in 1975. There is considerable—although far from overwhelming—support for a republican Australia with an elected Head of State, which would end this final constitutional link with the United Kingdom.

Parliament

The business of government is carried out through Commonwealth and state parliaments. In the Commonwealth, and all states except Queensland, parliament is bicameral; Queensland's upper house was abolished in 1922. Members of the Federal House of Representatives are currently elected in 148 single-member constituencies by preferential voting for maximum three-year terms. At normal Senate elections, half of each state's total of twelve senators are elected (both senators from each of the two territories are elected at each half Senate election), with each state or territory being a single electorate and senators elected by proportional representation using the Droop quota, single transferable vote system (Jaensch 1995, p. 30). Under exceptional circumstances, a double dissolution may be granted in which all seats in the House of Representatives and the Senate are contested. All state parliaments except Tasmania's have single-member constituencies using preferential voting for their lower houses; Tasmania uses its five federal electorates as seven-member constituencies using proportional voting. Both territories have only lower houses elected via proportional representation. State upper houses are variously constituted: New South Wales and South Australia have state-wide electorates and use proportional representation; Victoria has twenty-two provinces each electing two representatives; Western Australia has two electorates electing seven members and four others electing five by proportional representation; and Tasmania has nineteen single-member electorates (Jaensch 1995, p. 12).

Elected members of parliament in both houses normally divide as government (the party or coalition with a majority of members in the *lower* house) and opposition (the party or parties not having a lower house majority). The opposition may have a majority of members in the upper house, which is likely to make it difficult for the government to have its legislation passed there. Members of small parties, or those not affiliated with a political party (independents), sit on the cross-benches. A minority government is one which, although it does not have the majority of members in the lower house, generally has the support of sufficient members sitting on the cross-benches to have its legislation passed. If a minority government fails to have major legislation passed it will normally resign; the Governor-General or Governor may then request another member of the existing parliament to form a new government from existing members, or may call a new election.

Administration of parliament is effected by two departments headed by the 'presiding officers' of the parliament. In federal parliament, for example, the presiding officers are the Speaker of the House of Representatives and the President of the Senate. State parliaments are similarly presided over by a speaker in the lower house and a president in the upper. These presiding officers (or their deputies) chair parliamentary sessions in their respective houses except when that house is considering details of proposed legislation (i.e. when the house is 'in committee'), when they are replaced by a chairman of committees (or a deputy). Parliaments' activities include parliamentary and committee debate on legislation, receipt of reports, and parliamentary and backbench committees (cf. Tully 1995).

Some permanent statutory bodies are responsible, and report directly, to parliament. For example, the Auditor-General is responsible for validating the expenditure of executive government. Parliament may also establish temporary bodies for special purposes such as royal commissions and other investigations (cf. Weller 1994).

Voting for parliament

Australia prides itself as a democracy, and as one of the earliest nations to grant full adult suffrage. According to Jaensch (1995, p. 15) this 'applied for Federal elections from 1902, and Victoria, the last state to do so, granted women the right to vote in 1908'. Women also obtained the right to sit in Federal Parliament in 1902; again Victoria was the last state to grant this right in 1923. Despite Jaensch's claim, *full* adult suffrage did *not* apply in Australia from 1902. All Australians were only eligible to vote in federal elections from 1962, when Aborigines were fully enfranchised (Jaensch 1995, p. 15). Aborigines were eligible to vote in all states and territories by the end of the 1960s, but their voting status was only fully comparable with that of non-Aboriginal Australians in 1984 (Jaensch 1995, p. 25).

Government

The lower house of parliament provides the Head of Government (Prime Minister nationally, and Premier in the states). Ministers are drawn from both houses of parliament and, together with the Prime Minister or Premier and Governor-General or Governor respectively, comprise the executive government. The executive government is also assisted by the bureaucracy, which includes the departments of government and a plethora of statutory bodies that report to parliament via a Minister.

Judiciary

The third arm of government is the judiciary. The High Court is established under the Federal Constitution and there are also other federal courts (e.g. Federal Court, Family Law Court) and tribunals (e.g. Industrial Relations Commission, Administrative Appeals Tribunal) established under legislation. States have Supreme Courts (whose Chief Justice is usually the state's Lieutenant-Governor) and a variety of lower courts and tribunals.

Protection against the State

Unlike the USA, Australia does not have a Bill of Rights specifying the individual's rights against the State. There are some protections in the common law, which derive from English common law, and the High Court has recently explored the concept of an implied bill of rights, which it discovered in the Constitution (Galligan 1995, chapters 6–7). Some governments have also introduced partial protections for the individual. These include privacy legislation, various ombudsman offices to which individuals can appeal against inappropriate government action, and freedom of information (FOI) legislation by which individuals can gain access to information held concerning them. Queensland and New South Wales have permanent anti-corruption bodies, which are a hybrid of court and statutory body (the Criminal Justice Commission and Independent Commission Against Corruption, respectively).

One way of limiting the degree to which such bodies constrain the executive is for the latter to starve them of resources (e.g. Taylor 1995). Economic pressures can thus limit the degree to which individuals can protect themselves against the State. The New South Wales Ombudsman (1995, p. 3) noted the high fees and charges levied under NSW's freedom of information legislation.

3.2 ECONOMICS OF FEDERALISM

An economic theory of federalism

Unless there is log-rolling or other forms of vote trading, collective decision-making invariably involves 'winner takes all' decisions. In these cases, individuals could find that many, if not all, collective decisions were contrary to their wishes. Eventually, if this oppression became too onerous, an individual or group could choose to emigrate. In previous centuries, emigration—particularly European emigration—led to the emigrants establishing states more conducive to their aspirations. (Indigenous populations, however, usually became a newly-oppressed group.) Emigration is a particular form of 'voting with the feet' (Mueller 1989, pp. 154–7) where if an unsympathetic state could not be altered by an individual's vote, the individual could join another state. Effective 'foot-voting' requires that individuals have a variety of political systems from which to choose, that the chosen state will voluntarily accept the potential 'foot-voter', and that adjustment costs are not too high.

Galligan and Walsh (1992, p. 200) argued that the decentralisation of at least some public services provides a diversity of ranges and levels of services from which foot-voters could choose. They further argued that the Australian Federation, with differences in types and levels of services among the states, could be justified because it permitted foot-voting. Foot-voting benefits, however, merely justify the existence of some form of political system at the sub-national level with some services delegated from (or not arrogated by) the national government. Thus, the desirability of foot-voting might just as well justify eliminating the existing Australian states and retaining existing local government, or eliminating existing state and local government and replacing them with regional government.

Australia's inherited federalism

The six former colonies in Australia federated by an Act of the British Parliament on 1 January 1901. This Act became the Australian Constitution, which has since remained largely unaltered. Even many of the transitional arrangements, required to transfer powers from imperial and colonial governments to the Australian Government, remain as archaic legacies in the Constitution (cf. Winterton 1994). Other peculiarities of the Constitution include that, while the Prime Minister as Head of Government is the principal government officer of the Commonwealth, this office is not mentioned in the Constitution. The constitutional crisis of 1975 drew attention to the importance of unwritten conventions that modify the Constitution's operations.

Three broad constitutional areas of economic relevance (after Coper (1978, 1992).

Areas from which states are expressly excluded

i Excises and bounties

By s.90, states are precluded from levying customs and excise duties, and granting most forms of bounties. States are permitted by s.91 to grant bounties on mining metals, or for any production or export activity with the approval of both Houses of the Federal Parliament. Thus states cannot levy sales taxes (i.e. value added, consumption or goods and services taxes). Many possible state agricultural marketing schemes teeter on the brink of illegality because a classic agricultural marketing scheme rests on levying products sold on high priced markets (an excise), and returning these levies as an average payment on all sales (a bounty). States have sought to circumvent this proscription on excises via franchise fees (a tax related to sales in preceding years). An alternative avoidance mechanism is for state governments to demand dividends from state-owned enterprises (cf. Lloyd 1986, chapter 7 on 'public authority dividends'). When applied to widely-used services like electricity and water, these dividends have similar economic effects—if not legal form—to an excise. Such taxes are now threatened by the Hilmer reforms, which would severely limit the exercise of monopoly powers that make this taxation so valuable to the states (cf. Chapter 14).

ii Restraints on interstate trade

By s.92, the states (and the Commonwealth) are prevented from imposing restraints on interstate trade. This prohibition has made it more difficult for the states to implement unilateral marketing schemes for their producers, or to reserve a market for a state's own producers. Agricultural marketing schemes were extensively litigated from 1915 to the 1980s (Cullen 1988). By the mid-1980s, the general interpretation of s.92 had come to be that a law affecting interstate trade would have infringed s.92 'unless the burden imposed was indirect or merely regulatory'. The Cole and Whitfield case appears to have instituted an interpretation of s.92 which may endure, at least for a little while; under this interpretation, s.92 is infringed only by 'discriminatory laws of a protectionist nature' (Coper 1992, p. 136).

iii Other provisions

States are excluded from many other activities which have substantial economic effects. By s.117, residents of a state cannot be discriminated against in other states simply because of their state of residence. Sections 101–104 provide that 'There shall be an Inter-State Commission' to oversee regulation 'of the provisions of this Constitution relating to trade and commerce, and of all laws made thereunder.' The Inter-State Commission was moribund until revived by the 1970s Labor Government, and was ultimately merged into the Industry Commission in 1989. This merger—together with a more expansive interpretation of the corporations power (cf. below)—provided the Federal Government, through the Industry Commission, with a prima facie justification for extending the activities of the former Industries Assistance Commission into examining state-owned and state-regulated businesses which had previously been avoided.

The states cannot coin money nor declare anything other than gold or silver to be legal tender (s.115). This reinforces the Commonwealth's economic and financial power through control of the banking system (other than state banks) and currency (cf. s.51(xii)). Control of the defence forces is vested in the Commonwealth (s.114): this section has little direct implication for agriculture but, because of the economic importance of defence forces in labour force and procurement, the disposition of defence forces around Australia is clearly of economic significance to the states. Since defence forces reserve and use extensive land areas, which are often remote and little developed, defence force land has become of increasing environmental significance because of its wilderness values.

Commonwealth powers—but states permitted to legislate if Commonwealth has not

The Constitution specifies a variety of potential powers for the Commonwealth, most of which are contained in s.51 (table 3.2). The Commonwealth is not required to legislate in these areas, but it may do so. If the Commonwealth does legislate under these 'heads' of power, then s.109 of the Constitution precludes the states from legislating in

Table 3.2 Potential Legislative Powers of the Commonwealth

51. The Parliament shall, subject to this Constitution, have power to make laws . . . with respect to:

(i)	Trade and commerce with other countries, and among the States:
(ii)	Taxation . . .
(iii)	Bounties on the production or export of goods . . .
(iv)	Borrowing money on the public credit of the Commonwealth:
(v)	Postal, telegraphic, telephonic, and other like services:
(vi)	The naval and military defence . . .
(vii)	Lighthouses, lightships, beacons and buoys:
(viii)	Astronomical and meteorological observations:
(ix)	Quarantine:
(x)	Fisheries in Australian waters beyond territorial limits:
(xi)	Census and statistics:
(xii)	Currency, coinage, and legal tender:
(xiii)	Banking other than State banking . . . and the issue of paper money:
(xiv)	Insurance . . .
(xv)	Weights and measures:
(xvi)	Bills of exchange and promissory notes:
(xvii)	Bankruptcy and insolvency:
(xviii)	Copyrights, patents of inventions and designs, and trade marks:
(xix)	Naturalisation and aliens:
(xx)	Foreign corporations, and trading or financial corporations formed within the limits of the Commonwealth:
(xxi)	Marriage:
(xxii)	Divorce and matrimonial causes; and in relation thereto, parental rights, and the custody and guardianship of infants:
(xxiii)	Invalid and old-age pensions:
(xxiiiA)	The provision of maternity allowances, widows' pensions, child endowment, unemployment, pharmaceutical, sickness and hospital benefits, medical and dental services . . . benefits to students and family allowances:
(xxiv)	The service and execution . . . of the civil and criminal process and . . . judgments of the courts of the States:
(xxv)	. . . recognition . . . of the laws, the public Acts and records, and the judicial proceedings of the States:
(xxvi)	The people of any race [other than the aboriginal race in any State]# for whom it is deemed necessary to make special laws:
(xxvii)	Immigration and emigration:
(xxviii)	The influx of criminals:
(xxix)	External affairs:
(xxx)	The relations of the Commonwealth with the islands of the Pacific.
(xxxi)	The acquisition of property on just terms from any State or person for any purpose in respect of which the Parliament has power to make laws:
(xxxii)	The control of railways with respect to transport for the naval and military purposes of the Commonwealth:
(xxxiii)	The acquisition... of any railways of the State . . .
(xxxiv)	Railway construction and extension in any State with the consent of that State:

(xxxv) Conciliation and arbitration for the prevention and settlement of industrial disputes extending beyond the limits of any one State:

(xxxvi) Matters in respect of which this Constitution makes provision until the Parliament otherwise provides:

(xxxvii) Matters referred to the Parliament of the Commonwealth by the parliament or Parliaments of any State or States, but so that the law shall extend only to States by whose Parliaments the matter is referred, or which afterwards adopt the law:

(xxxviii) The exercise within the Commonwealth, at the request or with the concurrence of the Parliaments of all the States directly concerned, of any power which can at the establishment of this Constitution be exercised only by the Parliament of the United Kingdom or by the Federal Council of Australasia:

(xxxix) Matters incidental to the execution of any power vested by this Constitution in the Parliament or in either House thereof, or in the Government of the Commonwealth or in the Federal Judicature, or in any department or officer of the Commonwealth.

Note: # omitted by referendum of 1967; Winterton (1994, p.35) omitted s.127 of the Constitution which was deleted by the 1967 referendum: 'In reckoning the numbers of people of the Commonwealth, or a State or other part of the Commonwealth, aboriginal natives shall not be counted.'

a way that is inconsistent with the Commonwealth legislation. Section 51 bequeaths to the Commonwealth powers that are directly relevant to agricultural and resource policy:

- s.51(i) enables the Commonwealth to control exports and imports, and to regulate *inter*state but not *intra*state trade.
- s.51(iii) complements s.90 in enabling the Commonwealth to grant bounties on production or exports.
- s.51(ix) enables the Commonwealth to control the ingress of people and commodities for the control and prevention of potential agricultural and environmental scourges (pests, diseases and weeds)—i.e. quarantine.
- s.51(x) enables the Commonwealth to control fisheries beyond the states' territorial limits (5 kilometres) out to the 320 kilometre resources zone. Control of these fisheries is of major importance given the open access (i.e. no property rights) characteristics of ocean fisheries, the long migratory patterns of some species and the potential for exhaustion of renewable resource stocks.
- s.51(xxix), the 'external affairs' power, has little obvious connection with agriculture and natural resources. However, by entering into international agreements under the external affairs power, the Commonwealth obtains the power to implement international obligations.

Section 51 also grants powers to the Commonwealth which, if it chooses to legislate, may indirectly affect agriculture and resources:

- The taxation and borrowing powers (ss.51(ii) and (iv)) provide the financial resources for the Commonwealth to undertake major economic activities such as infrastructure funding.
- s.51(v) enables the Commonwealth to facilitate the implementation of national rather than state-based communication systems; the Constitutional framers pre-

sciently added 'and other like services' to this subject before the invention of television.

- Particularly for a largely atomistic industry like agriculture, the power to create a national information system via the census and statistics power (s.51 (xi)) is of major value to producers (cf. potential market failure with non-rival information).
- The currency and weights and measures powers (ss.51(xii) and (xv)) provide for national systems for describing trade (potentially reducing transactions costs of negotiation and revenue collection).
- Provisions for national banking and insurance systems and financial instruments (ss.51(xiii), (xiv) and (xvi)) facilitate lower transactions costs by permitting national markets to emerge in the finance industry.
- National systems for regulating corporations (s.51(xx)) and their dissolution (s.51(xvii)) similarly facilitated emergence of national markets.
- s.51(xxiiiA), inserted via a referendum in 1946, provides for national welfare services to a range of Australians, including farmers—although the thorny problem of income support to self-employed farmers remote from effective labour markets was only really addressed in the early 1990s (cf. Chapter 9).
- The 1967 referendum deleted the square bracketed material from s.51(xxvi), enabling the Commonwealth to legislate for Aborigines. This amendment permitted the Federal Government's response to the High Court's 1992 native title decision (cf. Chapter 10) despite the Commonwealth's lack of power over land (unless justly acquired under s.51(xxxi)).
- The conciliation and arbitration power (s.51(xxxv)) enabled the Commonwealth to regulate labour markets (Simms & Singleton 1993, pp. 289–90), both to create the highly centralised system in place by the 1980s, and then to start deregulating it (cf. Chapter 14).

Unspecified matters

Matters which were not expressly designated as legitimate areas for Commonwealth activity are theoretically reserved to the states by ss.106–7. However, in the *Engineers* case (1920), the High Court adopted a very broad interpretation of the Commonwealth's powers. Craven (1992, pp. 63–4) argued that, in the *Engineers* case, which was the High Court's 'basic interpretive yardstick for the constitutional division of powers':

> . . . the powers of the Commonwealth parliament were to be interpreted broadly, with all the liberality that the words [of the Constitution] would allow. Correspondingly . . . the High Court overturned the doctrine of reserved powers, whereby the powers of the Commonwealth were to be interpreted narrowly, leaving the greatest possible residue to the States. Only after the powers of the Commonwealth had been allowed their full and expansive operation was it possible to say anything of the residue left to the States.

Craven (1992, pp. 64, 55) argued that this decision tipped 'the balance of constitutional interpretation decisively and apparently permanently in favour of the Commonwealth' and 'it is increasingly difficult to think of important legislation which the Commonwealth could not pass were it absolutely determined to do so'. Expansion of the Commonwealth's powers was facilitated by the Labor Party's innovations, developed in the 1960s and implemented in the 1970s, of using the external affairs power (s.51(xxix)) to broaden the legitimate scope of Commonwealth power (see above) and tying grants to the states (s.96; see below). Craven (1992, p. 56) noted that 'The Commonwealth is at this moment in the press of grasping the previously forbidden fruit of uniform companies legislation, and there is no reason to suppose that the High Court will prove less acquiescent here than elsewhere in the past' (cf. Williams 1995).

In addition to the three areas outlined above, several other constitutional issues are important, or potentially important, to agriculture and resources:

i ***s.96 grants***

Under s.96, the Commonwealth parliament 'may make financial assistance to any State on such terms and financial conditions as the Parliament thinks fit.' As a consequence of High Court interpretation, there are 'practically no limits on the power of the Commonwealth to impose conditions on payments to the States under section 96' (Craven 1992, p. 78). The Federal Labor Government of the early 1970s, and subsequent governments, have extensively used s.96 grants to 'bribe' the states into implementing policies which the Federal Government wishes to apply nationally. If states want additional financial resources, and are prepared to accept tied grants from the Commonwealth in return for implementing federally-determined policies, the Commonwealth may dramatically expand its nominal Constitutional powers.

ii ***Constraints on Commonwealth's powers***

The Commonwealth has constraints on the ways it exercises its powers, as well having constraints on which powers it may exercise. For example, the Commonwealth cannot discriminate between states or parts of states on taxation (s.51(ii)); bounties must be uniform throughout the Commonwealth (s.51(iii)); and the Commonwealth cannot by law or regulation of trade, commerce or revenue give preference to a state or part of state (s.99). However, the wider use of s.96 grants enables the Commonwealth to escape the burden of s.99 if it can bribe states into implementing its policies rather than itself legislating for them directly, *and* if the policy incidentally happens to discriminate in favour of or against a particular state or states.

Whatever the contemporary interpretation of the 'little bit of layman's language' in s.92's injunction that 'trade, commerce, and intercourse among the States . . . shall be absolutely free', the Commonwealth shares this constraint with the states.

iii Changing the disposition of power

The Constitution provides two mechanisms by which the disposition of power between the Commonwealth and states may be altered. Section 51(xxxvii) provides that parliaments of states may 'refer' powers to the Commonwealth parliament. Alternatively, the Commonwealth may utilise s.128, which provides a mechanism to amend the Constitution by direct appeal to voters (cf. Chapter 2).

State governments

State governments' powers originate in the powers each colony was granted in the nineteenth century. The state constitutions are 'unwritten', like that of the UK, and comprise the powers they were initially granted as colonies and the common law, together with subsequent state legislation and any conventions adopted in a particular state. Each state's constitution was then modified by the enactment of the Federal Constitution, subsequent amendments to that Constitution, and constitutional case law (e.g. the *Engineers* case). Some states have recently taken to 'entrenching' constitutional change—i.e. requiring any future legislative changes to declared pieces of legislation to be approved by a state-wide referendum.

Constitutional games

The international growth in the role of national governments during the twentieth century was mirrored in Australia. Where the Commonwealth foresaw a need to expand its powers it could do so by persuading the states to relinquish powers, changing the Constitution by referendum, or by unilaterally enacting legislation to create circumstances in which the Commonwealth's powers could be tested in the High Court (e.g. the *Seas and Submerged Lands Act 1973*, cf. Haward 1991).

An example of Commonwealth unilateral action was the wheat marketing legislation of 1988. Previously it had been assumed that, because the Constitution only conferred power on the Commonwealth to regulate interstate and international trade, the regulation of *intra*state trade was the province of the states. Wheat marketing legislation from the 1940s to the 1980s involved complementary Commonwealth and state Acts. In 1988, to effect deregulation of wheat marketing, the Federal Government legislated unilaterally—effectively 'daring' the states to take action in the High Court to preserve their traditional right to legislate for intrastate trade.

Governments not only play games via their legislation against each other, but also against their citizens. Some key judicial decisions against agricultural marketing legislation resulted from aggrieved individuals seeking remedies against what they saw as oppressive government action—e.g. actions by James and Ubergang (cf. Coper 1985). Attempts by state governments to defend their producers' markets might have been clearly in breach of s.92, but court action was required by aggrieved individuals to assert their rights—e.g. NEDCO's 1975 action against the New South Wales Government which had attempted to prevent milk imports from Victoria (Coper 1978, Cullen 1988). Where individuals seek relief against government actions, the financial power of state or federal governments provides a powerful weapon which is likely to greatly disadvantage individuals and small to medium-size corporations.

Intergovernmental relations

Because of the federal system of government, and the often ill-determined boundaries between federal and state responsibilities, a system of formal interaction has evolved between Commonwealth and state governments. There are currently (March 1996) three bodies at the top level of this interaction:

1 The *Council of Australian Governments* (COAG) comprises the Prime Minister, premiers, chief ministers (of the two territories) and the President of the Local Government Association. It meets at least once a year. Its objectives are to increase cooperation among governments—especially relating to economic reform and structural reform of government—and consultation on general matters that might arise from more narrowly-focused ministerial councils, international treaties, and major initiatives of individual governments that might have spillover effects. COAG meetings are preceded by meetings of officials.

2 The *Premiers' Conference* comprises the Prime Minister, premiers and chief ministers, together with treasurers who are not also heads of government. The Premiers' Conference meets once each year in conjunction with the Loan Council. The focus of the Premiers' Conference is on intergovernmental financial arrangements, especially the level and distribution of general revenue assistance to the states and territories. Premiers' Conference meetings are usually preceded by a meeting of Treasury heads. Until the early 1990s, the Premiers' Conference combined the activities of COAG and the Premiers' Conference.

3 The *Loan Council* comprises the Commonwealth Treasurer and the premier or treasurer of each state. The Loan Council subjects the net borrowing requirements of Commonwealth, state and territory governments to scrutiny and possible restriction. It usually meets in conjunction with the financial Premiers' Conference, and is preceded by meetings of Treasury heads.

By the mid-1980s, there were more than forty ministerial councils dealing with specific issues like agriculture and resource policy, health, education, etc. (Sharman 1991, p. 28). Three reasons have been proposed for the growth in Commonwealth–state consultative mechanisms: growth of government, continued assertion of Commonwealth power, and a growing emphasis on 'managerialism' during the 1980s and 1990s (Sharman 1991, p. 30ff). There has been considerable rationalisation of ministerial councils up to the mid-1990s (cf. section 3.4).

Local government

The most visible forms of local government are the municipal, district or shire councils (terminology varies among states) but also included are special purpose local government authorities. Local government provides services including roads (other than major roads provided by the states), water, sewerage and drainage (including large-scale single purpose metropolitan authorities), electricity distribution, building and health inspection, libraries, pest disease and weed control (the Rural Lands Protection Boards in New South Wales). Local government varies from the very large scale, such as the Brisbane City Council, which provides all local government services in that city (unlike the Council of the City of Sydney which only provides services to the central business district), to large area but small population local government in rural areas. Local government is subordinate to state government: state government decides the powers that local government may have, how these powers may be exercised, audits the exercise of these powers and, if dissatisfied with the manner in which a local government body exercises its functions, may over-ride that body, with ultimate powers to dismiss the appointed or elected representatives and install an administrator or call new elections.

3.3 STRUCTURE OF GOVERNMENT

The formal structure of government comprises the parliaments themselves and the individual parliamentary representatives. The government comprises the members of the largest party or coalition in the lower house of parliament; the executive comprises the members of the governing party or parties with executive responsibilities and powers. At the local level there are elected or appointed councils or boards. Behind the elected parliamentary representatives are the political parties—with their bureaucracies and individual members—of whom the parliamentary representatives are members and whose programs they represent in parliament, and the staff of their parliamentary and electorate offices. Behind the executive members of government stand the bureaucracy —the public service—which implements government policy and supports the executive by providing information and advice.

Discovering details of government structure often presents a major problem, especially when searching for the body responsible for more obscure functions of government. There are some commercial directory services to government (e.g. International Public Relations (n.d.)) which may be available in libraries. Some cheaper alternatives worth investigating are the telephone directory's 'White Pages Index', which provides pointers to federal, state and local government services. Good directories for primary industries are provided by the Federal Government's *Rural Industry Directory*, the Federal Government's directory to its programs for rural people (DPIE 1994), the National Farmers' Federation's (1995) *Australian Agriculture*, and some state governments (e.g. New South Wales Agriculture's (1995) *Country Guide*). Future searching for government bodies and the services they offer should be assisted by electronic communications. The National Library of Australia currently offers an access point for government bodies and interest groups connected to the Internet. Future commercial and more widely available computer-linked services may offer improved access.

Instruments of government

The first and most obvious instrument of government is an Act of Parliament which 'is the way in which Parliament sets out its decisions on what the law is to be' (Gifford & Gifford 1982, p. 1). An Act of Parliament must be passed by the House(s) of Parliament and receive Vice-Regal Assent (or, occasionally, Regal Assent). These Acts, starting their lives as 'Bills', define what the government is permitted to do, extending from the broad-ranging money-raising and expenditure Appropriation Acts that constitute the Budget to the quite mundane.

The second government mechanism is a Regulation under an Act. Regulations, which are like subsidiary Acts, have been extensively used to implement the detail because the process by which they receive parliamentary assent is generally simpler than for Acts; thus Regulations are used to implement those parts of an Act which change frequently (e.g. rates of tax). Again, like their principal Act, Regulations specify what government may do.

The third government mechanism is the everyday administrative decisions that government makes. These decisions may be taken by the Governor-General in Executive Council, Cabinet, individual ministers, department heads, or individual bureaucrats right down to the most junior clerk. Such decisions range from composing and signing letters to constituents and lobbyists, to signing contracts potentially committing large government expenditure. In between is a plethora of decisions ranging from data collection to the prosecution of individuals for breaches of law. While parliamentary Acts and Regulations specify what *may* be done, both politicians and public servants frequently have a wide range of discretion with respect to what is done. Discretion is available at all levels of government, although politicians and public servants are increasingly scrutinised by 'independent' bodies (cf. 'Protection against the State' above).

Parliamentary process

In a Westminster system, parliament has three broad functions: to pass legislation that establishes the framework of government, to supply resources for the government to function, and to scrutinise the actions of government. A fourth important function is to continually pit government and opposition against each other so that the electorate can continually assess their performances. This gladiatorial context provides constant revision of the information electors may use to evaluate future voting options.

The process by which proposed legislation is enacted is illustrated in Turner (1993, figure 6.1). The key aspects are the origin of the Bill (as either government legislation proposed by the Cabinet, or a private member's Bill); the first, second and third readings, and the debate and committee stages following the second reading; and (vice-) regal assent and proclamation. Various authors describe a variety of linear, semi-linear or circular processes which underlie the essentially linear legislative process (e.g. Ward 1991, Braddock 1994, Turner 1993; cf. section 2.6).

High Court

Under the Constitution, the judicial power of the Commonwealth is vested in the High Court (and any other courts created by the federal parliament) (ss.71–76). The High Court is the final court of appeal from all Australian courts and tribunals, both federal and state. The High Court has 'original jurisdiction' in all matters arising under treaties, affecting other countries' representatives in Australia, where the Commonwealth is suing or being sued, and in legal suits between states or residents of different states, or where action is taken against officers of the Commonwealth. Parliament may grant original jurisdiction to the High Court for any matter arising from the Constitution, arising from federal laws, of maritime jurisdiction, or where laws of different states claim the same subject matter. The High Court currently comprises the Chief Justice, and six other judges.

The High Court played a major role in the past with respect to agricultural marketing schemes as those who wished to escape conscription into compulsory marketing schemes sought relief via s.92 of the Constitution. The Court was forced to rule on matters such as whether artificial schemes that might involve transport through one state and returning to the original state actually involved interstate trade and thus gained the protection of s.92.

Coper (1992, p. 138) extended the lessons of the High Court's deliberations on s.92 with reference to a more general issue concerning the High Court. He argued that 'while state laws of a non-discriminatory or non-protectionist kind are not necessarily in the national interest . . . it does not follow that the elimination of such laws is a job for section 92.' The 'proponents of deregulation' were criticised for using any available means for achieving the deregulatory goal. However, the courts cannot 'rove at large over the national landscape' and must have 'reasonable limits within which to work, and apply criteria suited to the judicial process'. Beyond the restricted judicial context in which the High Court ought to work, the resolution of social conflict was a role for politicians, not the judiciary (cf. Hilmer reforms, Chapter 14). The High Court's 1992 decision on native title, involving a common law appeal from the Queensland Supreme Court, is an even more contentious illustration of Coper's distinction between the appropriate spheres of action for politicians and judges (cf. Chapter 10).

Craven (1992, pp. 61–7) argued that the erosion of the states' powers with respect to the Commonwealth's were partly the consequence of the High Court's willingness to permit this erosion. Having reviewed the narrowing of the states' roles following the *Engineers* case, contemporary signs of change in the Court were noted, wherein the

Court determined 'the division of power boldly, essentially on the basis of whether a subject is or is not suitable for Commonwealth control . . . it would seem that the answer to this question will ordinarily be a resounding affirmative' (Craven 1992, p. 65). Craven argued that the reasons for the High Court's favouring the Commonwealth over the states were its belonging to the English common law tradition; reflecting the 'anti-federal tone [prevalent] in Australian intellectual circles'; and because the Court identified with its employer, the Commonwealth.

3.4 INFLUENCING GOVERNMENT—THE INTEREST GROUPS

If interest groups wish to have their preferences realised, they must be able to influence the process of government. 'Government' here is being interpreted broadly as encompassing parliament, the executive and the judiciary.

Political parties

Political parties are the 'firms' of the political market and its leaders, both parliamentary and organisational, are the political entrepreneurs. Political parties traditionally

Table 3.3 Political Parties Offering Candidates at 1993 Federal Election

Abolish Self Government Coalition	Cowper Greens	Pensioner and Citizen Initiated Referendum Alliance
ACT Green Democratic Alliance	Deadly Serious Party of Australia	Queensland Greens
ACT Referendum First Group	Defence and Ex-services Party of Australia	Republican Party of Australia
Advance Australia Party	Democratic Labor Party of Australia	Rex Connor (Snr) Labor Party
Australia First	Democratic Socialist Party	Richmond Clarence Greens
Australian Conservative Party	Eastern Suburbs Greens	Socialist Party of Australia
Australian Democrats	Green Alliance Senate—NSW	South Sydney Greens
Australian Family Movement	Greens in Lowe	Sydney Greens
Australian Indigenous Peoples Party	Grey Power	Tasmania Senate Team
Australian Labor Party	Illawarra Greens	Tasmanian Greens
Australian Shooters Party	Independent EFF	Tasmanian Independent Senator Brian Harradine Group
Australians Against Further Immigration	Irina Dunn Environment Independents	The Australian Ethnic Democrats
Call to Australia (Fred Nile) Group	Janet Powell Independents' Network	The Federal Party of Australia
Central Coast Green Party	Liberal Party of Australia	The Green Party South Australia
Citizens Electoral Councils of Australia Group	National Country Party of Australia (WA)	The Greens
Communist Party of Australia	National Party of Australia	The Greens (WA) Inc.
Confederate Action Party of Australia	Natural Law Party	Torres United Party
Conservative Party of Australia	Northern Territory Country Liberal Party	Unite Australia Party
	Nuclear Disarmament Party	Vallentine Peace Group
	One Australia Movement	Victorian Green Alliance
		Weekend Trading Party
		Western Suburbs Green

Source: *Sydney Morning Herald*, 15 March 1993, pp. 2, 23

have an ideology or mission statement that embodies the values which the party is supposed to represent. The party 'produces' a range of policies, which are the 'commodities' it markets through the political process.

If the number of parties contesting elections indicates the degree of democracy, Australia's democracy is very robust. At the 1993 federal elections, sixty-two registered parties contested seats in the House of Representatives and half the Senate (table 3.3). Many candidates (nominally) independent of any political party also offered themselves for election—but, while many were called, few were chosen. In the House of Representatives, eighty Labor members were elected as the Government; forty-nine Liberal members and sixteen National Party members comprised the Opposition; and there were two independents (for the electorates of North Sydney and Wills). In the half-Senate election, seventeen Labor members, nineteen Liberal/National members, two Democrats, one Western Australian Green, and one independent from Tasmania were elected (Jaensch 1995, p. 82).

While most voters have supported one of the three major parties at national elections since Federation, there has been a slow and irregular small trend of voter support away from these parties. Early in the twentieth century, nearly all voters supported one of the major parties; by the Second World War, less than 90 per cent did (arising from a collapse in support for the forerunner of the Liberal Party). After that war, close to 100 per cent of federal voters again supported a major party, but by the 1990s, only about 90 per cent did. From its heyday of 15 per cent support in the 1930s, the National Party has been quite stable around 10 per cent of the total vote but fell to 7 per cent in the 1993 election (Jaensch 1995, figure 5.2 and p. 82).[1] The Australian Democrats, a new party from the 1970s, peaked in support at about 10 per cent of the vote in 1990 and has since declined to about 5 per cent. From the late 1960s, there has been a reduction in voters 'very strongly' identified with the major parties, which has been largely off-set by increases in those 'fairly strongly' identifying with major parties; the reduction in 'very strong' identification was most marked for the Labor Party (Jaensch 1995, table 6.2).

Philosophies and platforms

In the heady days of the 1980s, Labor governments in Victoria, South Australia and Western Australia came seriously unstuck as a result of ill-considered ventures, a failure to control State Banks leading to spectacular debts in the two former states, and an excess of entrepreneurialism. Some elements of the last also affected the Labor government in New South Wales. In response to these excesses, governments throughout Australia became wedded to the notion that the first—and perhaps only—priority for government was to be a 'good manager'. One state premier went so far as to declare the 'end of ideology' (Greiner 1990).

If ideology has indeed ended, it is unclear what 'managerialism' implies about the role of government. Managerialism only dictates that governments should operate their activities well; it cannot reveal *which* activities governments ought to undertake. Perhaps economists who have adopted the assumptions of neoclassical economics as a religious rubric would know that governments ought to intervene only in the case of pure public goods—and then only by 'market' interventions. And through the 1980s, such religious economists were on the rise (cf. Pusey 1991). Mercifully, however, humans realise that life is about more than merely utility and profit maximisation. What governments ought to do is the result of a more or less considered view about how society should be organised, and the role that government should play in this society. Without an ideology or, at least, a practical political philosophy, it is unclear as to why government should be managing anything. Managerialism is therefore a means to an end, not an end in itself.

'Country-mindedness'

Verrall, Ward and Hay (1985, chapter 1) commented on 'the importance of the rural–urban cleavage which remains (despite frequent neglect) an essential and key notion in understanding Australian politics'. They argued that this cleavage explains why the Labor Party has disappeared as a major electoral competitor in rural areas, and why the Liberal Party is the National's key rural competitor. Verrall et al. argued that there are no socio-economic characteristics of the two conservative parties to distinguish the appeal each has for rural voters. A generally harmonious working relationship between the two conservative parties nationally has often not spilled over into the states. Verrall et al. noted the intense rivalry between the National and Liberal parties in Victoria (patched over prior to election of the coalition Victorian State Government in 1992) and Queensland (although there were calls in Queensland for the merger of the two conservative parties after the election of the Labor Government in 1989).

Aitkin (1985) presented seven key themes of 'country-mindedness' from its high point of the 1920s to the 1960s (Box 3.1). Verrall et al. (1985, pp. 21–2, emphasis in original) followed Aitkin in identifying country-mindedness as a key factor in maintaining the hegemony of the conservative parties in rural Australia, commenting that '[t]he ALP can make little headway against the ideological hegemony of "country-mindedness", precisely because it is seen as the electoral vehicle of the *enemies* of "country-mindedness"'. However, Aitkin (1985) also argued that country-mindedness had been dissipated by post-war immigration, reduction in the importance of agriculture to the national economy, increased national affluence and improvements in transport and, especially, communications. Recent major shifts in population have resulted in substantial increases in some rural populations made up of conservative urban retirees and professionals who have dropped out. Areas particularly affected by such changes are the south coast of Queensland, where the changes have tended to benefit the Liberal Party, and the north coast of New South Wales, where changes have tended to benefit the Labor Party. Improvements in transport and communications (especially television) may also be expected to affect the homogeneity of country-mindedness (see Rural

Media below). By the early 1980s there were signs of a trend that—at least in New South Wales—urban populations above 5000 inhabitants were beginning to consider themselves as 'town' or 'city' rather than 'country' (Lucy 1985, p. 92). Duncan and Epps (1992) suggested, however, that the demise of country-mindedness was still confined to rural areas of major population immigration.

Box 3.1 Country-mindedness at its Zenith, 1925–60

i Australian depends on its primary producers for its high standard of living, for only those who produce a physical good add to a country's wealth.

ii Therefore all Australians, from city and country alike, should in their own interest support policies aimed at improving the position of the primary industries.

iii Farming and grazing, and rural people generally, are virtuous, ennobling and co-operative; they bring out the best in people.

iv In contrast, city life is competitive and nasty, as well as parasitical.

v The characteristic Australian is a countryman, and the core elements of national character come from the struggles of country people to tame their environment and make it productive. City people are much the same the world over.

vi For these reasons, and others like defence, people should be encouraged to settle in the country, not in the city.

vii But power resides in the city, where politics is trapped in a sterile debate about classes. There has to be a separate political party for country people to articulate the true voice of a nation.

Source: Aitkin (1985, p.35)

National Party of Australia

The National Party—previously the Australian Country Party and the National Country Party of Australia—is almost unique in the world in being a specifically rural party; the only contemporary parallels are the Scandinavian agrarian parties (Costar & Woodward 1985b, p. 2). The party's key strengths are in Queensland, New South Wales and, to a lesser extent, Victoria; as Woodward (1985, p. 55) noted, it is 'a party of the eastern mainland states only'.[2] While the vote of the National Party (and its predecessors) has been much more stable than its two major rivals, there has been a slowly declining trend in its share of the total federal vote (Jaensch 1995, Figure 5.2). At a federal level, the National Party remains a major force because the Liberal Party normally requires its support to form a government; this dependence on coalition also characterises the National Party's role in New South Wales. In Queensland, by contrast, the

National Party was the dominant coalition partner from 1957, and governed alone from 1983 to 1989. In Victoria, serious conflict between the National and Liberal parties was a constant feature from early this century until the mid-1980s; the current Victorian Government is a coalition of the two previously-warring parties.

Woodward (1985, p. 61) argued that the (Federal) National Party 'is a strange blend of conservatism coupled with support for radical government intervention in certain economic and social areas', presenting itself 'as a party with the entire nation's interests at heart' with a matching policy platform, but where a 'fundamental concern for the welfare of rural dwellers' permeates all its policies. This preoccupation with rural dwellers includes emphases on special policies for the support of agricultural industries, and support for rural infrastructure (e.g. art and culture, education, communications); the National Party is also a strong exponent of 'balanced' development and decentralisation. The National Party sees no contradiction in its general support for private enterprise while arguing the specific case for intervention in the farming economy; this has been denoted 'agrarian socialism' (cf. Smith 1993b, p. 137) despite the Party's 'genuine and deeply felt' opposition to socialism (Woodward 1985, p. 62).

Liberal Party of Australia

In the colonial period to 1901, the combination of restricted franchises and the distribution of wealth in Australian society resulted in conservatives dominating colonial governments. The rise of the Australian Labor Party in the 1890s challenged this dominance, and led existing conservative parliamentarians to combine to counter the 'socialist' threat presented by the ALP. The contemporary Liberal Party was formed in 1944–45.

The formation of the Liberal Party as an anti-Labor party has had important implications for its success. Its principal belief is in the freedom of the individual, in contrast to the ALP's perceived subordination of the individual to the State. The Liberal Party's opposition to the centralist ideas of the ALP forged a party structure that was dominated by the state branches, which have tolerated only a weak national office. This structure works well for the state parliamentary parties. It also works satisfactorily while the Party is in national government when it can draw on the support of the bureaucracy; while important to all governing parties, being in government is of special importance to the Liberals because of the traditional lack of policy input from the Party's organisational wing. Dominating government gives party members a *raison d'être* which is difficult to sustain when the Party is in opposition;[3] and also allows tangible benefits to be offered to interest groups. Considerable internal difficulties characterised the New South Wales Liberal Party during its opposition years from defeat in 1976 to the capture of parliamentary leadership by N. Greiner in 1983; similar disarray generally characterised the Federal Liberal Party in opposition to the Hawke–Keating governments, except for the first three years under J. Hewson and the early years of J. Howard's second term as leader (1995–96). Belief in the freedom of the individual also sometimes creates difficulties in maintaining the parliamentary discipline required to combat the ALP's caucus solidarity.

Like most large political parties, the Liberal Party is a coalition of diverse interests. In recent decades a factional system has emerged of 'dries' (whose dominant interest is in an essentially market economy and small government) and 'wets' (who retain the Party's commitment to social support for individuals in a predominantly market economy).

The Liberal Party has governed in its own right in Victoria, Tasmania and Western Australia, and in South Australia as the Liberal–Country League. Federally, and in New South Wales, the Party governs only in coalition with the National Party. The 'modern' Liberal Party dominated Federal Government from 1949 to 1972 and 1976 to 1983, and also state politics for long periods (e.g. New South Wales 1965–76; Victoria 1955–82; in South Australia as the Liberal–Country League, 1938–65 under the same Premier, T. Playford; and Western Australia 1959–71 and 1974–83). The need for formal coalition to achieve government poses challenges for both Liberal and National parties. The Nationals' principal problem in opposition stems from its own stability: it can only regain government by the Liberal Party's winning enough seats to form a new coalition government, and it is effectively sidelined while the Liberals reinvent themselves sufficiently well to achieve this goal. An accompanying frustration for the National Party is that the seats the Liberal Party needs to regain coalition government are inevitably in the city, and to regain these seats the Liberal Party may need to articulate policies that are inconsistent with the National Party's own ideals. To overcome such problems, the merger of the two parties is sometimes proposed, but usually founders on the fear that the spectrum from Liberal 'wets' to the most conservative Nationals is so great that a merger would probably result in the emergence of a new rural party.

Australian Labor Party

The Australian Labor Party has international firsts in forming the first government of a labour party (Queensland, 1899) and the first national government of a labour party (1904). The ALP has always comprised a broad coalition of diverse interests, although the elements of this coalition have varied over time. Originally established as a coalition of the labour movement and liberal and radical reformers, the ALP has been transformed into a contemporary party of the labour movement and professional, middle-class reformers. Reflecting this broad and changing coalition, the ideology of the ALP has also evolved over time, although in its formal manifestation as the Party 'platform' the form of words was resilient over many decades, even if its implementation was not. The Party has identified itself, and has been identified by others, as 'socialist' although Woodward (1990, p. 151) argued that it has been more 'radical-liberal'. The Party has also been seen as the political wing of the labour movement, although this role has declined over time as the trade union movement itself has declined. Reflecting the nature of its coalition across a broad spectrum of supporters, the ALP has experienced three major splits—on military conscription (First World War), economic policy (Great Depression), and the role of Catholics and communists (1950s)—when the costs of maintaining the coalition have become prohibitive. Contemporary management of this broad coalition depends on three factions to carve up the spoils of party and government office. The Party appears to have become more resilient to periodic crises, as evidenced by its survival (without splitting) of the sacking of the Whitlam Labor Government in 1975, and the substantial deregulation of the economy by the Hawke–Keating Labor governments despite the Party's nominally socialist platform. The ALP federal governments between 1983 and 1996 provided the Party's longest continuous period in national office; apart from the Curtin–Chifley Governments of 1941–49, national ALP governments have been of short and generally tempestuous duration. The ALP has been more successful at state level, providing long-lived governments in Tasmania (1934–69 and 1972–82); Western Australia (1933–47 and 1983–93); South Australia (1970–79 and 1982–93); Victoria (1982–92); NSW (1941–65 and 1976–88).

The Party's early ideology stressed socialism, especially with respect to the more egalitarian sharing of wealth and income. The Party's early program stressed the role of direct government involvement in the economy through nationalisation, especially of financial institutions, to implement this ideology. Following the failure of bank nationalisation legislation—declared unconstitutional by the High Court—the Party has tended to stress egalitarian social programs, which it has achieved through legislation on education, health, and income support. The ALP has been characterised as a 'mildly reformist' party, more in the radical-liberal tradition than in a socialist one. The ALP tends to favour central over state governments, because it sees the national government as having the resources to achieve its program. Examples of this centralisation include the Whitlam Government's use of s.96 grants to induce state governments to implement ALP programs; the use of national government powers to stop the Franklin Dam in Tasmania, to implement national health support (through Medibank and Medicare) and native title legislation (cf. Chapter 10); and the use of the Hilmer report to attack state government monopolies (cf. Chapter 14). The Party attempts to control individual parliamentarians by requiring them to take a pledge to vote as decided by the caucus, which comprises all ALP parliamentarians of a particular parliament.

The bureaucracy of the ALP is organised at state level, with a federal bureaucracy whose power has grown in recent decades. Each state has a Conference, which is the policy-making organ for state-level policy. States select delegates to the biennial National Conference, which determines national policy. The smaller National Executive plays an important role between meetings of the National Conference. The composition and *modus operandi* of the National Conference and Executive have evolved substantially in recent decades, in a way that has loosened the control of Party members and its bureaucracy over parliamentary members (cf. Woodward 1990, pp. 147–9). Trade unions 'affiliated' with the ALP contribute substantial financial resources to the Party, and are rewarded by having large numbers of delegates, often a majority, at state conferences.

Support for the ALP is largely, although not exclusively, urban and it tends to achieve most of its electoral successes in metropolitan areas or large provincial cities. Its supporters' social beliefs tend to conflict with country-mindedness, and the ALP has difficulty winning parliamentary seats in rural and regional Australia except where such seats have large mining communities (e.g. the NSW state seat of Broken Hill), where political dynasties are established, or where there are parliamentarians who win despite being members of the ALP (e.g. G. Campbell, federal member for Kalgoorlie).[4]

Minor parties and independents

The Democratic Labor Party (DLP), formed from the ALP split of the 1950s, has essentially disappeared. It was of key importance to coalition governments from the mid-1950s to the mid-1970s because its preferences in lower houses provided a bridge for former Labor voters to transfer their vote to the Liberal or National (then Country) parties, and because the DLP attracted sufficient support to have senators elected who, by voting with the coalition, ensured coalition dominance of the Senate until the mid-1970s.

It is ironic that the Australian Democrats only qualify as a 'minor' party, since their share of the national vote is not dramatically different from that of the National Party. In the 1993 federal election, the Nationals obtained 7.2 per cent of the vote and the Democrats 3.8 per cent; however, the Democrats' Senate vote was somewhat higher at 5.3 per cent (Jaensch 1995, p. 82). The Democrats' peak voting support in 1990 was almost identical to the Nationals' peak voting support in the 1987 'Joh-for-Canberra' push (Jaensch 1995, figure 5.2). The Democrats' problem is that their vote is spread throughout Australia, rather than being geographically concentrated like the Nationals'—thus, the Democrats are unlikely ever to develop a significant lower house presence without a substantial increase in their vote which would require a collapse in the vote of either the Liberal or Labor party. Alternatively, abandoning single-member constituencies in favour of sufficiently large multi-member constituencies—about fifteen members given the Democrats' current level of support—could result in lower house success. The Democrats grew out of splits in the Liberal Party after the late 1960s; it is a highly participatory Party in which all members elect parliamentary leaders and policies are formulated by the Party as a whole via postal ballot (Shamsullah 1990, p. 171). The Democrats' ideological position is perhaps best summed up as the politics of caring (Shamsullah 1990, p. 173). They have held or shared the balance of power in the Senate since 1981, and have also held or shared upper house control in some states.

There is considerable contemporary discontent with the role of minor parties and independents because they control the Senate and, to a lesser extent, some state houses (e.g. both houses of the NSW Parliament 1991–95). Baxter (1994, p. 11) argued that:

> . . . the leader must have the authority to exercise his or her role and should have clear majority in both Houses of Parliament. I am doubtful if Australia can afford the indulgence of independents at such a critical time of its evolution. While they may make people feel warm and comfortable, they hold and wield power disproportionate to the number of people who elected them. It is often overlooked that government is elected to make decisions for the majority, for the 'common good'.

No doubt small parties or independents who exercise a balance of power frequently appear to be a nuisance to executive government and its bureaucracy. However, refusal to tolerate small parties or independents may result in the disenfranchisement of those who wish to vote for them. The argument that at this time Australia cannot afford independents could have been used at any time during Australia's history. Finally, the 'majority' is not necessarily the same as 'the common good'. In most Australian elections, the two-party preferred vote generally splits in the range 55:45. The majority that elects a government is thus rarely substantially greater than the minority who do not. Majorities thus frequently do not represent the *common* good; democracy might be better understood as a system in which representatives elected by the majority make decisions on behalf of society as a whole, while respecting the rights of a minority that might be almost as large as the majority. Since governments with untrammelled power may not respect the wishes of the (large) minority, democracy may well be better served by parliaments in which minor parties or independents *do* have the balance of power.

Barriers to entry

Funding

A recent innovation in Australia is the public funding of political parties. The principal justification used by political parties for public funding is that it reduces the need for them to seek funding from private bodies—e.g. individuals, firms or interest groups—and thus reduces the extent to which the winning party is beholden to particular interests after the election. However, it is also in the interests of the established, especially major, parties to limit the entry of new political parties. Public funding that is tied to past performance establishes an entry barrier which it is difficult for new parties to surmount. The public funds paid to political parties after the 1990 federal election are illustrated in table 3.4. As shown in table 3.5, public funding has come to represent a significant source of funding for the established political parties. In 1995, legislation passed by the Federal Parliament doubled the value of public funding for political parties.

Table 3.4 Public Funding of Political Parties, 1990

Party	($'000)
Australian Democrats	1 495
Australian Labor Party	5 301
Call to Australia	19
Citizens Electoral Councils	3
Greens in Lowe	3
Illawarra Greens	7
Independent EFF	3
Irina Dunn Environment	4
Liberal Party	4 612
National Party	1 140
NT Country Liberal Party	38
Rex Connor (Snr) Labor Party	7
South Sydney Greens	4
Sydney Greens	5
The Greens	10
United Tasmania Group	12
WA Green Party	87
Independents (22 candidates)	129
Total	12 879

Source: Jaensch (1995, table 3.1)

Table 3.5 Party Income and Expenditure, 1990 Federal Election ($'000)

	Income		*Expenditure*			*Deficit*
	Public funding	*Gifts*	*Broadcasting*	*Publishing*	*Other*	
Labor	5 300	3 978	8 337	2 880	3 539	5 477
Liberal	4 612	4 373	5 119	3 251	3 746	3 131
National	1 140	2 039	1 672	696	837	27
Democrat	1 495	83	348	216	599	415
Other	332	1 448	1 970	3 950	1 708	5 850
Total	12 879	11 921	17 446	10 993	10 429	14 068

Source: Jaensch (1995, p. 51, with corrections for apparent arithmetic errors)

Political parties are required to provide detailed reports of their funding and expenditure. A traditional source of funding is donations from individuals and organisations. Donations to the major political parties are shown in table 3.6. Part (a) of the table neatly illustrates the concept of 'induced institutional innovation'. Each of the large parties receives most of its large donations from a single source: the ALP's national secretariat received $1.6 million from John Curtin House Ltd; the Liberal Party's national secretariat received $7.2 million from the Free Enterprise Foundation; the National Party's federal secretariat received $40 000 from the Free Enterprise Foundation. By contrast the Western Australian and South Australian divisions of the Liberal Party, with long lists of large donations from individuals and companies (cf. table 3.6(b)), do not appear to receive substantial funds from single companies, foundations or trusts.

Table 3.6 Donations to Australian Political Parties, 1993–94
(a) Summary

Party	*Donations < $1500 ($)*	*'Individuals' donations over $1500*	
		Total	Some individual donations ($'000 except where indicated otherwise)
Australian Democrats			
National level	1 300	27 270	
NSW	23 800	4 000	
Victoria	11 900	18 200	Parliamentarians and candidates only
Queensland	5 400	10 650	Parliamentarians and candidates only
WA	800	10 650	
SA	32 900	38 300	
Tasmania	400	2 200	Parliamentarians only
ACT	470	—	
Australian Labor Party			
National level	12 400	1 963 400	Amcor 30, BIL 25, Coles Myer 34, John Curtin House Ltd $1.63m, Leighton 20, Lend Lease 25, Lustig and Moar 25, McDonalds 25, Mercantile Mutual 20, MIM 50, SA Brewing 25, Wesfarmers 20
NSW	50 600	98 900	McDonalds 10, Randwick Labor Club 29
Victoria	92 500	24 000	
Queensland	110 400	906 700	ACTU 22, AWU 43, Labor Holdings P/L 150, Lakes Corp. 10, Reana Developments 11, Brisbane Futures Ltd 271
WA	31 359	118 618	National Secretariat 20, Connell Lighbody 20, Transfield 10
SA	53 484	301 479	Santos 20, Shop Distributive Assoc. 10, Western Mining 20, Australian Liquor Union 16
Tasmania	29 358	62 182	
ACT	3 584	85 986	Canberra Labor Club 86; Other income: Canberra Labor Club 110
NT	20 277	266 638	Nabalco 10, Zapopan 20, MIM 25, National Secretariat 170
Liberal Party of Australia			
National level	550	7 472 390	Amcor 60, Australian Reinsurance 15, Free Enterprise Foundation $7.2m, Lend Lease 25, McDonalds 25, MIM 25, Santos 50, Toyota 29; Other income: International Democratic Union (UK) 70
NSW	287 381	247 090	Amcor 10, 500 Club 29, Free Enterprise Foundation 101, Stable Coalition P/L 25
Victoria	134 903	370 708	Amcor 10, Phillip Morris 25, Pratt & Co. 50, 500 Club 220; Other income: 500 Club 472
Queensland	114 654	118 716	Free Enterprise Foundation 15, Feez Ruthning Trust 25
WA	357 600	621 301	see part (b) of Table
SA	372 909	931 358	see part (b) of Table
National Party of Australia			
National level	2 719	77 500	Amcor 10, Free Enterprise Foundation 40, MIM 25
NSW	39 407	9 675	
Victoria	88 680		

Table 3.6 cont'd

Queensland	106 500	255 000	Mineralogy P/L 10, Kaldeal P/L 50, Townsville Trade Waste, 10, Towners Cleaning Service 20, MIM 25, John McEwan [sic] House P/L 25
WA	1 600	51 000	Wesfarmers 20

Source: Australian Electoral Commission

(b) Liberal Party in Western Australia and South Australia, Individual Donations exceeding $1500 ($'000)

Western Australia	*South Australia*
Ardross Estates 10, Austal Ships 17, Cape Bouvard Investments 20, Clements Consulting 36, Gold Mines of Australia 10, Griffin Coal 20, Hancock Prospecting 80, Hogan and Partners 10, Paladian Securities 25, Eyres Reed 12, Robe River Iron 25, Sansom and Associates 10, St Barbara Mines 52, Titan Ford, 40, Wavemaster International 50, Furama P/L (accommodation as gift in kind) 156	AAMI 10, Adelaide Brighton Cement 30, Arrowcrest Group 13, Catch Tim Ltd (in Hong Kong) 100, Cavill Power 10, C. E. Heath International 20, CIC Insurance 10, F. H. Faulding 35, Gerard Industries (printing as gift in kind) 68, G. H. Michell & Sons 10, Impeach P/L 10, Independent Holdings (cocktail party as gift in kind) 49, Kangaroo Island Sealink 10, KS Freighter 10, MMI 10, Mercantile Mutual 10, Motor Trade Electoral Action Committee 71, Santos 75, South Corp. Holdings 50, WMC 20, Wolf Blass and Associates 11

Source: Australian Electoral Commission

Malapportionment

Jaensch (1995, tables 3.1 and 3.2) presented indicators of malapportionment in Australia. The first of these tables, on zonal malapportionment, relates to those states which classify their electorates into zones, and have different mean numbers of electorates by zones. Data for Queensland and Western Australia relating to the 1980s reveals ratios between electorate sizes in the unfavoured and favoured zones of lower house electorates of around 2.0—that is, it takes twice as many voters to elect a member in the unfavoured seats. Conversely, a voter in an unfavoured seat has a vote worth only half that of a favoured voter. In the Western Australian Legislative Council, the ratio between unfavoured and favoured seats was about 4.0.

Bureaucracy

The ostensible purpose of the bureaucracy (or public service) is to implement the decisions of executive government. In particular, the bureaucracy administers those Acts and Regulations which the executive government proposes and parliament passes. The bureaucracy also provides continuity across changes of government. Such change is often accompanied not only by substantial policy change but also by the cleaning out of those (particularly senior) public servants seen to be too closely aligned to the outgoing government.

The bureaucracy can be divided into several sectors. In examining the Australian federal bureaucracy, Pusey (1991) identified an 'inner triangle' of departments. Unfortunately, his research coincided with an extensive agglomeration of federal government departments after the 1987 election; the list below thus refers to the new mega-departments rather than the departments Pusey identified.

- Core, or 'Central Agencies'—Prime Minister and Cabinet, Treasury, and Finance—which support the Prime Minister as the principal minister, the Treasurer, who is responsible for setting (and possibly controlling) the economic agenda, and the Minister of Finance, who is responsible for monitoring the spending of Commonwealth government departments.
- 'Market' departments, which support the business sector—e.g. Foreign Affairs and Trade; Industry, Technology and Commerce; Primary Industries and Energy.
- 'Program and Service' departments, which primarily provide welfare services to individuals—e.g. Human Services and Health, Social Security, Veterans' Affairs.

Many other federal government departments did not fall within Pusey's 'Inner Triangle'. These included Foreign Affairs, now included with Trade as part of the 'Inner Triangle', and Defence. Some of the 'outer' departments primarily provide support to other government departments (e.g. Attorney-General, Administrative Services), while others deal with public infrastructure (e.g. Employment, Education and Training; Transport and Communications). Other departments are grab-bags—e.g. Environment, Sport and Territories.

Because Pusey concentrated on departments delivering services, he omitted the activities of many statutory authorities. These include economic authorities such as the Reserve Bank and the Industry Commission, various statutory authorities delivering marketable services (e.g. QANTAS, Federal Airports Corporation, Commonwealth Bank, Australian National Railways, Telstra), and statutory authorities delivering public goods (e.g. Australian Bureau of Statistics) (cf. Box 3.2). The former Department of Aboriginal and Islander Affairs has since become the Aboriginal and Torres Strait Islander Commission.

Box 3.2 Some Australian Government Statutory Authorities and Officers

Administrative Appeals Tribunal
Australian Broadcasting Authority
Australian Competition and Consumer Council
Australian Government Publishing Service
Australian Quarantine and Inspection Service
Australian Security Intelligence Organisation
Australian Taxation Office
Austel
Austrade
Australia Council for the Arts
Australia Post
Australian Broadcasting Corporation
Australian Industrial Property Organisation
Bureau of Air Safety Investigation
Bureau of Consumer Affairs

Bureau of Meteorology
Civil Aviation Authority
Commonwealth Employment Service
Child Support Agency
Coal Industry Tribunal
CSIRO
Customs Service
Defence Housing Authority
Director of Public Prosecutions
Electoral Commission
Export Finance and Insurance Corporation
Film Australia
Film Commission
Health Insurance Commission
Human Rights and Equal Opportunity Commission
Immigration Review Tribunal
Insurance and Superannuation Commission
International Development Assistance Bureau
International Training Institute
Industrial Relations Commission
Joint Coal Board
Landcare Australia
Law Reform Commission
Maritime Safety Authority
Materials Testing Laboratories
Medicare
National Acoustic Laboratories
National Audit Office
National Crime Authority
National Film and Sound Archive
National Health Occupational and Safety Commission
National Library
National Maritime Museum
National Media Liaison Service (service to government)
National Office of Overseas Skills Recognition
National Science and Technology Centre
National Standards Commission
Nuclear Science & Technology Organisation
Office of Film and Literature Classification
Ombudsman
Patents Office
Pharmaceutical Benefits Scheme
Prices Surveillance Authority
Privacy Commissioner
R&D Councils
Race Discrimination Commissioner
Sea Safety Centre
Securities Commission
Sex Discrimination Commissioner
Statutory Marketing Authorities
Special Broadcasting Service
Telecom
Tourist Commission
Weather Bureau
Worksafe Australia

The general form of government departments and statutory authorities is mirrored in the states and territories (e.g. Reynolds 1991 on Queensland). The roles and functions of state and federal departments reflect the powers they exercise.

Lobby groups

A wide variety of other interest groups have access to, and seek to influence, the decisions of parliament, the executive, the government, and the bureaucracy. These interest groups include:

- 'summit' employer groups—e.g. national groups such as the Business Council of Australia, Australian Chamber of Commerce and Industry (formerly Confederation of Australian Industry), Australian Mining Industry Council, National Farmers' Federation, Metal Trades Industry Association, and corresponding state-level summit groups;
- associations representing particular industries (e.g. education, manufacturing, and sub-groups such as vehicle manufacturing or component parts) or products (e.g. farmers, forestry, fishing, mining) at national or state level;
- individual firms—some firms are so large (e.g. in mining, some manufacturing, retailing) that they exercise influence independently of corresponding employer associations;
- professional associations and societies—e.g. the Australian Medical Association and the specialists' colleges, the lawyers' Bar Councils and Law Societies, Institution of Engineers, and corresponding groups for accountants, auditors, economists and agricultural economists, agricultural scientists etc.;
- 'summit' employee groups—the Australian Council of Trade Unions;
- individual trade unions;
- service organisations (e.g. motorists' associations), cultural associations (e.g. the RSL, and arts and sports clubs), ethnic associations, religious groups, secular and religious welfare organisations (e.g. philanthropic such as Smith Family and Salvation Army, consumer such as Australian Federation of Consumer Organisations and Australian Consumers Association, pensioners, overseas aid) and social organisations for non-human welfare (e.g. RSPCA and WIRES for animals; and Landcare groups, Australian Conservation Foundation, Greenpeace and Wilderness Society for conservation generally).

Interest groups' mechanisms of influence include:

- donations to political parties (with or without the expectation of specific government decisions);
- collecting and analysing information, and presenting it to politicians, bureaucrats or inquiries;
- disseminating public information, advertising and spreading propaganda; and
- providing future politicians (e.g. current politicians S. Crean and M. Ferguson from the ACTU, I. McLachlan from NFF, and B. Nelson from the AMA).

Policy consultants and think tanks

Some of the functions of departments and authorities are increasingly being undertaken by private firms—some examples are services such as cleaning, some legal services and staff training. Policy analysis, policy reviews and policy development are also increasingly being privately provided (cf. table 3.7). Interesting economic analysis could be undertaken on the private provision of policy advice: in particular, what are the demand conditions for the services of particular firms and how do they relate to past activities of a particular firm? In the economic policy field, private policy providers include Access Economics, ACIL, Hassall and Associates and the Centre for International Economics.

Table 3.7 Commonwealth Government Consultancies

	1987–88	*1988–89*	*1989–90*	*1990–91*	*1991–92*	*1992–93*	*1993–94*
(i) Expenditure ($m)							
DFAT–AIDAB	..	122.1	405.4	575.3	194.6	116.2	182.4
PIE	1.8	3.3	4.8	8.3	8.4	8.4	7.3
Admin Services	70.3	65.6	58	47	30.9	33.9	33.2
Defence	6.6	18.9	23.1	31.9	20.9	40	33.2
Total	91.8	246.7	547.7	728.4	324.4	269.7	342.4
Consultancies (number)							
DFAT–AIDAB	..	411	429	470	390	230	194
PIE	136	162	229	293	450	482	288
Admin Services	4142	3988	2583	2053	1041	1344	1426
Defence	..	437	524	571	782	1068	1710
Total	4873	6468	5721	5702	5309	5584	6832
Average cost ($ per consultancy)							
DFAT–AIDAB	..	297 000	944 900	1 224 000	498 900	505 200	940 200
PIE	13 200	20 300	20 900	28 300	18 600	17 400	25 300
Admin Services	16 900	16 400	22 400	22 800	29 600	25 200	23 200
Defence	..	43 200	44 000	55 800	26 700	37 400	19 400
Total	18 800	38 100	95 700	127 700	61 100	48 200	50 100

Source: Howard (1995, Tables 4 and 5)

The market place for ideas is one of society's great formative influences, although one person's educational and rational debate is another's propaganda and ideology. The universities—traditionally public but now including private universities—have historically been major, although not the exclusive, centres of intellectual exploration. In recent decades, universities have established internal institutes to harness the collective efforts of like-minded academics. In the policy field, such bodies include the National Institute of Economic and Industry Research (Melbourne), National Institute of Labour Studies (Flinders) and the Centre of [sic] Policy Studies (Monash). In scientific fields, Australia has also had state-run intellectual organisations such as CSIRO and some parts of the bureaucracy (e.g. ABARE).

Australia is increasingly following the American lead of developing private sector or non-government organisations which operate as centres of intellectual enquiry. Established think tanks in Australia are generally on the political right, but have been joined in recent years by institutes from the political left.

The Institute of Public Affairs (IPA) was established in 1943 and claims to have become one of Australia's most important sources of ideas alternative to socialism. The Institute merged with the Australian Institute for Public Policy in 1991. The IPA sees its role as being 'to identify the means that will enable Australians most effectively to realise their values and to argue publicly for those means against objections inspired by opposing ideologies or vested interests' and it shares the values and aspirations of the vast majority of the [Australian] population: the rule of law; parliamentary democracy; a prosperous economy with full employment; high standards in education; stable family life; sound environmental management; security from the threat of crime and invasion; the freedom to associate, express opinions, own property and practise one's religion; and a tolerant, peaceful society (*IPA Annual Review 1993–1994*, p. 4).

The Institute's main interests are government, economics, education and the environment. The IPA undertakes 'tasks' from federal or state governments, but does not accept commissions or donations from political parties or grants from governments.

The purpose of the Centre for Independent Studies 'is to preserve and enhance the individual freedoms, including the rights to life, liberty and property, of Australians and New Zealanders'. It has a 'bias towards classical liberalism and social and economic freedoms within a framework of autonomous civil institutions'. The Centre believes among other things that 'the best possible way to defend and promote [a free, democratic society under limited government] is by a continuing process of education based on rigorous, public, intellectual debate about major social and economic issues' (Centre for Independent Studies n.d.). The Centre is independent in that it receives no government funding, but relies on funding from a 'broad mix of public and private companies, individuals, subscribers' plus those who purchase its publications and services. As suggested in a recent review by one of its Academic Advisory Council, the Centre's philosophy might also be described as 'libertarian' (cf. Arndt 1994–95). Among other activities, the Centre's Economics Education Resource Centre claims to be 'one of Australia's largest providers of up-to-date resources and information for secondary economics teachers.'[5]

The Tasman Institute, founded in 1990, undertakes projects directed at improving well-being in the Australian region. Its approach to policy rests upon its 'research and evidence that prosperity requires both competitive markets and well defined property rights'. Without these attributes, 'trade may not be beneficial, and political freedom may not be sustained' (*Tasman Institute Annual Review 1992*, p. 4). The Tasman Institute has an associated commercial research arm, Tasman Economic Research. The Tasman group's Advisory Council comprises major Australian business leaders, and its Research Council comprises conservative, especially economist, academics from Australia, North America and the UK. Tasman appears to have been closely involved with the policy changes of the Kennett Coalition government in Victoria, and has collaborated with the Institute of Public Affairs in 'Project Victoria'. Tasman also fosters connections with conservative US think tanks such as the American Enterprise Institute.

There is some overlap among the various boards of the Institute of Public Affairs, the Centre for Independent Studies, and the Tasman Institute (table 3.8).

Table 3.8 Think Tanks—Common Board, Council or Trustee Members and Research Advisers

Institute of Public Affairs	Tasman Institute	Centre for Independent Studies	Menzies Foundation
Institute of Public Affairs	W. Bailey, K. Gosper, E. Mayer, H. Morgan, R. Murdoch	H. Morgan, I. Roach, B. Watson	
	Prof. J. Freebairn, Prof. R. Snape	Prof. K. Clements, Prof. R. Officer, Prof. W. Kasper	Prof. J. Rose
Tasman Institute		Prof. R. Blandy, Prof. M. Hartwell, Prof. K. Minogue, Prof. R. Parish	

Sources: *Tasman Institute Annual Review 1994*, Centre for Independent Studies (n.d.), Institute of Public Affairs *Annual Review 1993–1994*

Initially established as the New South Wales Branch of the Institute of Public Affairs, the Sydney Institute became an independent and privately funded organisation in 1989. It is 'dedicated to the principle of policy debate', and its longer term aims are 'to promote human and commercial freedoms and to contribute to a community awareness of Australia's strategic positioning in the world.' The Institute has established links with the American Enterprise Institute, the Manhattan Institute, the European Policy Forum, Keidanren (in Japan) and the Centre for Strategic and International Studies (Indonesia).

The Menzies Research Centre, named after R. G. Menzies who was Prime Minister from 1939–41 and 1949–65, was established by the Liberal Party in October 1994 but 'it will be an independent organisation' funded by business and other supporters. The Party's Federal President said, on announcing the Centre's establishment, that the 'Liberal Party was built on the power of ideas' and that the Menzies Research Centre 'is a major initiative to reinforce the importance of ideas as the foundation of our Party'.

The Evatt Foundation was named after H. V. Evatt, a former judge, Labor politician, and President of the United Nations' General Assembly. The objectives of the Foundation are to promote education, ideas and the discussion of public issues; to develop academic research for the labour movement; to promote human rights for disadvantaged and minority groups; to promote youth education and training; and to promote debate about the arts and cultural policy in Australia.

The Foundation is incorporated as a non-profit organisation, and was established with grants from a New South Wales Labor government and a Tasmanian government, and donations from organisations including trade unions and business. From 1984, the Foundation has received an annual federal grant of $0.25m (indexed from 1993).[6] A similar organisation, the Whitlam Institute for Social and Economic Research, was established in 1992 by politicians, trade unionists and academics as an additional counter to the 'hijacking' of the policy debate by economic rationalist arguments. The Victoria Foundation was established in 1993 with the similar aim of challenging economic rational orthodoxy.

The purpose of the Australia Institute 'is to help create a vision of a more just, sustainable and peaceful Australian society'. The Institute recognises that, while private markets are often effective at encouraging efficiency, they frequently fail to reflect the community's economic, social and environmental priorities. The Institute rejects the belief that market outcomes are value free. The Directors of the Institute are drawn from parliament, bureaucracy, non-government organisations, the trade union and university sectors, and private sector consulting.

Media and information

Information is a key commodity in the political market place. Rational policy making demands that the implications of policy decisions are known as far as possible. Information creators, such as the education sector and lobby groups, also disseminate information privately (i.e. directly to decision makers) and publicly via the media. Public information dissemination has begun to include electronic information services such as the Internet.

The media are not only means for disseminating information but also players in the political process. The way that 'news' is reported can influence the public's perception of events. In the commercial media, senior executives who control the way news is reported know that their appointment and continued employment depends on satisfying the requirements of proprietors. Where proprietors have political as well as commercial objectives, it would be naive to think that these objectives do not affect news content, presentation and editorial tone either explicitly or implicitly. Since politicians' futures depend on voters' perceptions, they have an interest in attempting to ensure that the structure of the media is not inimical to politicians in general, and governments may attempt to achieve 'balanced' media coverage (i.e. favourable to their own party). There is therefore likely to be continual tension between politicians and the commercial media. Different issues affect publicly-funded broadcasters, who depend on politicians for resources.

The media are highly concentrated in Australia. The publicly-funded Australian Broadcasting Corporation, together with SBS, has two national television networks, three metropolitan radio networks, and two regional radio networks. Following aggregation of regional television in the 1980s, non-remote rural Australia has access to three commercial television networks: one of these is owned by K. Packer's Consolidated Press, one was sold by R. Murdoch's News Corporation because of the introduction of restrictions on cross-media ownership (of print and electronic media) in the 1980s. R. Murdoch's News Corporation is reported to own 60 per cent of the newspaper titles in Australia, including its purchase of the Herald and Weekly Times group in the late 1980s. The other major newspaper group is Fairfax, currently with K. Packer's Consolidated Press and C. Black as major shareholders.

The cross-media restrictions of the 1980s are an interesting example of the Commonwealth extending its powers. Under the Constitution, the Commonwealth has no power to control print media. However, because the Commonwealth did have the power to control the electronic media, it implicitly extended its control to print media by restricting media owners from having dominant positions in both print and electronic domains.

3.5 RURAL POLICY PROCESS

The rural policy process is a microcosm of the Australian policy process described above. An understanding of the rural policy process therefore principally demands an understanding of the general policy process in Australia. Discussion in this section focuses narrowly on those parts of the policy process that are especially important to the rural sector.

Commonwealth and states, and ministerial councils

There is no specific mention of agriculture and little about resource industries in the Australian Constitution. A policy role for the Federal government in the agricultural and resource industries thus arises primarily from the more general powers that the Constitution confers (cf. section 3.2). The Commonwealth's key powers directly affecting agriculture and resources include (ss.90, 51(i), (iii), (ix), (x), (xxix)), and the Commonwealth's indirect powers come via ss.51(ii) and (iv) and 96.

The agriculturally-oriented programs that the Commonwealth incorporates under its heads of power are extensive (Box 3.3; see also Wonder 1995). State government's powers are technically the 'residual' of the Commonwealth's specified powers (cf. section 3.2); as an example of state legislation, New South Wales Acts which affect the agricultural industries directly are detailed in Box 3.4.

Box 3.3 Commonwealth Department of Primary Industries and Energy Programs on Agriculture and Resources, 1994

1. Commercial Linkages
   - Agri-Food Industries Growth Strategy
   - Marketing Skills Program
   - Rural Industries Business Extension Service
   - World Best Practice Scheme
   - Rural Enterprise Network Program
   - Rural Development Incentive
   - Innovative Agricultural Marketing Program
   - Agribusiness Programs Board & Business Development Programs
   - Food Standards and Quality Assurance
   - Ausindustry

2.
   - Farm Productivity
   - Rural Adjustment Scheme
   - Farm Household Support Scheme
   - Job Search Allowance
   - Rural Financial Counselling
   - National Drought Policy
   - Farm Management Bonds
   - Income taxation incentives
   - Rural R&D Corporations
   - Landcare

3. Rural Communities
   - Rural Communities Access Program
     – Countrylink
     – Rural Access Program
     – Telecentres
     – Rural counselling Program
     – Business Advice in Rural Areas
   - Integrated Regional Database
   - Services in Rural Areas

4. Broader Linkages
   - ARMCANZ/SCARM
   - AgriFood Council
   - RAS Advisory Council

Source: Department of Primary Industries and Energy

Box 3.4 New South Wales Agricultural Acts

Agricultural Scientific Collections Trust 1983
Agricultural Holdings 1941
Agricultural Tenancies 1990
Animal Research 1985
Apiaries 1985
Banana Industry 1987
Biological Control 1985
CB Alexander Foundation Incorporation 1969
Cattle Compensation 1951
Commonwealth Powers (Meat Inspection) 1983
Dairy Industry 1979
Dried Fruits 1939
Exhibited Animals Protection 1986
Exotic Diseases of Animals 1991
Farm Produce 1983
Farrer Memorial Research Scholarship Fund 1930
Fertilisers 1985
Grain Marketing 1991
Homing Pigeons Protection 1909
Horticultural Stock and Nurseries 1969
Institute of Rural Studies 1973
McGarvie Smith Institute Incorporation 1928
Marketing of Primary Products 1983
Meat Industry 1978
Murray Valley Citrus Marketing 1989
Non-indigenous Animals 1987
Noxious Weeds 1993
Pesticides 1978
Plant Diseases 1924
Poultry Meat Industry 1986
Poultry Processing 1969
Prevention of Cruelty to Animals 1979
Prickly Pear 1987
Rural Lands Protection 1989
Rural Assistance 1989
Seeds 1982
Stock (Artificial Breeding) 1985
Stock (Chemical Residues) 1975
Stock Diseases 1923
Stock Foods 1940
Stock Medicines 1989
Swine Compensation 1928
Sydney Market Authority 1968
Tobacco Leaf Stabilisation 1967
Veterinary Surgeons 1986
Wheat Marketing 1989
Wine Grape Processing Industry 1979
Wool and Hide Dealers 1992

Source: Mortensen (1994)

Until recently, the Commonwealth tended to exercise legislative power concerning agriculture and resources in cooperation with the states. Until the late 1980s, for example, the Commonwealth implemented statutory agricultural marketing schemes cooperatively with the states through complementary Commonwealth and state legislation, and quarantine activities were jointly undertaken. The mechanism for achieving such agreements was the agricultural ministerial council. The Agricultural and Resource

Management Council of Australia and New Zealand (ARMCANZ) comprises ministers of the Commonwealth, states, territories and New Zealand responsible for agriculture, soil conservation, water resources and rural adjustment. The heads of the corresponding government departments comprise a working group—the Standing Committee on Agricultural and Resource Management (SCARM)—which negotiates agreement where possible for formal ratification by ARMCANZ.[7] The Council and its Standing Committee meet biannually. The Council's objective is to 'develop integrated and sustainable agricultural and natural resource management policies, strategies and practices for the benefit of the community'. It does so by reference to the following issues in the agricultural and water industries, and natural resource management:

- research and development (for efficient, sustainable and equitable development), education, training and extension;
- regional and rural economic and community development;
- enhancing effectiveness and efficiency of rural industries;
- providing for efficient water supply, sewerage, and waste management in the water industry;
- adopting uniform policy on external marketing and water pricing;
- ensuring natural resource practices provide a basis for food and water of high safety and quality standards; and
- monitoring natural resource 'health'.

ARMCANZ and SCARM are highly visible within the bureaucracy as preparation for the biannual meetings consumes considerable bureaucratic resources. However, ARMCANZ is only externally visible through the publication of resolutions that follow its meetings, and SCARM is practically invisible.

Corresponding to ARMCANZ and SCARM, there are also ministerial councils for:

- the environment—Australia and New Zealand Environment and Conservation Council (ANZECC) (incorporating the former Australian Environment Council and Council of Nature Conservation Ministers) with separate standing committees on Environment Protection, and Conservation;
- fishing and forestry—Ministerial Council on Forestry, Fisheries and Aquaculture, with separate standing committees for forestry, and fisheries and aquaculture;
- food—National Food Standards Council.

Because of the complexity of river basin management, the Murray–Darling Basin Council was established by the Commonwealth, New South Wales, Victorian and South Australian governments to coordinate land use management in the Murray–Darling Basin. The most recent Murray–Darling Basin Agreement was signed by the Commonwealth, New South Wales, Victoria and South Australia in 1992 and subsequently by Queensland. The purpose of the Agreement is 'to promote and coordinate effective planning and management for the equitable efficient and sustainable use of the water, land and other environmental resources of the Murray–Darling Basin.' The Murray–Darling Basin Commission (corresponding to the standing committees of other ministerial councils) was established under the agreement to:

- advise the Ministerial Council in relation to the planning, development and management of the water, land and other environmental resources of the Murray–Darling Basin;
- assist the Ministerial Council in developing measures for the equitable efficient and sustainable use of the water, land and environmental resources of the Murray–Darling Basin; and
- coordinate and or implement measures authorised by, or policies or decisions of, the Council (Murray–Darling Basin Agreement 1992, section 17(1)).

There are also less formal Commonwealth/state cooperative arrangements. For example, the Australian National Committee on Large Dams coordinates common arrangements relating to the physical integrity of large water storages, with corresponding state bodies such as the New South Wales Dams Safety Committee.

The Federal Labor government of 1983–96 also strongly encouraged industry policy councils to advise government on the form and implementation of policy (cf. Martin 1990, p. 172).

Bureaucracy

The principal government departments dealing with agriculture and natural resources are the Commonwealth Department of Primary Industries and Energy (DPIE) (agriculture, water, forestry, fisheries, as well as mining and energy) and state agriculture and natural resource departments. DPIE primarily establishes a national policy framework for agricultural and natural resource production in Australia, and delivers or coordinates external quarantine services. DPIE also has research arms—the Australian Bureau of Agricultural and Resource Economics (ABARE), and the Bureau of Resource Sciences. Agricultural trade matters may be dealt with by the Department of Foreign Affairs and Trade in concert with DPIE, or by DPIE separately.

State government departments are involved in delivering agricultural extension, research and veterinary services; irrigation water supply and drainage; fisheries management and research; timber production and policy; and soil conservation research, extension and contracting services. State governments' statutory authorities include those administering the Rural Adjustment Scheme, statutory marketing authorities, and product handling and marketing facilities, and have frequently regulated the transport of agricultural products. Pest, disease and weed control activities may be undertaken by state agriculture departments, or by specialist bodies. The latter include the Australian Plague Locust Commission (the Commonwealth jointly with Queensland and New South Wales), the Agricultural Protection Board for Western Australia, and the Rural Lands Protection Boards (RLPBs) in New South Wales. The RLPBs are a form of local government providing animal health, stock identification and documentation for travel, noxious animal and insect control, management of travelling stock reserves, and administration of natural disaster relief. Local government may also provide market services (e.g. saleyards), and has provided processing facilities such as abattoirs.

Statutory bodies

At both federal and state levels there are non-departmental bodies which have a substantial impact on agricultural and resource policy. Some federal bodies have an economy-wide remit, and thus affect agriculture and resources along with other sectors. As well as these statutory bodies with a general remit, there are many specifically agricultural statutory bodies, and some relating specifically to resources.

- Industry Commission (IC). Established under the *Industry Commission Act 1989*, and was formed out of the Industries Assistance Commission (established 1974) and the Inter-State Commission (see section 3.2). The functions of the Industry Commission are to hold public inquiries and report on the matters referred to it by the Federal Government; undertake activities incidental to its inquiry function, such as advising on business regulation and publishing information papers to enhance community understanding of industry policy issues; and report annually on the economic importance of industry and the effects of assistance and regulation on industry and the economy.

  In undertaking its activities, the IC must have regard to the government's policy framework which includes to:
  - encourage the development and growth of Australian industries that are efficient in their use of resources, self-reliant, enterprising, innovative and internationally competitive;
  - facilitate adjustment to structural changes in the economy and ease social and economic hardships arising from those changes;
  - reduce regulation of industry (including regulation by the states and territories) where this is consistent with the social and economic goals of the Commonwealth government; and
  - recognise the interests of industries, consumers, and the community, likely to be affected by measures proposed by the Commission.

    The IC has been important for agriculture for two reasons. First, some of its early inquiries investigated cherished agricultural institutions such as fertiliser bounties and statutory marketing schemes, and it has continued to rigorously scrutinise agricultural arrangements. Second, because it has been consistently 'economic rationalist', it has help succour the climate in which there have been major attitudinal shifts to economic regulation, and has had some successes in decreasing regulation in the agricultural sector and generally. In recent years, the

IC has also investigated non-agricultural rural and related issues such as fisheries, water resources and waste water disposal, energy, and greenhouse gas emissions. In mid-1996, the Productivity Commission will be formed from the IC, the Bureau of Industry Economics and the Economic Planning Advisory Council.

- Trade Practices Act and Australian Competition and Consumer Commission. The basic approach of the Trade Practices Act of 1974 is to prohibit and to provide for judicial enforcement against anti-competitive arrangements. Agriculture was formerly extensively exempted from the provisions of trade practices legislation while benefiting from the effects of the Act on other sectors of the economy, but these exemptions have expired. The Australian Competition and Consumer Commission was formed in 1995 from the Trade Practices Commission and the Prices Surveillance Authority with a primary focus of furthering the Commonwealth–state agreement on competition policy arising from the Hilmer report (cf. chapter 14).
- Environment Protection Authorities (Commonwealth and state). These bodies are responsible for environmental policy, investigations, and monitoring and enforcement. The Commonwealth's EPA also has powers previously exercised by the Resource Assessment Commission (abolished in December 1993).
- Statutory Marketing Authorities (SMAs). Statutory marketing authorities have a long history in Australia of managing the marketing process for agricultural commodities (cf. Industry Commission 1991b, Chapter 2 for a brief overview). SMAs may be established by either Commonwealth or state governments to market particular products, or groups of products. They may be established primarily to market commodities (e.g. Australian Wheat Board), to facilitate marketing (e.g. Australian Meat and Live-stock Corporation), to run price stabilisation schemes (e.g. the former Australian Wool Commission), or to run price discrimination schemes (e.g. market milk authorities). SMAs may have a primarily export focus, a primarily domestic focus, or operate across both markets. Objectives of statutory marketing arrangements include increasing returns to producers; stabilising prices, production and incomes; reducing marketing costs and stimulating demand; and providing or delivering assistance to primary producers (Industry Commission 1991b).
- Research and Development Corporations. The principal economic function of the agricultural Research and Development Corporations (RDCs, formerly known as Rural Industry Research Funds or RIRFs) is to circumvent potential excludability and non-rivalry problems in the agricultural research process (cf. Chapter 5). The RDCs enable the taxing of primary producers (as s.90 excises) to create funds to support agricultural research; these funds are also augmented by direct Commonwealth subsidy. In the resources area, there are specialist resource management bodies such as the Australian Fisheries Management Authority.

Rural interest groups

As well as the public service and statutory authority bodies noted above, there are many rural interest groups involved in the agricultural and resources sector. Lobby groups frequently wield considerable influence, partly arising from their financial resources. For example, the National Farmers' Federation established the Australian Farmers' Fighting Fund in the mid-1980s during the Mudginberri dispute with public and corporate donations exceeding $10 million. Its annual investment income is believed to be $1 million. The original purpose of the fund was to finance a challenge to trade union power using Mudginberri's dispute with the Australasian Meat Industry Employees' Union.

Political parties

In addition to the general political issues previously raised (cf. section 3.3), the role of political parties at the state level provides further opportunities for political trading because the states are responsible for providing specific services. Thus, for example, for agricultural and related resource industries it matters whether, when and where infrastructure is provided, who pays for the investment, and the price that is charged for the services provided by this infrastructure. Important investment issues include the construction or modernisation of airports, dams and roads, and whether or not railways remain open. The provision of agricultural support services—such as provided by departments of agriculture, water, soil conservation—is important in terms of scale and type of services, location, accessibility and the price charged. More generally, the provision of community rural infrastructure—such as schools and other educational institutions, hospitals and associated medical services, police and justice—are important issues in rural areas, just as they are in major cities. Politicians have an opportunity to provide benefits for their constituents through such services.

The New South Wales National Party, when in government, has generally sought to control the Department of Public Works because of its large spending budget and consequent opportunities for 'pork barrelling'. According to a former Director-General of the New South Wales Cabinet Office, who had served under the Coalition government, this department's 'construction bias' is partly the consequence of an 'engineering' culture which had been dictated to it by the National Party, A reported consequence of these tendencies was substantial damage to the quality of coastal waters and the subsequent threat to the fishing industry (Morris 1995).

Farmers' trade unions

These are the National Farmers' Federation at national level with commodity subgroups such as the Grains Council of Australia, and various state affiliates of farmers' organisations (e.g. Victorian Farmers' Union, South Australian Farmers and Stockowners, New South Wales Farmers' Association) and constituent commodity groups.

Business

Individual agribusinesses and their trade associations: processors (meat, dairy, wheat, stockfeed, brewers, sugar millers and refiners, canners), shippers, pastoral and finance houses.

Labour trade unions

Those affecting the rural sector especially are the Construction Forestry Mining and Energy Union, the Australian Workers' Union, Waterside Workers' Federation, and the Australasian Meat Industry Employees' Union.

Welfare groups and citizens

Welfare groups include church and philanthropic organisations, and rural counselling services. Citizens may be consumers (with concerns, for example, about food quality and product prices), service users (e.g. motorists' associations lobbying for better roads), and community welfare groups (with concerns, for example, about animal welfare). Citizens may also form environment groups with concerns about erosion, wilderness, scenery, water quality, native flora and fauna, feral animals and weeds (e.g. Landcare groups).

Rural media

Until recently, discussion of Australia's rural media would have emphasised local monopolies in newspapers, television and radio, in addition to ABC television and radio (including FM radio, and Radio National, JJJ and SBS Television from the 1990s). Rapid structural and policy changes in the Australian economy during the 1980s did not, however, leave the rural media unscathed. By the early 1990s, two media groups dominated regional daily newspaper circulation: Australian Provincial Newspapers had 29 per cent and News Limited 21 per cent (Henningham 1993, p. 66). Prior to aggregation of regional television in the second half of the 1980s, regional television monopolies were the most profitable sector of the Australian television industry. This was achieved through a combination of local monopolies in local television advertising and regional monopsonistic program buying from the three major metropolitan stations. Aggregation of licences into larger geographic areas created larger television markets and introduced regional competition.

The more complete integration of rural and regional Australia into the national media networks is likely to have a substantial impact on rural communities. The dissemination of local and regional information is likely to be squeezed by national and international programming (especially news and sport), and rural businesses and interests are likely to have greater difficulty in obtaining access to electronic media, especially television, as national advertising competes with local material.

As well as the general press there is also specialist rural media disseminating information to regional Australia. This specialist media includes state-based rural newspapers such as *Land* (NSW), *Weekly Times* (Victoria), *Stock Journal* (South Australia); and farmers' organisations' trade journals such as *Queensland Graingrower*, New South Wales Farmers' Association's newspaper and newsletter, National Farmers' Federation's *In Brief*; national news services such as *Primary Industry Newsletter* and magazines such as *Australian Farm Journal*. There are also specialised programs on the electronic media, such as the Country Hour on ABC Radio National, and Countrywide and its predecessors on ABC television (cf. Black 1995). Providing television and other communications services to 250 000 people in remote rural Australia, and especially to remote Aboriginal communities, created special problems.

3.6 CONCLUSION

Understanding the institutional structure of Australian agricultural and resource policy demands knowledge of the nation's general government structure. Appreciation of this policy framework also demands, more importantly, an appreciation of the processes that cause evolution in these structures. Integrating efficiency and public choice approaches is useful in explaining policy structures using economic analysis. Explaining changes in these structures as the consequences of changes in efficiency and policy choice variables is a valuable tool in understanding and explaining the evolution of Australian government policy in general, and of agricultural and resource policy in particular.

ISSUES FOR DISCUSSION

1 Two referendum questions were carried at the same time as the 1995 New South Wales general election. Evaluate these comments on the results, published three days after the vote:

> 'These [referendums] were . . . on entrenchment (that is, the requirement that any further change will need a referendum) in the New South Wales Constitution Act of the independence of the judiciary and a fixed four-year term of parliament. On the early votes, both of these will be passed with overwhelmingly majorities. This is despite the fact that the referendum questions were barely publicised at all until the last week of the campaign, all the major parties ran dead on them, and many voters were amazed when they entered the polling booths and were presented with the questions.
>
> . . .
>
> Now the ordinary voter is unlikely to spell all this [arguments in favour of the two referendums] out, or even have a very clear idea of what voting Yes to the fixed term will mean. Nor have they been helped by the media, which with very few exceptions failed to make any serious attempt to inform the electorate of even the fact of

the referendums, let alone seriously debate their pros and cons until the very last week. Even then the arguments were mainly confined to editorials.'

'It was in the end a triumph for good, instinctive judgment' (McGuinness, P. P., 'Electorate's good judgment', *Sydney Morning Herald*, 28 March 1995, p. 12.).

You might also like to consider the following comment, published four days earlier:

'While I support in principle the idea of judicial independence and tenure which can be broken only by the most difficult of means, I am beginning to think it might not be a bad thing at all if referendum question 1, which proposes to entrench tenure in the New South Wales Constitution Act, were defeated tomorrow' (McGuinness, P. P., 'Reject fixed judges' tenure', *Sydney Morning Herald*, 24 March 1995, p. 18.).

2 Describe how the Australian Constitution distributes powers with respect to agriculture between the Commonwealth and the states. What problems for agricultural policy result from this distribution, and how could these problems be overcome?

3 Are the origins and activities of Australian (agricultural) policy institutions better explained by market failure or public choice theory?

4 Should the High Court have made a determination on native title in the *Mabo* case, or should this decision have been left to the parliaments of the individual states?

Endnotes

1 Jaensch (1995, figure 5.3) purported to show support for parties at the state level from 1946 but the legend to this figure is clearly erroneous; what is labelled in the figure as votes for 'other' parties appears to be the Liberal Party; the votes labelled 'Liberal' appear to be for the National Party; and those labelled 'National' appear to be votes for 'other' parties. It is also ambiguous as to whether this figure reports voting at the state level in federal elections, or voting in state elections.

2 Although the Western Australian National Party is now in coalition with the Liberals, and its leader is Deputy Premier.

3 Difficulties in coping with Opposition are not unique to the Liberal Party but appear to be especially difficult for it because of its party structure. Partly because of its more stable electoral support, the National Party probably survives Opposition best of the three major parties.

4 Eventually the costs of retaining Campbell within the Labor Party became too great and he was expelled in late 1995. Campbell successfully contested this seat at the March 1996 election as an independent.

5 Quoted material in this paragraph is taken from publications and correspondence of the Centre for Independent Studies.

6 This material is taken from publications of the Evatt Foundation.

7 ARMCANZ was previously the Australian Agricultural Council (the first Ministerial Council, established in 1934) and subsequently the Agricultural Council of Australia and New Zealand (ACANZ). ARMCANZ was formed in 1992 from ACANZ, the Australian Soil Conservation Council and the Australian Water Resources Council. ARMCANZ amalgamated with the Rural Adjustment Ministers' Meeting in 1993. SCARM was formerly the Standing Committee on Agriculture.

Part II

Intrasectoral Issues: Input and Product Markets

The flow of product through the agricultural and resources sector provides one framework for structuring evaluation of intrasectoral policy issues. Topics considered in Part II include policies affecting the provision of natural resources to agricultural production, research activities affecting the technical efficiency of farm production, and policies modifying the marketing of agricultural products. In considering natural resources, attention is focused on water provision and degradation of the landscape by land-using activities like agriculture (Chapter 4). The role of government in promoting technological change in agriculture through support for research and development is evaluated in Chapter 5. Governments have intervened extensively in the marketing of agricultural produce, and the general framework of such interventions is evaluated in Chapter 6. Recent reductions in government intervention in the marketing of wheat, wool and sugar are evaluated in Chapter 7. The contemporary policy favourite of 'value adding' is evaluated in Chapter 8, together with an investigation of food safety as a value-adding activity.

4 Farm inputs: water and externalities

Agriculture uses both natural resources (land, water and air) and produced inputs (machinery; agricultural chemicals, fertilisers and other farm supplies; services such as machinery repair, accounting and legal, farm technical advice; and contracting services such as shearing and pest control). Some farm output, such as on-farm conserved feed; grain for seed and other planting material; and feedstuffs for the intensive livestock industries like pigs, poultry, dairy, beef feedlots and horses, is consumed within the agricultural sector. Most output is marketed into the food and fibre industries, although some is distributed in the form of the externalities. The objective of this chapter is to investigate how government policy interacts with agriculture's use of natural and produced resources. The principal issues discussed are water provision and water pricing (section 4.1) and natural resource degradation (section 4.2).

4.1 WATER

Water is the lifeblood of agriculture. Without water, there would be no crops or pastures, and thus no crop or livestock products—however, it is not the only scarce resource in agriculture, and there are partial (although not absolute) substitutes. Most crops and pastures are rain-fed, and most livestock is supplied by on-farm watering systems supplied by farm dams, but discussions about water in agriculture—especially concerning the Murray–Darling basin—are almost invariably discussions about large-scale public irrigation schemes. Sometimes they are about smaller-scale schemes based on pumping artesian or sub-artesian water although in parts is being increasingly privatised. The economics of irrigation are fascinating because few if any Australian irrigation schemes were ever economically viable—'Much of the past public investment in irrigation would not have proceeded had irrigators been required to meet full costs, including capital costs' (Industry Commission 1992b, p. 85, cf. Davidson 1969).

An individual farmer's analysis of on-farm water supply is an investment decision. Would the benefits of the improved water supply exceed the required costs, and would the rate of return on this investment exceed the rate of return in any other use to which these funds could be put? Similarly, decisions about public irrigation schemes are generally investment decisions, requiring the comparison of future returns with costs borne both now and in the future.

Public irrigation investment decisions have become increasingly complicated because of the higher values now often ascribed to the environment. Water withdrawn from river systems has a deleterious impact on the environment, either through the effect of reduced flows on water quality, threats to habitat, or through the return of waste water. Decisions about irrigation water supplies are, therefore, not only investment decisions about marketable farm outputs, they also have environmental opportunity costs.

The water policy debate in Australia has been 'informed' by, among other things, the knowledge that Australia is the driest continent; consequently there has been an almost religious belief in the virtues of impounded water.[1] A former federal Liberal politician is reported to have said that whenever he heard of a dam being announced he felt an election coming on—the damming of unregulated rivers continues to be urged.

Policy issue

The key policy issue is the efficient use of both water resources, and the capital and labour required to harvest and manage water provision. Important components are the setting of water prices; the optimal level of water storage and the optimal level of water recycling. The following discussion focuses on the first of these issues.

Policy intention

Until very recently, the intention of policy has been to provide water to nearly all users regardless of the cost, and irrespective of the price users have been prepared to pay. Cheap water has been provided to both urban and agricultural users, implicitly justified on market failure grounds. Public health issues have underlined the cheap supply of high quality water to urban households. The flow-on effects to regional development have been used to justify increased irrigation.

Water restrictions in Sydney during 1994–95 provide an interesting comment on policy intention. Much of New South Wales was severely affected by drought, although after several drought years Warragamba Dam, Sydney's major water supply, never fell below 45 per cent of its capacity. In October 1994, water restrictions were introduced; they were tightened in early 1995 and subsequently relaxed in March 1995. The state government faced an election on 25 March 1995, when rural New South Wales had been suffering severe drought for up to four years. Not only was agriculture severely affected but also, by mid-1994, rural towns and cities, particularly in northern New South Wales, were starting to run out of water. Thus, despite Sydney still having about half its storage capacity remaining, nominal water restrictions were introduced. The Coalition government apparently wanted to demonstrate to its rural constituency that metropolitan residents were sharing the pain, although that pain was voluntary, minuscule, unenforced, and unjustified by the metropolitan area's water storage situation. Reinforcing this perception, the incoming Labor government announced, following rain in 1995, that Sydney's water restrictions would only be lifted after discussion with rural organisations *inter alia* (Sharp 1995).

Area of agricultural sector

The principal beneficiaries of irrigation water subsidies have been farmers on land to which water has been delivered. Where water has been made available for irrigation, there are substantial 'upstream' economic effects on industries supplying inputs complementary to water. There are even more substantial 'downstream' economic effects. The change from dryland to irrigation farming reduces the output of some enterprises (e.g. extensive wool growing) and dramatically increases irrigated production (e.g. rice, cotton, orchard fruit, vegetables, grapes, dairy). Irrigation expansion creates opportunities for downstream processing (e.g. rice milling, cotton ginning, juice extraction, fruit canning and vegetable processing, vine fruit drying and wine production, and milk products).

Diverting water to irrigation and urban uses, and disposing of wastewater, has also created significant externalities, especially in the return of saline water to streams, and reductions in river flow which have contributed to recent problems with blue-green algae. Consequential effects on the quality of downstream drinking water, especially affect Adelaide, which draws its water from the Murray–Darling system. Recreational aspects of water quality are also increasingly important.

Policy orientation

The policy orientation of water supplies is complicated by Australia's major eastern river (the Murray) forming the border between Victoria and New South Wales, and being the principal water supply for Adelaide. Construction of major storage and associated works on the Murray thus required agreement between the states. The Snowy Mountains Scheme required inter-state agreement because some water flowing through two states was impounded in New South Wales and subsequently released for the benefit of all three states. The financial magnitude of the Snowy scheme, and especially its national myth-making potential, resulted in the Federal Government being a key partner. Some subsequent irrigation schemes were relatively large compared to the economies of especially smaller states that the Federal Government became a major partner (e.g. Ord and Burdekin schemes). On one occasion, the Federal Government acted for environmental reasons to prevent development of a major hydro-electric water storage scheme (the Gordon-below-Franklin in Tasmania).

Policy mechanism

The Council of Australian Governments (COAG) has established a Working Group on Water Resource Policy to 'develop a strategic framework for efficient and sustainable reform of the water industry [taking into account] the diversity that exists across states and territories' (Office of State Owned Enterprises 1993, p. 4). Water harvesting in the Murray–Darling Basin is approaching its limits and this, together with the increasing recognition of water's environmental value, has given urgency to coordinating the environmental management of this resource, and led to the increasing importance of the Murray–Darling Basin Ministerial Council and the Murray–Darling Basin Commission.

Construction of particular water storages generally proceeds via an Act of Parliament. In New South Wales, the government recently sought finance from the potential beneficiaries of new irrigation schemes to construct storages; the unwillingness of potential beneficiaries to commit funds precluded the development. Because capital investment in water storage is so large, decisions to proceed with construction are taken at high levels—in New South Wales, for example, the Capital Works Subcommittee of Cabinet has a primary role in recommending new investment. Smaller modification of capital structures might be funded from annual budgets, although special appropriations might be required together with approval at a high level of government. The continuing provision of water requires operating and maintenance funds that would normally be approved in a government's annual budget.

Once storage is provided, key decisions on providing irrigation water include allocating access to water (water rights, quotas, licences or entitlements) and setting the price at which this water is provided. If irrigators are required to contribute to water provision, the level of their contribution might be specified in particular legislation or, for easy revision, regulations under more general legislation. However, the policy principles underlying these charges would probably require Cabinet level decision-making. The allocation of water rights has become a serious problem, at least in New South Wales, because of the variable nature of water harvesting, exuberant assessments of likely water availability from new storages, past policies which have implied that water has a zero value in environmental use, and increasing contemporary demands for water for environmental purposes. This problem became so critical in New South Wales that a former government appointed a Catchment Assessment Commissioner to make recommendations on revising water allocations, although the arrangement did not survive the 1995 change of government.

Traditionally, irrigation water was allocated with a nominal licence cost, and an annual water price that was a small percentage of the operating and maintenance cost of distributing water from the main storages (Industry Commission 1992b). Water entitlements were often tied to land, and thus there was no explicit expression of the value of the water allocation rights except through land prices. Trading of water entitlements has recently been permitted and has created an explicit price for water via its opportunity cost. In the case of metropolitan water in Sydney, the Government Pricing Tribunal now recommends on water pricing (Government Pricing Tribunal 1993).

Over-allocation of water rights becomes particularly problematical where decisions are taken to reduce water allocation to individuals. Water entitlements may be either a formal or a *de facto* property right. Modifications to water allocations reduce the capital value of the right where a formal market exists for the transfer of entitlements, if water rights are tied to land, or even if there is no explicit market related to the right. Changes to entitlements are conceptually simple if the rights to water were originally sold to water users—it seems equitable and efficient for the government to adjust the aggregate level of water entitlements by entering the market and purchasing just sufficient rights to meet its policy objectives. Even where water allocation was originally simply granted, efficient adjustment to aggregate water entitlements might still be best administered by the purchase of rights, since those who value the rights least would be prepared to sell them at the lowest prices. However, on distributional grounds, it may seem unreasonable to compensate beneficiaries of government largesse.[2]

Apart from the Snowy Scheme, the delivery of water is principally a state matter. In Victoria, for example, agriculture uses about 77 per cent of the state's regulated water; Melbourne uses some 9 per cent; and other urban industrial, stock and domestic users each consume about 7 per cent.

Metropolitan water is supplied by the Melbourne Water Corporation which has functions combining essentially private goods (water supply), goods with significant externality attributes (sewerage disposal), and essentially public goods (stormwater disposal). Rural water in Victoria was supplied by some 120 different water authorities. The Rural Water Corporation provided the strategic overview of rural water services, and owned the headworks and delivery network (Office of State Owned Enterprises 1993). Other states have similar delivery structures.

Institutional structure

Past decisions to construct water storages have traditionally been related to state and national politics. The declining relative power of rural Australia, and tighter macroeconomic constraints on governments in recent decades, have substantially limited governments' ability to provide additional water storage. Increasing pressure from environmental groups is contributing to a re-evaluation of water for urban, agricultural and environmental uses.

The key non-government institutional players are the water users. In agriculture these groups are relatively small, identified with particular watercourses and easily organised. Potential water users may in some cases be easily organised, particularly when there are major proposals to augment water supplies such as by a new dam. Alternative users, particularly environmental users, are diffuse, difficult to organise, and may not be geographically closely related to the resource. There are clearly problems with identifying and obtaining the views of future generations who have an obvious interest in current environmental decisions concerning water.

In urban areas, consumers, industry and commerce are the major users. In most cases, water comprises a small proportion of their budgets, and water delivery issues do not play a major role in their decisions—except, perhaps, when the water market is used for taxation purposes (e.g. 'environmental' levies on water bills in New South Wales).

Policy performance evaluation

Until recently, there has been very little public evaluation of the water industry. The major exception was the fierce warfare between economists and engineers in the late 1960s and early 1970s as to whether or not irrigation was an efficient use of social resources (cf. Davidson 1969). This situation began to change with the Industry Commission's 1992b report. This report has subsequently been followed by economic analyses of water pricing (e.g. New South Wales Government Pricing Tribunal 1993) and analysis of the structure of the Victorian water market (Office of State Owned Enterprises 1993).

Economic issues

The principal current economic issues are:

- *multiple uses*—water is used directly by domestic, commercial, industrial and agricultural users; water may also have environmental values for direct consumption in recreation or passive environmental (including existence) values for the ecosystems which depend on aquatic environments; regulation of river systems may also be important for flood mitigation;
- *multiple attributes*—in urban uses, water is an important input for direct consumption, and is a principal input in existing sewerage systems; there may also be complementarities with stormwater disposal, although there may be difficulties in maintaining the integrity of sewerage and stormwater disposal in older systems;
- *mixed characteristics*—some uses of water are rival, but there are significant externalities in all uses, and some uses may be complementary (e.g. water storage and recreation); even where water is rival at a particular point in time, it is quasi non-rival over time because of recycling possibilities;
- *natural monopoly*—water supply, sewerage and stormwater disposal, and irrigation water and flood mitigation are natural monopolies; and
- *water harvesting is stochastic*—rainfall is uncontrollable and, unless storage and use is carefully related to rainfall distribution, water availability may also be conjectural.

There are clearly opportunities for market failure, and these possibilities have (implicitly) justified past policies. Economic analysis has both single period dimensions (e.g. optimal pricing given existing investment) and dynamic aspects (e.g. optimal investment in water storage).

Analytic techniques

Suppose a river system has been dammed, and an amount Q0 of water (figure 4.1) can be reliably delivered from that system in any one year—thus annual supply (S) equals water availability Q0. Suppose the agricultural demand for water from that river system is Da. The efficient price to charge for that water is P0. At price P0, users have an incentive to use all the water available, but there is no incentive to attempt to extract additional water from the system.

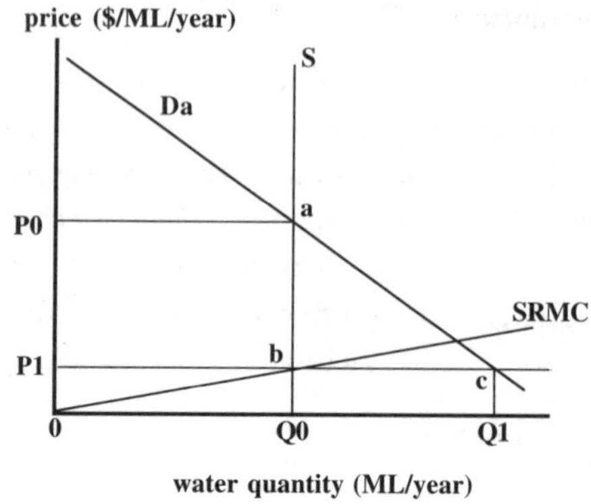

Figure 4.1 Simple Water Market

However, public irrigation schemes are generally highly capital intensive (the costs of constructing dams, weirs and supply channels are high) but the annual operating costs (short run marginal costs) are relatively low (e.g. SRMC in figure 4.1). If there is any economic rationale to irrigation water pricing in Australia—and, until recently, irrigation water has generally been supplied to farmers as a (nearly) free good—this rationale has been to recover operating costs. For example, the Industry Commission (1992b, p. 85) argued that, if 'the intention is to let [irrigation] systems run-down' then 'efficiency would demand only that prices be sufficient to at least cover operating costs'. Consider, however, the consequences of only charging water at its operating, or short run marginal, cost (cf. figure 4.1). If the water supply authority charged the short run marginal cost corresponding to water availability Q0 (i.e. P1), then water users would want to access a quantity Q1, far in excess of Q0. There are at least three effects of making water available at its short run marginal cost (cf. figure 4.1):

- the water supply authority is making a gift of P0abP1 to those who are supplied with the water;
- some mechanism is required to ration Q0 water when, at price P1, (potential) users are demanding Q1; and
- water users would be likely to agitate to increase the supply of water from that system (e.g. by constructing more storages, or diverting water from other uses such as environmental flows) *even if it was intended to disinvest in the water supply.*

Figure 4.1 can also be used to illustrate economic analysis of increasing the water supply in this system. Suppose water supply could be increased to Q1, at a capital cost of K dollars. The *gross annual* value of the *additional* water is given by area acQ1Q0 in figure 4.1 (= V, say), and this gross annual value will accrue for every year of the life of the new water supply facility. Using conventional investment criteria, the discounted net present value (NPV) of the stream of future benefits of the additional water is:

$$\text{(4.i)} \qquad NPV = \sum_t \{V - C\}/(1+r)^t - K \qquad \text{for } t = 1, \ldots, N$$

where C is the annual maintenance cost of the additional water supply facilities which last for N years, and r is the discount rate. Only if NPV > 0 would it be worthwhile investing in the additional water supply facilities; even if NPV > 0, there may be other projects that yield higher returns.

If only the short run marginal cost (P1) was charged rather than the efficient price (P0), water rationing would be required because the quantity of water demanded at P1—i.e. Q1—exceeds the available supply. An efficient water rationing mechanism would result in the allocation of Q0 water to those users who would have used Q0 water at price P0. A bureaucratic water rationing mechanism that ignored economic values would be unlikely to allocate the water efficiently. Where water licences have been allocated historically, there is no necessity that efficient water users would be able to access water. A contemporary solution to this problem is to enable trade in water rights. Suppose a water user had an historical water allocation but would only use this water if its price was at P1, and there were potential users without water who would be prepared to pay P0. A user with an historical allocation but a low valued current use would be financially better off by renting the licence to a user prepared to pay P0. More permanent solutions might involve the sale of the right itself, rather than just its annual lease.

Suppose, however, that there are potential urban water users (with demand curve Du) in addition to agricultural users (demand Da) (figure 4.2). The aggregate demand for water is now Dt. The efficient water supply price is P2, with Q* water being used in urban uses, and q* (= Q0 – Q*) water being used by agriculture (figure 4.2). If a price below P2 is charged, there will again be an excess demand for water, a need to ration available water, and a demand for additional investment in the water supply.

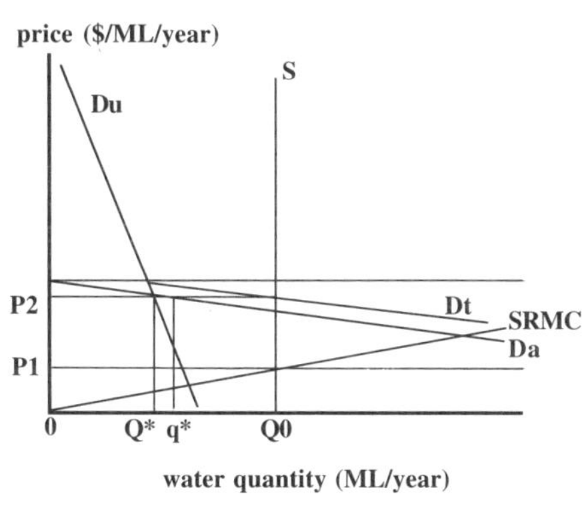

Source: Randall (1987, Figure 4.9)

Figure 4.2 Water Market with Two Demands

In the water market represented by figure 4.2, suppose the efficient price (P2) was not charged. It is possible to design a differential water pricing scheme so that all the available water is used, but different prices are charged to agricultural and urban users. In figure 4.3, urban users are charged Pu, and use Q#; agricultural users are charged Pa, and use q—where (Q# + q) = Q0. The social loss from differential pricing is the loss from reduced urban use (area abQ*Q#) less the value of the increased agricultural use (area cdqq*) where (Q* – Q#) = (q – q*). In the case illustrated, the urban loss from higher prices is considerably greater than the value of the additional water used in agriculture.

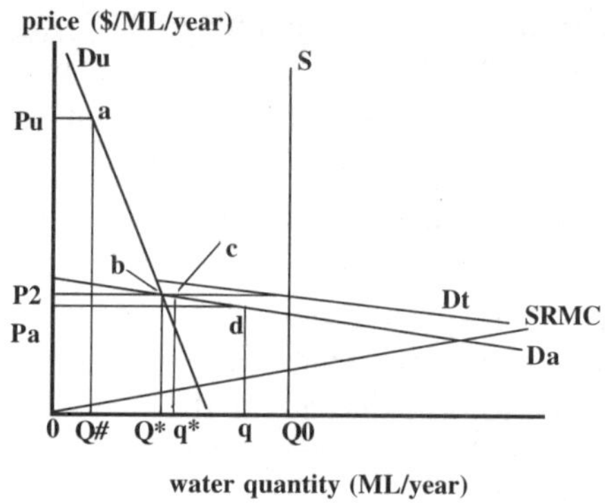

Source: after Randall (1987, Figure 4.10)

Figure 4.3 Simplified Water Market with Two Demands and Price Discrimination

4.2 EXTERNALITIES AND NATURAL RESOURCE DEGRADATION

Resource degradation is widespread. In releasing a study into water quality in the Hunter Valley, the New South Wales Director-General of Water Resources noted, 'While all sampling sites in the valley failed the guidelines for drinking water supply *it would be unreasonable to expect any natural streams in New South Wales to meet the guidelines* because of inputs of bacteria from various sources' and 'the river system meets the guidelines for stock watering, irrigation, boating and other recreation *which did not expose people to prolonged contact with the water*' (Millington 1994, pp. 2,1, emphases added).

Natural resources, including those used by agriculture, may be degraded by agricultural activity. Failure to maintain sufficient ground cover to prevent wind or water erosion may cause soil erosion. Other forms of degradation peculiar to agriculture include damage to soil structure and fertility, soil acidification, soil and water salinity, and scourges (pests, diseases and weeds). Agricultural resources may also be degraded by non-agricultural activities—e.g. release of saline water from coal mines. Blue-green algal blooms in Australian rivers appear to be caused both by soluble phosphates in urban wastewater and phosphates originating as phosphatic fertilisers or wastes from intensive animal industries. Globally, changes to solar radiation through degradation of the ozone layer, or climate change resulting from increasing atmospheric concentrations of greenhouse gases, may significantly change the environmental conditions for agriculture.

Policy issue

Natural resource degradation does not necessarily imply a role for government intervention. All produced capital assets deteriorate with time and use, but there is not a general rationale for government intervention to prevent this. Some natural resources must degrade because of their use—e.g. mineral resources are necessarily depleted by mining. Whether or not resource availability changes with the exhaustion of a mine depends on exploration, recycling, and substitution possibilities with other natural resources and produced commodities. There will be an allocative *efficiency* rationale for government intervention to modify natural resource degradation if there is market failure in the use of resources, or the markets for resource extraction or use are imperfect. There may a *benign distributional* rationale for government intervention in the use or degradation of natural resources if particular social groups are adversely affected or especially benefited by such degradation. Finally, government may intervene to modify natural resource degradation for *public choice* reasons.

Degradation of natural resources may occur where there are satisfactory property rights and no externalities—e.g. on-site gully erosion or soil acidification. In such cases, there may be little efficiency justification for government intervention unless there are particular problems of ensuring adequate bequests of natural resources to future generations or particular problems associated with information markets. Some resource degradation problems are, however, clearly associated with property rights failures, especially in the case of water. Such problems include dryland and irrigation salinity, where the creator of the problem—e.g. the tree feller or the irrigator respectively—does not suffer major damage from the externality. Other externality problems associated with water include reduced stream flow causing environmental damage such as blue-green algae.

Natural resource degradation problems in agriculture associated with property rights failures range from local problems—e.g. spray drift or weed seeds blowing across a boundary fence, or dryland salinity in localised catchments—through regional salinity, or pest and disease problems to global problems such as the enhanced greenhouse effect.

Policy intention

Social control of natural resources has traditionally been a state government matter. While control of natural resources remained with the states, and the states were dominated by 'developmentalism' (Bennett 1986, Galligan 1986), the intent of natural resource policy was exploitative. In that fine Australian aphorism: if it moves, shoot it; if it doesn't, chop it down. These attitudes began to change by 1970, with the battle over flooding Lake Pedder in Tasmania as the first major controversy. As these debates accelerated over time,

> ... the fight was waged by often-uncomprehending state governments against a new type of political opponent who usually ignored party lines, was often linked by informal networks with other states, and whose tactics were often quite sophisticated. It was clear that many people were no longer prepared to leave the development of Australia's resources to state governments, having become more aware of 'people and ideas' than before (Bennett 1986, p. 215).

Greater environmental awareness in society generally has slowly been reflected in greater environmental awareness by government. Economic analysis that accounts for non-market environmental values will better identify allocative efficiency. But government decision-making that reflects environmental values does not necessarily improve the efficiency of resource allocation. Such decision-making may also be distributional —for example, improving social welfare by increasing the environmental amenity of a particular group even if allocative efficiency is unchanged or even diminished. It may also occur within a public choice model—for example, Senator G. Richardson's deliberate courting of the 'green' vote for the Labor Party during the 1990 federal election (Richardson 1994, chapters 13, 15).

Area of sector

The principal focus of environmental decision-making as it affects agriculture is on agriculture's use of natural resources. Superficially, this focuses attention on on-farm natural resources—principally soil and its degradation (e.g. erosion, dryland salinity, acidification) and other land-based problems (e.g. feral and native animals, 'woody weeds' and introduced plant pests)—and the supply of irrigation water. The wider issue of natural resource degradation and agriculture includes its effects on biodiversity and the production and absorption of greenhouse gases.

Policy orientation and policy mechanism

All levels of government have been involved in combating environmental degradation associated with agriculture. The National Landcare Program involves a Land and Water Program administered by the Department of Primary Industries and Energy; a Natural Resources Management Strategy administered by the Murray–Darling Basin Commission; and the community components of Save the Bush, One Billion Trees, and Waterwatch Programs funded through the Australian National Conservation Agency. These programs are cooperative with the states, which have their own programs.

State agencies involved in protecting natural resources include departments of agriculture, soil conservation, water resources, forestry, conservation and land management, national parks, environmental protection authorities, and agencies involved with control of scourges (pest, disease and weed control).

Local government has traditionally been involved in the control or eradication of agricultural weeds, pests and diseases (e.g. the Rural Lands Protection Boards in New South Wales). Local government is also involved in the abstraction of water for domestic, industrial and commercial uses (e.g. metropolitan water and sewerage authorities, and rural councils) and the return of water to the natural environment via sewage; local government is also generally responsible for urban water runoff. Both the abstraction of water, and its return as sewage or runoff, have potential major implications for the downstream availability of water and for its quality, because of the nutrient load of water returned to the river system.

Institutional structure

Because environmental degradation is such a catch-all, there is a wide range of groups interested in various related aspects. Farmers have traditionally considered themselves to be the best stewards of the land, although the Landcare movement appears to have revealed to many that, despite their best intentions, their knowledge of environmental processes and 'best practice' farming technologies may have been less than perfect. Farmers have been increasingly pressured by environmental lobby groups, especially the Australian Conservation Foundation. Farmers' initial responses to this pressure were to deny environmentalists' claims. However, a fascinating institutional development was the joint agreement between the National Farmers' Federation and the Australian Conservation Foundation preceding the Prime Minister's 1989 environment statement (Australian Nature Conservation Agency et al. 1993, p. 3).

Policy performance evaluation

The evaluation of environmental degradation has been essentially piecemeal. This lack of an integrated approach partly reflects the diffuse nature of the problem. The piecemeal nature of the evaluation response also reflects the multi-disciplinary nature of environmental studies. While economists' interests in natural resources long pre-date the modern environment movement, economists are only one player in policy performance evaluation. Many branches of the sciences are key players in environmental monitoring and evaluation, as are other social sciences (e.g. geography) and, increasingly, the humanities (e.g. philosophy). There have been recent Australian attempts to codify the 'state of the environment' (e.g. Australian Bureau of Statistics 1992, Environment Protection Authority 1993). The Murray–Darling Basin Commission has reported extensively on the environmental status of the Basin (e.g. Murray–Darling Basin Ministerial Council 1989). Individual authorities have also begun to report regularly on the state of the environment for which they are responsible (e.g. environmental reviews in the New South Wales Department of Water Resources' annual reports). There are also reviews of specific problems directly relating to the state of the environment per se (e.g. Blue-green Algae Task Force 1992), or to government policy affecting the environment (e.g. Office of State Owned Enterprises 1993, NSW Government 1994). The failure of the Resources Assessment Commission to survive the Canberra maelstrom means there is no environmental equivalent of the Industry Commission to continuously monitor the environment nor to undertake regular reviews—although the IC itself has investigated economic policy aspects of environmental problems such as greenhouse gas emissions and the water industry. In the absence of a widely-accepted methodology for evaluating environmental issues, much of the evaluation of existing and alternative policy proposals relating to land and water degradation is at the disciplinary level.

Economic issues and analytic techniques

The environmental degradation issues confronting agriculture may be categorised in two dimensions: property rights and time (table 4.1). Some of the most interesting and socially vital problems are those in the 'dynamic' row, but the analytic techniques required to evaluate these problems are very difficult. Further, many 'environmental' problems are largely issues for the individual decision maker. For example, where problems of spray drift and chemical contamination, dryland salinity, or scourges do not extend beyond the boundaries of a firm, that firm has appropriate incentives—as long as there are no significant information transactions costs—to optimise its own output of degradation and wastes.

Table 4.1 Agricultural Environmental Problems—time and property rights

Time	*Property Rights*	
	Private Goods (excludable and rival)	*Public (non-excludable and non-rival)*
Static	• dryland salinity • spray drift • pest, diseases and weeds • erosion, acidification, soil structure • chemical contamination	• dryland salinity • spray drift • pest, diseases and weeds • erosion • irrigation salinity • chemical contamination
Dynamic	• erosion, acidification, soil structure • chemical contamination	• erosion • chemical contamination • ozone layer • enhanced greenhouse effect and climate change • biodiversity • blue-green algae

However, where resource degradation effects extend beyond the boundaries of an individual firm, and other firms cannot affect the degrader's activities through markets—i.e. there are 'externalities'—there are significant opportunities for allocative inefficiencies. The focus of this section is, therefore, on (static) problems of environmental inefficiencies.

Simple static efficiency analysis

A simple analysis of externality was presented in section 1.2. To recapitulate: a firm produces some commodity Q and sells this output in a perfectly competitive market (and so its output price P is constant). A waste (W) is also produced proportionally to the commodity:

(4.ii) $$W = \Omega.Q$$

The waste is also produced deterministically—that is, in every production period, if Q output is produced then so is W (= Ω.Q) waste. The key economic characteristics of this waste are:

- the producing firm disposes of the waste without regard for costs imposed on others;
- the waste 'disappears' at the end of the current production period because of the 'assimilative capacity' of the environment—i.e. the waste does not accumulate over time (it is a 'flow' rather than a 'stock' pollutant); and
- the waste affects the profitability of other firms or the government (they have to clean it up), or individuals' utility, and these other economic agents have no market in which they can pay the waste emitter to reduce or eliminate its wastes.

A standard representation of this problem was given in figure 1.4, where—for a perfectly competitive firm—the efficient solution is to equate the firm's marginal revenue (commodity price) with the marginal *social* cost of its activities. An equivalent solution may be derived by equating the firm's marginal net profit (MNP = P – MPC) with the marginal external cost (MEC) of its activity (figure 4.4). The optimal social solution is then expressed as:

(4.iii) $$MNP = P - MPC = MEC$$

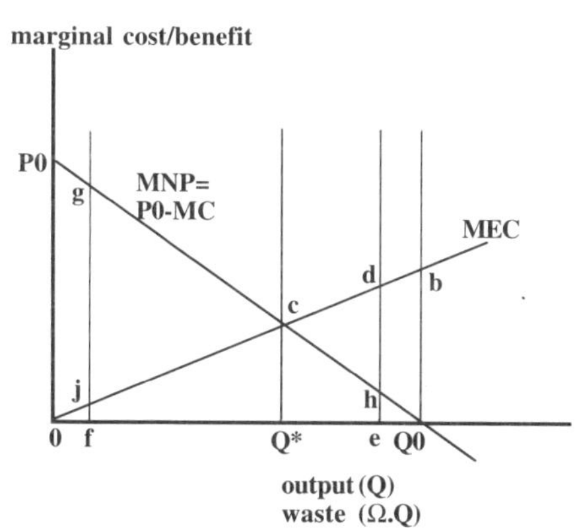

Figure 4.4 Alternative Representation of the Pollution Problem

A firm's private optimum is Q0 : MNP = 0 is equivalent to MR = MC. If those affected by pollution had access to finance, they could offer the firm the amount heQ0 on condition that it reduced its output and therefore the associated wastes. The firm would be indifferent between producing at Q0, and producing at e with this 'bribe' or 'side-payment'. Those affected by pollution could offer the firm to produce the socially optimal level Q* with a 'side-payment' Q*cQ0. The pollution they suffered would be reduced, and the net value of this reduced pollution is the difference between cbQ0Q* (the gross benefit from reduced pollution) and Q*cQ0 (the 'side-payment' to the firm). There would be no incentive for those affected by pollution to induce the firm to reduce its output and pollution below Q* because the costs of obtaining the firm's cooperation exceed the benefits of reduced pollution.

Alternatively, suppose the firm initially was banned from producing because of its pollutant. If the firm offered those who would be polluted an amount 0jf; they would be indifferent between no pollution and pollution corresponding to f plus a side-payment 0jf. By analogy with the previous argument, the firm could bribe those potentially affected by the pollutant to accept pollution corresponding to output Q* together with a side-payment 0cQ*.

Regardless of the starting point, output level Q* in Figure 4.4 is Pareto efficient *ceteris paribus* because no economic agent could be made better off by voluntary exchange without some other agent becoming worse off.

Underlying assumptions

Additional to previously-noted assumptions, the optimality of Q* in figure 4.4 depends upon the following:

1 Perfect knowledge by all affected parties as to the MNP and MEC curves—i.e. there are no (or 'sufficiently low') information transactions costs about the relationship between production and pollution, and the effects of pollution on those affected by it. But, particularly for medical or subtle environmental changes, years or even generations may elapse before the effects of pollutants become either known in theory or detected in practice.
2 No (or 'sufficiently low') negotiation transactions costs—i.e. the net benefits from negotiating a reduction in pollution are greater than the costs of organising the parties to negotiate. However, where a pollutant affects many individuals, the costs of organising a bargaining group may be so high that organisational costs exceed the potential benefits of reduced pollution. Further, group organisation requires leaders: if the individual effect of pollution is low, an individual's benefit from pollution reduction may be less than the individual's costs of organising negotiations to reduce pollution, even if the aggregate benefits of pollution reduction exceed negotiation costs. Unless those benefiting from pollution reduction offer an *ex ante* side-payment to an organiser, there may be insufficient incentive for pollution reduction to be organised. A firm seeking to expand production to Q* may also face negoti-

ation transactions costs as it seeks to identify those who would require bribing to enable production and pollution to proceed, and such costs reduce the net benefits of pollution reduction. Finally, monitoring and enforcement costs would also be required to ensure that each of the bargaining parties adhered to the terms of the bargain.

3 Sufficient disposable income to enable a bribe to be offered—pollution reduction from Q0 to Q* in figure 4.4 requires that those wanting the pollution reduction have discretionary income corresponding to area cQ0Q*. Without this income, they could not offer a side-payment sufficient to induce the firm to reduce its output to Q*. A firm wishing to expand output to Q* requires, before production commences, income equivalent to P0c0 to enable it to bribe those potentially affected by the pollution to permit production Q*.

Policy options

Identifying the optimal level of pollution only solves a small part of the pollution policy problem. This problem also involves identifying a policy mechanism to achieve the optimal level of pollution. Possible mechanisms include:

1 Moral Suasion

Misallocation of resources from pollution occurs because the polluter ignores the cost imposed by the externality. If the polluter could be induced to recognise the cost imposed by the pollutant, and act as if it were affected by it, then the decision maker would have 'internalised' the pollutant. Advertising campaigns, such as litter reduction, attempt to induce individuals to consider externalities, and act as if the activity affected them. Much of the contemporary environmental movement, including rural-oriented activities such as Landcare, attempts to change individuals' attitudes to the environment and hopefully modify decision making accordingly. Economists frequently overlook such behaviour-modification activities, even though market economies are riddled with similar preference-altering activities such as product advertising. Moral suasion will only optimise pollution if polluters acted as if they equated marginal social costs and benefits of pollution.

2 Command and control

The standard 'command and control' or regulatory response to pollution is to seek to regulate associated behaviour—e.g. by identifying a 'standard' which indicates the pollution emissions that will be tolerated, and by instituting appropriate monitoring and enforcement activities to ensure polluters adhere to this standard. For example, control of motor vehicle emissions has proceeded by Australian governments decreeing that all new vehicles must use lead-free petrol. In agriculture, there are controls on land-clearing close to watercourses or on steep slopes to reduce erosion. Agricultural (and other) activities are regulated in the water catchments of dams storing water for human consumption. Command and control will only result in the optimal level of pollution as identified in figure 4.4 if the regulator chooses a standard corresponding to Q*, and undertakes monitoring and enforcement activities that ensure producers emit the corresponding level of pollution.

3 Market-based instruments

These instruments induce firms to identify and adhere to the socially optimal level of pollution:

- The above bargaining procedure, by which the optimal level of pollution was identified, could actually be a mechanism that achieved the optimal level of pollution, as long as transactions costs were sufficiently low. Such solutions are often known as 'Coasian' bargains after their originator (R. H. Coase).[3]
- A tax on each unit of pollutant, or a tax on each corresponding unit of output, could force a firm to produce at the socially optimal level of pollution or output, rather than at its private optimum. The required tax is equivalent to the marginal external cost evaluated at the socially optimal output level. This tax creates an 'after-tax' MNP curve which intersects the horizontal axis at the socially optimal point Q* (figure 4.4). Such taxes are known as 'Pigouvian' taxes after their originator (A.C. Pigou). Section 90 of the Australian Constitution appears to preclude states' operation of Pigouvian taxes as they would be excises.
- The aggregate amount of pollution identified as socially optimal could be identified, as in a standard, but—rather than require producers to emit no more than this level—the corresponding *total* level of emissions could be declared as social property, and individuals could be allocated rights to emit some part of this total. If a producer had rights to emit W# units of pollution they could do so, but they could not emit more than this level. Prospective producers would either be required to have pollution-free production processes, or purchase pollution permits from existing producers who could reduce their pollution at a lower cost than the price at which they could sell pollution permits.

Pollution permits imply that the relationship between output and pollution is not necessarily the same for every producer. Generalising this argument identifies the importance of pollution abatement—at some cost, the production process can be modified so that a different level of pollution emission occurs per unit of output. Emission levels at any given output level therefore depend on the particular production and waste processing technologies. In figure 4.5, for example, a production cost curve MC1 (relating to marginal net profit curve MNP1) might be associated with a marginal external cost curve MEC1. If, however, abatement technology were included as part of the production process, marginal production cost (including abatement costs) would rise to MC2 (corresponding to MNP2) but marginal external costs would fall to MEC2. Marginal external costs would fall because, with the abatement technology, each unit of output Q would be associated with a lower level of pollution emission (e.g. Q = μ.Q for the production process corresponding to MEC2, where μ < Ω).

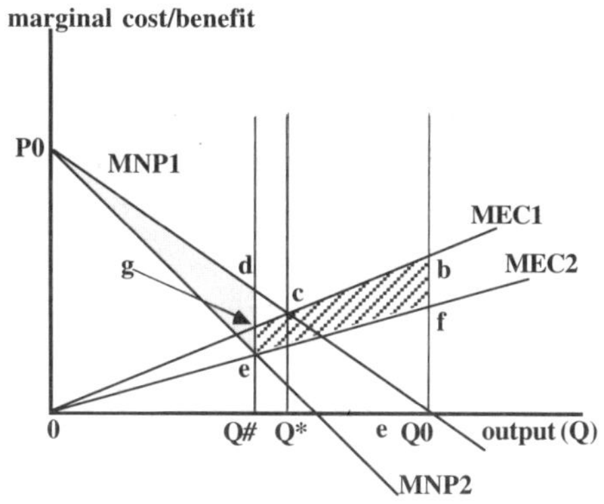

Figure 4.5 Pollution Problem with Abatement

The new socially optimal level of output is Q# with pollution W# = μ.Q#—if the reduced profit of using the abatement technology (area P0de in figure 4.5) is less than the reduced damage from the lower emission rates of the pollutant (area bgef). The new production level with lower pollution costs would probably require capital investment. Hence the socially optimal decision requires comparison of the investment cost and the present value of the future stream of net benefits (increased operating costs or reduced profits, and decreased pollution costs). It will be an empirical matter as to whether or not investment in a particular pollution-abatement technology is worthwhile.

Because the profitability of production falls with the pollution-abatement technology installed, producers are unlikely to install such technologies voluntarily. Unless there are indirect mechanisms operating—e.g. producers are identified as 'bad corporate citizens' because they pollute, and thereby have greater difficulty selling their produce (a form of 'moral suasion')—it would be necessary to compel producers to install pollution abatement. In facing 'command and control' pollution standards, polluting firms could choose output reduction to reduce pollution, or implement pollution-abatement technology. Which strategy the firm chooses will depend on whether profit reduction from reduced output is greater than profit reduction from the increased costs of operating pollution-abatement technology.

Similarly, the possibility of Coasian bargains, Pigouvian taxes, or emission permit systems will force firms to confront the effects on their profitabilities of reducing output to reduce pollution, or implementing pollution-abatement technologies to reduce pollution. The appropriate strategy for an individual firm will depend on that firm's particular circumstances.

Limitations of the general pollution model

The above model assumes a form of pollution where there is a simple, continuous, non-stochastic relationship between output and environmental degradation, and the pollutant does not accumulate in the environment. While this model is appropriate for problems like car exhaust emissions, it must be modified for many resource problems linked with agricultural production. For example, even simple forms of land degradation like soil erosion are not directly relevant to the assumptions of the preceding model, because damage caused by soil erosion in the current period remains in all subsequent periods, or until rectified. Soil erosion is not necessarily directly related to production, but to external events such as rainfall, and this relationship may not be a simple a continuous one: there may be thresholds below which erosion does not occur, and there may be discontinuities when damage crosses a threshold and the nature of the event–damage relationship changes. Similarly, damage may be associated with stochastic events like intense rainfall, or lack of river flow (e.g. drought), and may also be rectified by such natural events (e.g. flooding, which cleans out blue-green algae or phosphate deposition in rivers). Other important examples of degradation falling outside this model include spray drift, where degradation may be caused by the way agricultural practices are carried out rather than by the level of output. Thus, while the model provides a useful first introduction to economic analysis of environmental degradation, it cannot be used uncritically to evaluate real policy problems.

National Food Levy

Economic environmental analysis can be nicely illustrated using a call by an irrigation researcher for 'a special levy to be added to domestic food prices to help tackle the [salinity] problem' where this levy 'should be up to 0.5 per cent of the $40 billion annual value of retail food sales'. Part of the justification for this policy was 'That it would be affordable . . . and would bring home to urban consumers the environmental price they pay for cheap food, as well as raising up to $400 million for groundwater programs' (Beale 1995).[4] This levy would be passed back to domestic consumers, which was a principal objective of its proponents. But irrigation is not only used to produce food—why should other irrigation enterprises such as cotton escape the net of this tax? Further, since produce exported from irrigation enterprises is similar in its environmental effects to domestically-consumed products, why should domestic consumption attract the tax but exports escape? Additionally, much domestically-produced food is not produced using irrigation but would attract the tax: is this desirable or efficient?

Because the proposed levy is not a Pigouvian tax, and has no impact on private or social costs of production, it would have no direct impact on land-degrading activities. Consider the aggregate market for food in Australia (figure 4.6). With supply (S), domestic demand (Dd) export demand (Dx, with associated price Px) and no food levy, total output would be Qt with Q consumed domestically. If a domestic food levy

(P′ – Px) were imposed, total output would remain at Qt (there is no change to farmers' marginal revenue) but domestic price would rise to P′ and domestic consumption would fall to Q′. Because output remains unchanged, exports rise by Q – Q′. As production conditions (given by the supply curve S) and total output remain unchanged, the levy has no effect on the way agricultural output—especially agricultural output of a land-degrading kind in irrigation—is produced. Only if the funds raised from the tax (the dotted area in figure 4.6) result in reductions to the social cost of land degradation from agricultural production would the proposed levy have any beneficial environmental impact on agricultural production.

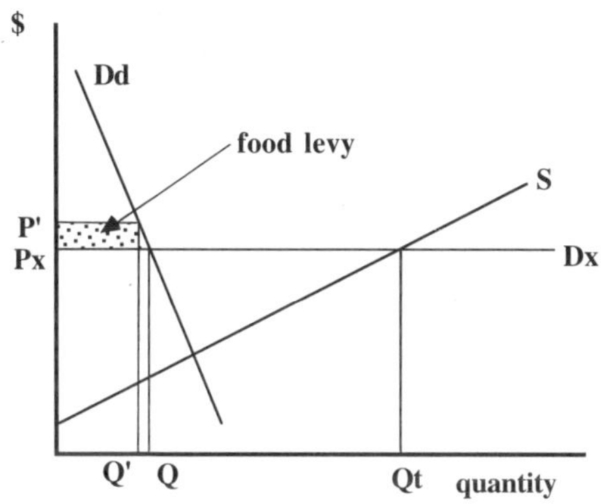

Figure 4.6 National Food Levy for Sustainable Agriculture

The principal effect of the proposed levy would be a reduction in domestic consumer surplus. Because the aggregate demand for food is likely to be highly inelastic, the reduction in consumer surplus and the amount of the tax raised will approximate the expenditure times the tax rate. As the levy would not discriminate between food of irrigation and non-irrigation origins, nor discriminate between food produced with land-degrading and non-land-degrading practices, the reduction in domestic consumption would be determined principally by the elasticity of domestic demand for individual food products. It would be ironic if the reduction in domestic consumption were greatest for *non*-irrigated food produced with *non*-land-degrading husbandry practices!

ISSUES FOR DISCUSSION

1 Consider the following ways in which government might intervene in input or product markets:
   i flat rate bounty on an input;
   ii flat rate bounty for an output; and
   iii supporting a minimum domestic price for an output.

Graphically represent the markets for these commodities, and the economic impact of these interventions.

What are the effects of the following qualifications:

i whether markets are open or closed to external trade; and

ii whether markets are perfectly competitive or monopolised?

What are the distributional impacts of these policies? Whom would you expect to gain, and whom to lose, from these policies? What would be the possible effects of these gains and losses on interest groups' participation in political action? How secure do you think these policies would be—and what would affect their degrees of security?

2 'After the bore'.

'Irrigation bores are now driven deeper and deeper down hundreds of metres to access water created and stored thousands of years ago.

This water is wasted via evaporation on dry low-rainfall lands, now stripped of all vegetation except for the rapidly expanding areas of irrigated crops.

The future deserts of our Western Plains! The drought is a precursor of its destiny—poisoned silted rivers, a dry wind-blown empty earth, its topsoil gone forever into a brown sky. Our meagre rainfall can no longer maintain river flow while greedy irrigation pumps take its life-water.

When the soil is gone, so are the crops, the soil, the farmers and the economy.'

Jacq Jackson, Woollahra, 15 January (*Sydney Morning Herald,* 23/1/95, p. 10)

What water will replace our rapidly depleting artesian water once it is used up?

3 Why might the National Farmers' Federation and the Australian Conservation Foundation have reached agreement on a joint approach on (some) environmental issues?

4 'The Murray–Darling Basin produces 40 per cent of Australia's agricultural output and supplies water to Adelaide and many rural towns and cities. However, despite extensive water storage schemes in the Basin, it is still vulnerable to drought. The waters of all rivers in the Basin, and nearby rivers that flow to waste directly into the sea, should be harvested and used to maximise production and drought-proof all farms in the Basin.' Discuss.

5 In 1996, second-hand harvesting machines imported from parts of the USA with the serious fungal disease Karnel Bunt were found to contain grain residues. How should Australian farmers, machinery importers, quarantine authorities and grain exporters respond to this problem?

6 Describe and evaluate the effects of a per unit input subsidy for a particular agricultural industry. Compare these effects to those of a per unit output subsidy, and a subsidised output price. Discuss the circumstances where such subsidies might increase economic welfare.

7 A New South Wales livestock researcher, commenting on possible outbreaks of exotic diseases in Australia, was reported as saying that if such an outbreak occurred 'this country's standard of living would crash' and that 'People don't realise the knife-edge on which this country has walked as far as exotic disease is concerned. It would take the most minor slip and Australia would be thrown into chaos.' Comment, particularly in the context of the New South Wales Government's attempt to substantially reduce the size of the state's agriculture department, and the way potential pest and disease problems were used—by departmental workers, the farming community, the veterinarians association among others—to oppose the cuts, especially the closure of two regional veterinary laboratories.

Endnotes

1 Technically, Antarctica is a drier continent than Australia (House of Representatives Standing Committee 1994, p. 46) and on the measure of water runoff per head of population, Australia is actually one of the world's *better-endowed* countries (House of Representatives Standing Committee 1994, p. 64). Brownhill (1994) supported another Snowy-type scheme to divert coastal waters to inland Australia; and the Watering Australia Foundation proposed diverting the Burdekin, Herbert, Tully and Johnston Rivers in Queensland, the Daly in the Northern Territory, and the Ord and Fitzroy Rivers in Western Australia, as well as the Clarence (McGeough 1995, cf. Davidson 1969, 1984). Australian politicians are not unique in their belief in the virtues of impounded water. Marglin (1990, p. 128) described an ideal Hindu kingdom as follows:

> Converting uncultivable into cultivable land is the task of the king. . . . the ideal king conquers a kingdom, builds a temple, and then constructs reservoirs to irrigate the land. Then he proceeds to give these territories, newly valuable because of the drainage, to the deity of the temple and to the Brahmins.

2 In the case of hen quotas in New South Wales, the State government compensated all growers at a generous rate ($15 per bird) when the state's egg marketing scheme was terminated (Strong et al. 1990).

3 Reported Coasian bargains are few. One example is:

> A British company has agreed to compensate a farmer for turning his white sheep black.
> Farmer Joe Hardwick is upset about the effect that smoke from a nearby factory chimney has on the fleeces of his 150-strong flock.
>
> Mr Hardwick, 54, of Avonmouth, Bristol, said: 'I can leave a new lamb out overnight and it'll be grey in the morning. Within three weeks it'll be black.
>
> 'Now I get heckled at market because people think I'm trying to sell duff animals.'

The company, Sevalco, which produces carbon black for the ink and rubber industry, admits its chimneys produce a certain amount of harmless soot. It has given Mr Hardwick a dishwasher, a tumble dryer and compensation of £3 (about $6.50) for each lamb.

Its deputy general manager, Cliff McCloskey, said: 'We have had no complaints at all, but we are very concerned with the environment and pollution control.'

'And never darken my sheep again', Stay in Touch, *Sydney Morning Herald,* 9 May 1990, p. 28.

4 The origin of the levy proposal appears to be Meyer (1992, pp. v, 12) although there is no economic analysis of the proposal. Hopefully the proponents of this levy are better at irrigation science than mathematics: 0.5 per cent of $40 000 million is $200 million not $400 million.

5 Research and development

Agriculture is one of the most innovative sectors of the Australian economy.[1] Australian farmers and scientists have a proud record as developers and innovators of new technologies, and as adopters of relevant overseas technologies. Australian agriculture has successfully undergone dramatic structural changes in response to changing economic conditions within the sector and the Australian economy generally, and also in response to the changing world economy. Structural change has also occurred as a consequence of applying new technologies to agricultural production (cf. Chapter 9). Australia has been institutionally innovative in creating research and development (R&D) structures that recognise deficiencies in the provision of private sector R&D, and has developed institutions which challenge market failures.

'Information' provides an ideal illustration of market failure. It is often not easy to define property rights in information and, even where they are defined and price excludability is technically feasible, providing sufficient information about an information market in order to have well-informed consumers causes problems. Information is also typically non-rival, meaning that even if price-excludability problems are solved, there may still be barriers to its efficient production. Further, information comes in discrete bundles and is a lumpy, non-homogeneous good; negotiating market transactions in information is qualitatively different from negotiating over relatively standardised and essentially continuous private goods. Because of transactions costs problems, new technological information, produced by the R&D process, perfectly illustrates the difficulties in ensuring the existence of efficient information markets. Market failure in R&D also arises because of the riskiness of the R&D process, and the likely differences between individuals' and society's attitudes to risk. Further, the riskiness of the R&D process may mean that large-scale R&D firms are optimal to enable risk-spreading across projects, but this may mean unbalanced market power between R&D providers and users, possibly leading to imperfect markets. Additionally, there may be economies of scale in R&D, especially as it becomes more specialised, reinforcing the trend to large R&D providers and potential market power conflicts. The long lead times between research investment and the development of commercial technologies may also result in sub-optimal resource allocation to R&D, especially if private and social discount rates differ. However, R&D in the resources sector is not simply a matter of efficiency. How it is funded, and how its activities are undertaken, have significant distributional consequences.

Policy issue

Given the likely difficulties with providing efficient markets in information, the efficiency policy issue that arises concerning R&D in the agricultural sector is how to ensure the efficient production and distribution of new technologies for Australian agriculture and resources. This issue has two dimensions. The first concerns the type of decisions necessary in developing R&D policy. Initially these are about the optimal level of funds for R&D, and the optimal allocation of this funding between the public and private sectors. If the optimal level of publicly-funded research can be determined, a subsequent issue is how much of this publicly-funded research should be carried out in the public sector as opposed to contracted out to the private sector. Given determination of the optimal level of publicly-undertaken R&D, choices are required as to which particular projects should be funded. A complementary issue affecting R&D funding allocation between public and private sectors is whether the public sector can encourage additional private sector R&D funding via other instruments such as property rights, education policy or competition policy.

These efficiency issues must be resolved not only in aggregate, but with respect to various areas of the agricultural and resources sector (input industries and natural resources, farm production, post-farmgate transformation), the type of new technology (embodied, disembodied), and the type of R&D activity (on a basic to development research spectrum). Embodied technologies are those that can only be acquired by purchasing a commodity or service. For example, new agricultural chemical technology is embodied because farmers must purchase a product 'containing' the technology in order to apply the technology. Disembodied technologies are, conversely, those whose application does not require the purchase of a commodity—examples include the knowledge and skills related to plant and animal husbandry practices. Embodied technologies are price-excludable and the commodity in which they are embodied is rival although the technology itself remains non-rival. Private markets are therefore likely to exist in these technologies, as they do in the agricultural chemical and agricultural machinery markets. Conversely, disembodied technologies are frequently likely to be non-price-excludable and non-rival, and private markets in such services non-existent or operating at sub-optimal levels.

The research and development spectrum (figure 5.1) classifies R&D into a range of related activities. At one end of this spectrum are laboratory and theoretical activities concerned with the production and dissemination of knowledge (basic research); at the other are new commodities or services that have been demonstrated to improve economic efficiency (development and marketing). Of particular importance in agriculture are the publicly-funded 'extension' or 'advisory' services, which provide the link between the development of new products and processes and their widespread implementation in agricultural production. Between basic research and development are research activities directed towards using knowledge to find better products and practices (invention), and proving the commercial value of these technologies (innovation). Much of the total cost of commercial R&D occurs in development (especially testing and marketing) rather than in research. Since basic research primarily produces knowledge and is disembodied, its output is generally non-price-excludable and non-rival. Intellectual property rights (e.g. patents) generally do not permit claims to private property in knowledge itself—although, since legal systems are not static, firms are continually pushing to expand the boundaries of what is patentable. Intellectual property rights do, however, protect new inventions (both products and processes), particularly against commercial competitors.

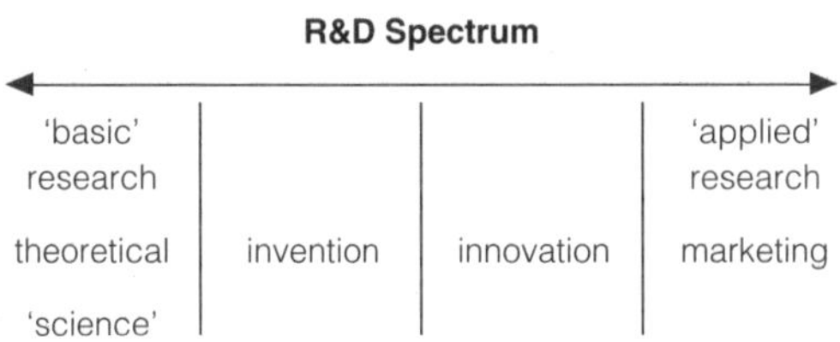

Figure 5.1 The R&D Spectrum

In figure 5.1 the R&D process is depicted as a linear spectrum. However, it does not necessarily follow that technological change commences with discoveries at the 'basic' end of the R&D spectrum, and proceeds by transforming this new knowledge into commercial products or services. New technologies may be developed before an understanding of the corresponding basic science. The research process is better conceptualised as a continuous process which is constantly evolving, and also being affected by outside influences. Figure 5.2 depicts a development of this idea (cf. de Janvry 1978). The top left-hand box labelled 'perceived latent supply of innovations' represents the potential technologies that, based on the current state of knowledge, could become commercial technologies by being transformed into new products or processes. Whether or not it is worth committing resources to create these new products or processes partly depends on the 'socio-economic structure', which influences the potential profitability of these new technologies should they become available. The 'expected pay-off matrix' represents the potential profitability of each of the possible new technologies, and these pay-offs determine the 'latent demand for innovations'—how much would the key economic agents in the technology process be prepared to pay to have these new innovations developed. The latent demand for innovations may be transformed into an actual demand by two possible pathways.

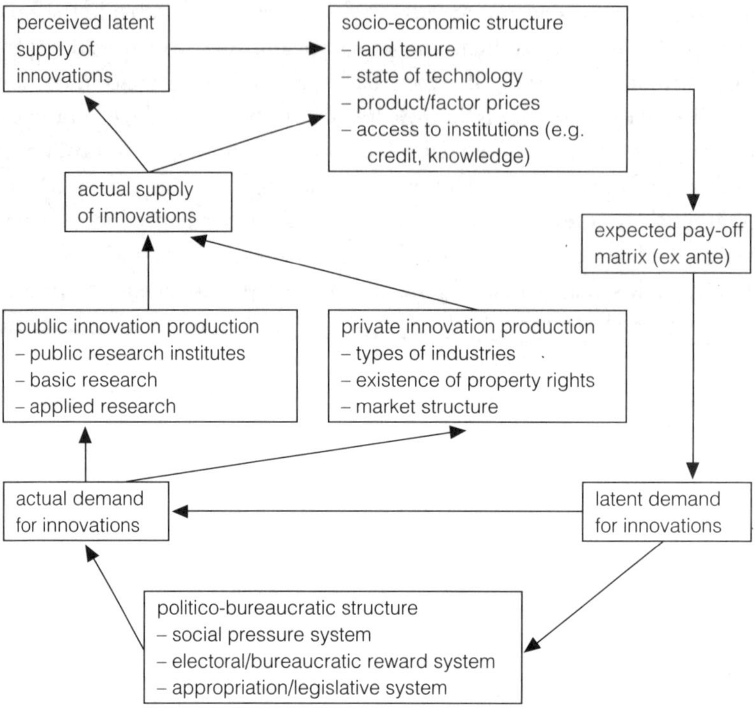

Source: after de Janvry (1978, figure 11.1)

Figure 5.2 Supply of and Demand for Technological and Institutional Innovations

De Janvry emphasised public research: the key determinants of whether or not potential innovations are actually developed by public research institutes is the 'politico-bureaucratic structure', which determines the rewards to those agents who control the public research system. Depending on these rewards, some potential innovations are transformed into new innovations via 'public innovation production' in the public research institutes (e.g CSIRO, state agriculture departments, universities). De Janvry ignored a corresponding private research production process: the profitability of research to firms engaged in 'private innovation production' determines which research is undertaken, and thus which new private technologies get developed. The new public and private technologies become the 'actual supply of innovations'; which innovations are successful depends on their profitability, or on whether users are forced to adopt these technologies (e.g. by regulation). The actual supply of innovations modifies the new 'perceived latent supply of innovations' in several ways. If new technologies are successfully innovated, these technologies reduce the perceived latent supply of innovations. However, development of these technologies is likely to have expanded the knowledge base, and this extended knowledge base may expand the set of technologies that is perceived possible to develop.

Policy intention

The intention of public policy towards agricultural R&D is nominally to improve economic efficiency, although it is not always directly expressed in those terms. Public decision makers have recognised that the private sector appeared to under-invest in agricultural R&D, or not invest at all. Without necessarily understanding why this under-investment occurred, governments have committed extensive public funds to agricultural research and extension. This preparedness was made easier through the relatively large size of the agricultural sector until after the Second World War and, in Australia's case, the key importance of agricultural exports to aggregate economic activity. Since the large agricultural sector had many voters, the rural sector could effectively demand assistance in this area. The key role of the Country/National Party in conservative coalitions in the period 1949–72 reinforced the ability of the Australian agricultural sector to have funds committed for research and extension.

Developments in economic thought regarding market failure, including market failure in the R&D process, reinforced the rationale for public sector involvement in agricultural R&D. Quantitative economic analysis of agricultural R&D productivity, commonly showing annual rates of return to agricultural research in the range 20–50 per cent, and often as high as 100 per cent, strongly suggested that continued investment in agricultural R&D was well justified (Hayami & Ruttan 1985, Table 3-A1). These economic nostrums were enthusiastically adopted by farmer organisations:

> Many of the benefits of [agricultural] research flow to other industries and sectors, often in a way which limits the ability of innovators to preserve a significant portion of the total benefits . . . For these reasons compulsory industry levies are collected from all farmers to fund research. In addition, because there is a remaining likelihood of under-investment in research and because benefits from rural research accrue in part to people and groups other than farmers, a high level of Government funding of [agricultural] research is both justified and necessary (National Farmers' Federation 1981, p. 61; cf. Industry Commission 1994).

Compulsory industry levies are not economically justified simply because some of the benefits of agricultural research are captured by non-farm groups. Further, under-investment in research and the spillover of benefits to other groups may justify government intervention, but not necessarily of a pecuniary kind.

Emphasising the origins of market failure rather than its consequences in sub-optimal resource allocation to R&D activities suggests that policy mechanisms other than direct government funding may provide solutions. Government has become increasingly interested in institutional innovations such as creating new property rights (e.g. in new plant varieties) to encourage greater private sector involvement in R&D. The Australian government has also increasingly distinguished between the existence of market failure in R&D—e.g. non-rivalry and non-price-excludability—and the need for governments to directly fund R&D.

While it is easy to find efficiency justifications for their intervention in R&D provision, this is not the only rationale that has appealed to governments. Publicly funded R&D provides a mechanism for shifting resources directly into the rural sector (by providing research facilities and advisory services).

The intention of government policy is also expressed through the choice of research priorities. Historically, the bodies controlling the choice of research activities by the Rural Industry Research Funds (RIRFs)—now the Research and Development Corporations (RDCs)—were farmer-dominated and most (but not all) funds were directed to farm-level research (Jarrett & Lindner 1982, p. 93). This domination occurred despite only 50 per cent of RIRFs' funding being farmer levies and the rest being general taxpayer funds. As government funding for research gradually tightened through the later 1970s and 1980s, funding for public agricultural research became increasingly dependent on funding from the RIRFs. The priorities of the RIRFs increasingly dominated the priorities of the government research providers (Jarrett & Lindner 1982, p. 92). By the mid-1980s, the Federal Government decided that public research priorities required reorienting, particularly towards activities beyond the farm gate (Commonwealth of Australia 1986, p. 45).

Area of agricultural sector

The areas of the agricultural and resources sector in which R&D may be undertaken include natural resource inputs; produced inputs and services; the farm and resource production; the transport, storage, handling and processing industries; and the wholesale, export, import and retail distribution systems. In the input industries, extensive private research is undertaken in the non-farm sector, particularly in agricultural chemicals and farm machinery, but also in some other inputs such as prepared farm feedstuffs, fertiliser (especially to improve the efficiency of the fertiliser production process) and farm materials (e.g building material and fencing). Private research output is most likely to be embodied in a product (to facilitate price excludability against purchasers) and patent protected (to facilitate price excludability against commercial competitors). Process improvements allow firms to capture the benefits of cost reductions in their own activities. Private non-farm research also produces considerable disembodied technology—e.g. information on how to use agricultural chemicals—which is tied to products through company representatives. Considerable technological change in the non-farm sector spills over into the farm sector, primarily through reductions in input costs; for example, use of computers in the accounting industry provides lower cost and improved farm management and tax accounting services.

Some private farm-level research into inputs may also occur: farmers are particularly engaged in animal breeding, and may also undertake some plant selection. Some farmers also invent machinery and tinker with existing machinery to improve its capabilities or performance. The public sector also undertakes extensive research into farm inputs, particularly in disembodied technologies; this may involve testing the use of private farm inputs (e.g. optimal application rates for fertilisers or chemicals) or more holistic cultural practices (e.g. the optimal conditions for growing a particular plant variety, or good livestock handling skills). The public sector also produces embodied technologies: it has a major rôle in plant breeding (partly because plant breeders' rights are a relatively recent intellectual property right—e.g. Godden 1994b), and it has also undertaken research into animal vaccines.

Improved technologies in the farm sector (as distinct from the farm input sector) occur through improved management, which combines purchased and natural inputs through the knowledge and skills of the farm operator and farm labour, taking into account information about the markets for farm products. Technological improvements such as improved computer technology and improved communications technology enable farmers to better match their production technologies with available inputs and product opportunities. Improved management is generally innovated over a much longer time scale than new input technologies, as each succeeding generation of farmers and workers obtains a better general and specialist education—partly through formal education, and some through lifetime education offered via state agriculture departments' extension services, and other sources such as Landcare groups. Some farmers themselves improve management technologies such as better methods of farm record keeping, or ways of using farm records.

Improved technologies in the post-farmgate sector occur both within and for that sector. The machinery industry develops better machines for storage, handling, transport and processing firms, and these firms may themselves be directly engaged in such R&D. Processing firms, especially, may be engaged in developing new or improved products that use farm output, or in reducing the costs of their processing. Storage firms may develop ways of protecting agricultural products against pests and diseases, or rely on other industries such as chemicals to improve them.

Policy orientation

Agricultural research appears to be principally a domestic operation, because agricultural environments tend to be geographically unique. Federal and state governments share in funding and carrying out agricultural research, partly because there is no mention of agriculture—let alone agricultural research—in the Commonwealth Constitution. This Constitutional silence suggests that, under the 'reserve powers' doctrine, it would be the domain of the states, and they do undertake considerable agricultural research through their agriculture departments (table 5.1). The Commonwealth also provides considerable R&D funding through its financial support of the universities, and its financing of CSIRO, although there is no direct constitutional power for the Commonwealth's participation in these activities. The Commonwealth has also previously funded state activities extension (s.96 grants). The Commonwealth provides a mechanism for collecting levies on farmers to support agricultural research, since s.90 of the Constitution prevents states from raising funds in this way. The Federal Government also provides taxpayers' funds to match the levies. The Commonwealth shapes the direction of research by its policy, particularly as enacted through the decisions of the RDCs, and through its policy directions to and funding of CSIRO. The states implement their agricultural policies primarily through the funding of their agriculture departments.[2]

Table 5.1 Total Agricultural Research Funding ($m)

	Various dates[a]	*1990–91*
NSW	118	101
Victoria	52	44
Queensland	100	78
South Australia	35	37
Western Australia	53	57
Tasmania	13	13
Northern Territory		7
ACT		0
Total		337

Source: Industry Commission (1994, pp. E.6–E.9)

Note: a: NSW (1993–94), Victoria (1991–92), Queensland (1992–93, probably underestimate), South Australia (1990–91), Western Australia (1993–94), Tasmania (1990–91).

There are also important international dimensions to agricultural research and research policy. Australian researchers have drawn on considerable international research in the past and will continue to do so in the future. For example, Brennan and Fox (1995, p. 9) estimated that 86 per cent of Australia's wheat crop is based on CIMMYT varieties. Since Australian agriculture is primarily based on exotic plants and animals—with the exception of native pastures, macadamia nuts, some kangaroo and emu harvesting and a small crocodile farming industry—the international dimension is crucial to the genetic improvement of commercial plants and animals. Australia could never afford to undertake all the basic agricultural research that it requires; thus, for example, considerable basic research in plant and animal molecular biology—'genetic engineering' —has been, and will continue to be, borrowed from overseas. This borrowing occurs through the international research literature, overseas travel, overseas research training, and overseas researchers visiting Australia.

The large proportion of exports for many agricultural commodities also means that Australian industries must understand overseas market requirements, and shape their products to be compatible with these markets. The Australian wool industry has helped fund considerable processing research in overseas countries, especially through the International Wool Secretariat (Vines et al. 1991). The Australian wheat industry is undertaking increasing research into the characteristics of overseas consumers' wheaten products to ensure that exported Australian wheat is used effectively (McMaster & McKay 1991).

Policy mechanism

Most of the RDCs were established under the *Primary Industries and Energy Research and Development Act 1989*: those for cotton, dairy, fisheries, forest and wood products, grains, grape and wine, pig and sugar. Also established under this Act were RDCs for Energy, Land and Water Resources, and rural industries generally. The latter, an umbrella for five R&D councils (chicken meat, dried fruits, egg, honeybee, tobacco), serves small or emerging rural industries, and funds research generic to the agricultural sector. The corresponding bodies for meat, horticulture and wool have their own Acts. There is also a Forest and Wood Products Research and Development Corporation and a Fisheries Research and Development Corporation. The above corporations and councils fund research providers (cf. table 5.2). Legislation also establishes the organisations that undertake agricultural R&D—Commonwealth legislation for CSIRO, Commonwealth and state legislation for the universities, and state legislation for the agriculture departments providing research and extension services, and related providers (e.g. soil conservation, irrigation water supply).

Table 5.2 Allocation of RDC Funding, 1992–93 (% unless specified)

Corporation/ Council	*State Agriculture Departments*	*CSIRO*	*Universities*	*Other*	*Funds ($m) 1993–94 (approx.)*
Grains	54	13	25	8	50
Meat	16	20	11	54	50
Wool	15	60	18	7	40
Horticulture	60	7	9	24	18
Dairy	33	14	19	34	15
Fish	58	26	6	10	10
Sugar	3	9	5	83	8
Pig	35	12	41	12	8
Cotton	29	48	19	4	8
Grape & Wine	23	7	6	65	4
Chicken Meat	20	22	52	5	1
Tobacco	97	0	0	3	1
Dried Fruit	44	38	16	1	1
Egg	18	19	48	15	1
Honeybee	30	10	51	15	0
RIRDC					15
Total					230

Sources: Industry Commission (1994, figure E1.3; appendix F, table F1)

RDCs are funded via tax-deductible levies collected by the Commonwealth (as s.90 excises), and matched by Commonwealth Consolidated Revenue, generally with a dollar-for-dollar subsidy up to a maximum of 0.5 per cent of an industry's Gross Value of Production (GVP) for most of the commodity RDCs. As a proportion of GVP, levies for nine commodities were in the range 0.3–0.5 per cent for 1992–93 and 1993–94, except for grains in the drought year 1993–94, when it was 0.7 per cent (Industry

Commission 1994, figure E2.1). The matching of levy funds with Consolidated Revenue indicates that the Appropriation Acts comprising the annual Budget are also important for implementing the RDCs. Similarly, federal Budget legislation funds the CSIRO and the universities, and corresponding state legislation funds research and extension/advisory functions in the agriculture and related departments. Despite the existence of industry levies, the Industry Commission (1994, pp. E.2–E.3) estimated that Australian governments funded at least 82 per cent of total agricultural R&D.

Administrative arrangements for agricultural R&D affect two main areas. First, there are administrative decisions that implement major or minor policy decisions of government. For example, the Federal Government's choice of members on the boards of the RDCs, and any explicit or implicit guidelines these members and boards are expected to follow, has a profound impact on the ways in which funds are disbursed. These decisions may affect shares of research funding across farm and non-farm industries; shares across enterprises (since most of the RDCs are responsible for several major industries); allocation of funds between basic and applied research, and between scientific and socio-economic (including marketing) research; allocation of funds between private and public research; and between CSIRO, state agriculture departments and the universities.

Second, there are administrative decisions that affect the funding of agricultural research. For example, decisions by the RDCs as to how much funding will be allocated each year to research projects, and which particular projects will be funded, are made by the Boards of the RDCs. While previously RDCs primarily funded public research in the CSIRO, state agriculture departments and the universities, in recent years there has been an increasing trend towards using private research organisations. Similarly, research funding within the state agriculture and related natural resource departments is essentially a matter of negotiation between those departments and their state's Treasury in drawing up the departments' budgets. Decisions as to how Consolidated Revenue funds are ultimately allocated to projects are taken within the agriculture and related departments, although tighter budgets in recent years have given RDC funding great leverage as these marginal funds are often necessary for research to proceed (cf. Jarrett & Lindner 1982, p. 92).

Public funding also provides governments, particularly at the state level, with the means to favour individual electorates.

There has been a dramatic change in the organisation of Australian agricultural R&D in recent years. Partly in response to previous policy enquiries, CSIRO has made a major shift away from basic research towards applied or 'mission' research.[3] Theoretically, CSIRO's previously-undertaken basic research was to be taken up by the universities, but simultaneous major changes in the structure and funding of the universities have constrained their ability to respond in the intended manner. CSIRO is now also required to raise 30 per cent of its research funding outside its Consolidated

Revenue grant from the Commonwealth (Industry Commission 1994, p. B.25). In agricultural R&D this has placed heavy reliance on RDC funding. State agriculture departments' research is primarily applied; the state provides the basic research infrastructure—i.e. land (critical in applied agricultural research), most capital structures and buildings, most salaries of permanent staff, and most operating expenses. The extension activities of state agriculture departments are a considerable portion of the development part of R&D, and are largely taxpayer funded. These extension activities assist the adoption process for new technologies, and recognise potential information transactions costs in a largely atomistic industry like agriculture. But not all potential extension activities are 'public' goods; there is extensive potential for private good 'consulting' activities in agricultural extension (cf. Lloyd 1986). Like state agriculture departments, universities provide most of the basic research infrastructure (land, most capital structures and buildings, most salaries of permanent staff, most operating expenses). The universities undertake research along the research spectrum from basic to applied, with a traditional emphasis at the basic end.

Institutional structure

The key interest groups in the funding and performance of agricultural R&D in Australia are:

- farmers and their trades union—who are interested in the development of new technologies that enable them to increase their profitability, and are also the levy payers who partly fund the R&D corporations;
- corporate R&D providers—CSIRO, universities and state departments, who undertake research using funds provided by taxpayers and farmers;
- agricultural researchers and their trades union and professional societies—who are employed via the research funds provided by taxpayers and farmers;
- R&D corporations—statutory bodies who distribute the funds collected via farmer levies and matching Commonwealth grants;
- taxpayers—who provide the bulk of the R&D funding for agricultural and related research; and
- non-farm private sector—which benefits from the provision of publicly funded research (input suppliers, and product processors, handlers, distributors), and also funds some of its own research.

The objectives of these interest groups are not necessarily mutually consistent. Each group of funders may wish to reduce its own level of funding as long as the aggregate level remains unchanged. Agricultural researchers would prefer the funding largesse without the increasingly onerous bureaucratic burden being demanded by government and the RDCs. Taxpayers are probably unaware that they provide the bulk of the funding for agricultural R&D in Australia.

Policy performance evaluation

Because of its importance to general Australian R&D activity, CSIRO has been repeatedly evaluated over the last two decades (cf. Industry Commission 1994, pp. B.12–B.13; Senate Economics References Committee 1994). Agricultural and resource R&D has been included in these reviews because of CSIRO's extensive research in these fields. The Industry Commission's 1994 draft report discussed agricultural and resource R&D, and included radical proposals to dramatically reduce taxpayer funding of the RDCs. Victoria has also reviewed its field-based extension and regulatory services (Watson et al. 1992).

Economic issues

Research organisation[4]

The key issues of agricultural research policy were identified above as concerning the optimal funding and undertaking of agricultural research. These questions arise because of the economic characteristics of research in general and agricultural research in particular. Two dimensions of the background to research policy are illustrated in table 5.3 as research objective and form of research organisation. Two standard explanations of research objectives are efficiency and rent-seeking. Research may be undertaken by private profit firms, private not-for-profit firms, and government bureaus. Private firms can and do undertake research but they must solve problems arising from non-excludability and non-rivalry. Private research firms solve non-excludability against users of their output by embodying their research in products which users must purchase. Embodiment also solves non-rivalry because users are unable to apply research results per se in production. Private firms must also solve non-excludability against commercial rivals, and attempt to do so via intellectual property rights (patents, PBR/PVR, copyright, trademarks etc.) and trade secrets. These routes do not entirely solve non-excludability: patenting, for example, provides valuable information to commercial rivals to assist 'inventing around', and trade secrets may 'escape'. Since it is difficult to obtain intellectual property rights for disembodied technologies and, even where this is possible, non-rivalry limits their value, private firms tend to concentrate their research on products such as agricultural chemicals, machinery and new plant varieties. Private profit research is thus most likely to occur as an adjunct to firms producing inputs for agricultural production, or those engaged in post-farmgate activities transforming agricultural products. Because private research solutions are likely to be concentrated in the production of embodied technologies, a purely private solution is unlikely to be allocatively efficient *across the sector*.

A second form of organisation to attack the research problem is voluntary, profit or 'not-for-profit' associations. Individuals in a similar industry band together in some corporate form to supply services that would otherwise not be adequately supplied. Cooperatives have formed in the past, particularly in agricultural marketing, with these characteristics. The modern economic terminology is 'club'.

Table 5.3 Rationales for Government Intervention in Research

Objectives	*Organisational form*		
	private, profit-oriented, research output selling firms	*voluntary, profit or not-for-profit, research output distributing associations*	*statutory, research output distributing or selling bureaus*
Economic efficiency:			
positive aspects	beneficiaries pay; embodiment avoids non-rivalry	beneficiaries pay	domestic excludability solved (e.g. levies)
negative aspects	private solution not *necessarily* optimal—activity concentrated where excludability possible	transactions costs; non-rivalry remains; private solution not *necessarily* optimal	non-rivalry remains; solution not *necessarily* optimal
Rent seeking:			
positive aspects	strengthen property rights	strengthen property rights	strengthen property rights
negative aspects	over-strengthen property rights; shift costs (e.g property rights defence) to public sector		maximise bureau objectives rather than optimise R&D output

Through the club, individuals supply capital to produce the required product or service, and membership of the club controls the distribution of the output. For example, farmers could band together in a particular industry or region to produce appropriate technologies for their products or environments. The R&D products could range across the entire spectrum (cf. figure 5.1). Were clubs to produce embodied technologies, the allocative outcome would be similar to that of private profit firms, with one possible exception. The private firm is interested in maximising its profit, but not necessarily those of its users. There may be some technologies that would return high returns to users but not to the firm; in this case, although users might 'demand' such research, it may not be supplied. Conversely a club would be interested in maximising the aggregate returns to research providers and users since they are indistinguishable. A club may also be able to undertake research in disembodied technologies despite the inherent problems of excludability and rivalry. If the club can persuade all potential beneficiaries of new technologies to contribute to the club's finances, it has partly solved non-excludability—particularly if it can relate the level of membership fees to potential benefits (e.g. through fees scaled on farm or enterprise size). Non-rivalry is implicitly solved because, while a technology may still be technically non-rival, non-rivalry does not jeopardise the club's finances if all potential users belong to the club (except for international leakage of new technologies). With a fully-encompassing membership, the club's problem reduces to charging membership fees which reflect the benefits the members obtain. If the fees are too high for some members relative to the benefits they receive, they may opt out of the club and re-create the non-excludability problem. Thus, research clubs may solve some of the problems facing commercial research, but significant transactions costs remain. A bias towards embodied research may also remain. Like the private profit research firm, the club must also solve the principal-agent problem to ensure that its research output is consistent with the demands of the club's owners.

One solution to the problem of voluntary club membership is to have a 'compulsory' club. This is the solution adopted with Australia's agricultural R&D corporations. Compulsory membership is obtained via compulsory levies on product, backed by the Commonwealth's constitutional power to collect excises. Compulsion solves non-excludability and, as with a fully-encompassing club, non-rivalry is also essentially solved except for international leakage. A compulsory club is less concerned with whether or not the new technologies are embodied or disembodied since it is not reliant on selling its research output. Indeed, the potential for sales of research output may create problems for a compulsory club, since it creates financing tensions between the club owners—should research be fully funded by membership fees (i.e. compulsory levies) or should research output sales (e.g. via licences on new vaccines or plant varieties) be used to fund research? While non-excludability may be solved explicitly and non-rivalry implicitly, the optimality of the solution depends on the level of membership fees. In the case of Australia's RDCs, the optimality of the solution also depends on the level of additional government funding. And, of course, the optimality of the research process depends not only on the optimal level of research funding, but on the appropriate selection of research projects. There are likely to be substantial principal-agent problems in research funding and project selection because there are many 'principals' (taxpayers, levy payers, governments, RDCs) and many research 'agents' spread over disparate institutions (e.g. CSIRO, state agriculture departments, universities, individual researchers) all of whom—and especially university researchers who also teach—have objectives that both compete with and are complementary to the objectives of the research principals.

If potential market failure problems could be successfully solved by institutional means, such as via research organisation or directly addressing excludability and rivalry, the allocative problems of research funding and project choice would also be solved. Where institutional problems remain, however, government will remain a major participant in the agricultural R&D process.

The potential form of the research organisation affects not only the efficiency of the research process but also distributional aspects. From the private profit research firm's perspective, anything that enhances its profit is clearly desirable. Thus, for example, private firms (and not just in agriculturally-related industries) have sought and obtained special consideration from government, especially in the form of tax concessions for research funding (Industry Commission 1994, chapter D2). Similarly, clubs may seek government intervention to support their operations, ranging from the type of support that private firms may seek, through to support that would be peculiar to clubs—for example, clubs may seek government assistance in identifying the potential beneficiaries of research so that they can target membership drives (particularly important in a geographically-dispersed Australian rural sector), or seek special tax or financial consideration for research clubs. The capacity for rent-seeking behaviour in the statutory research bureaus is also high because there are many competing interest groups involved. From the government's perspective, research funding is likely to contain elements of PERTs (political economic resource transaction) which reduces market failure or transactions costs, and PESTs (political economic seeking transfer) which are rent-seeking activities (cf. Rausser 1982). PESTs and PERTs could be viewed in at least two ways: (i) the PERT is the objective, and the PEST is the sweetener to win passage for the PERT (i.e. a Pareto-safe change); or (ii) the PEST is the objective, and the PERT is the sweetener, or camouflage, for the PEST.

Extension services

The Industry Commission (1994, pp. E.51–E.57) briefly discussed the rationales for public provision of extension services in contrast to private sector provision as part of the agricultural R&D process. These extension services have been an important component of the delivery (dissemination) of new technologies to farmers, particularly in the state agriculture departments. Lloyd (1986, p. 141) argued that advisory services 'have both public and private good aspects and benefits', and noted that 'Information provided by the extension services has one characteristic of a public good: one person's use does not diminish its value in either quantity or quality. However, the method of supplying information may be private, thus enabling a user-fee to be charged.'

Some extension services are purely private in nature, and have specific private benefits for individuals that are equal to social benefits. Highly specific financial and technical advice on individual situations is generally of this type. The key issue for the public provision of extension services is not, therefore, whether they should be provided but what is their appropriate form. Key issues include the industries in which such services should be provided, the types of information that should be publicly provided, how it should be delivered, and the extent to which users should pay for it.

Publicly provided extension services are also a key political good. They are a vehicle for shifting resources to the farm sector and a means of providing employment, especially in small country towns.

Managing the R&D process

Research and development is expensive. Because of government's current and likely future extensive funding of agricultural R&D in Australia, there is considerable interest in improving the efficiency of the research process. Allocative efficiency has been addressed via evaluating the institutional framework for research and the level of research funding (cf. Industry Commission 1994). Economists have also contributed by evaluating pay-offs to agricultural R&D. The technical efficiency of the agricultural R&D process is under the control of the research providers, but little external evaluation of technical efficiency has been undertaken (cf. Baker 1990).

There are several possible approaches to improving the technical efficiency of R&D:

- intra-organisation search for improved management structures;
- improved incentives for managers and researchers (a 'market' solution); and
- bureaucratisation (a 'command and control' solution).

Recent Federal Government initiatives suggest that bureaucratisation has been chosen as a principal mechanism for delivering these improvements. There are two strands to this bureaucratisation—competency standards and performance indicators.

Bureaucratisation—competency standards

The Federal Government's enthusiastic promotion of 'competency standards' has been accepted by the Australian Institute of Agricultural Science, which developed standards for 'agriculturists'— i.e workers in agricultural and related resource industries (Steering Group for the Development of Competency Standards for Professional Agriculturists 1993; see also Steering Committee on National Agricultural Research Strategy for Australia 1995, p. 5). The economic objectives appear to be to encourage workers to maintain or increase their investment in human capital by establishing a formal list of competencies against which individual workers might be assessed. A worker's assessed competencies provide information for potential employers.

Competency standards have been adopted for agriculturists despite there being no evidence of significant market failure in the labour market for agricultural professionals. Such standards might be economically useful if there were a demonstrable failure and the proposed standards reduced the costs of this failure by more than the cost of a competency accreditation scheme. However, employers are unlikely to abandon their own assessment procedures for recruitment or promotion in favour of the competency standards proposed, if only because they need to discriminate between candidates who may have similar competencies. Hence, procedures for determining competency are likely to run parallel to those of employers. Proposed competencies will therefore increase the cost of demonstrating professional standards, but with no appreciable benefits to be gained. Individuals may also exhibit perverse responses to competency standards by aggrandising qualifications rather than actually producing valuable output.

The proposed competency standards require administration by a 'Standards for Professionals in Agriculture and Agri-industry Board'. This Board would comprise representatives of interest groups with particular agendas in agricultural and related sectors. The most serious potential consequence of the proposed competency standards is that, if they were successfully implemented, they would be likely to stifle the existing plurality of views in agriculturally-related professions. Individuals would be less likely to critically evaluate proposals, policies or attitudes of key interest groups or powerful individuals in those interest groups if they knew that representatives of those same groups could wield decisive power over their professional status. Especially in policy-related fields such as agricultural economics, vigorous public and professional debate might be stifled. For example, previous vigorous debates about the economics of irrigation developments in Australia, or the merits of protection all round for Australian agriculture and industry, would be less likely to occur if participants feared that their involvement in such debates would prejudice their future careers. The stifling of such debates could seriously disadvantage the economic development of Australia as a whole, and the agricultural sector in particular.

Bureaucratisation—performance indicators and benchmarking

There is increasing emphasis on measurable performance outcomes in research and other economic activities. For example, the Steering Committee on National Agricultural Research Strategy for Australia (1995, p. 4) proposed, as one strategy, to 'Ensure that Australian agriculture and related industries remain internationally competitive by focusing applied research on adopting techniques of benchmarking and seeking to establish international best practice'.[5] Emphasis on benchmarking reflects an expectation that research is just another industrial production process, and the production of research can be evaluated in the same way as the performance of machines.[6] Research does have outcomes and these outcomes can be measured in a variety of ways; however, several characteristics of measurement systems for research should be noted:

- Performance indicators establish an incentive mechanism. Unlike machines, humans respond to being measured (cf. Boulding 1970, pp. 120–1). If performance indicators are inappropriately chosen, inappropriate responses will be elicited.
- Performance indicators are not a substitute for management. Management requires information, and performance indicators are one potential source of such information. But performance indicators must be relevant to a management problem. If the wrong dimension of activity is measured, or inappropriate measurements are taken, or if the activity is inherently difficult to measure, selected performance indicators may be at best irrelevant to the research management problem—and at worst counter-productive.
- Performance indicators are not costless. The more researchers are evaluated, and the more that is required of them in this evaluation process, the less time is available for research itself. There is, therefore, an opportunity cost of research management and, at least conceptually, an optimal level of performance evaluation and an appropriate set of performance indicators.

Performance indicators exist under a variety of aliases such as 'benchmarking' and especially 'international benchmarking'. The valid use of an indicator requires that the decision problem has been appropriately specified, and that the indicators chosen are relevant to the specified problem. Performance indicators are only valid management tools if the performance indicator(s) is (are) are derived from an appropriate model of the management unit, its goals, its internal and external constraints, and the particular problem it is facing. The technical conditions under which 'benchmarking' is appropriate are—for partial productivity measures—that technologies (including, in agriculture, environmental conditions) are identical for compared firms, and that output:input price ratios are identical. For 'benchmarks' that use relative input ratios, the production functions of the compared firms need to be identical except for a Hicks-neutral technical progress scale factor. None of these conditions is likely to hold in a comparison of Australian agriculture with agricultures in other developed countries. 'Benchmarking' is therefore highly unlikely to offer much to advancing R&D for Australian agriculture.

Because performance indicators are a 'reduced form' of a unit's decision problem, it does not necessarily follow that performance indicators should be maximised. In most cases there will be an optimal level of the performance indicator that is less than its possible maximum value—beyond this optimum, the costs of obtaining higher performance indicator values will exceed the benefits. As performance indicators are specific to a firm's problem, it will generally be inappropriate to compare them across firms; such comparisons across different decision-making entities are only valid if the entities are sufficiently similar.

The current political vogue for attaining 'international best practice' and using 'benchmarking' to do so, is not necessarily to Australia's advantage, especially in the case of agriculture. 'Benchmarking' is usually based on partial productivity measures—either a ratio of output to input (e.g. labour productivity, output per unit of labour input), or input ratios (e.g. stocking rates, number of head per hectare). International 'best practice' agriculture—i.e. high partial productivity measures or given input ratios—is likely to be irrelevant to Australian production. A simple example may be drawn from wheat production: in Western Europe, farmers' wheat yields average around 8 tonnes per hectare, about four times higher than average Australian yields. European yields are attained through a benign temperate climate, and high input use supported by heavy subsidisation of agricultural product prices. It would simply be unprofitable for Australian farmers to attempt to attain such yields using current—or even likely future—technologies, and inefficient for the aggregate Australian economy for such resources to be devoted to agriculture. Not only would such attempts probably be unprofitable, they are also likely to be ecologically unsustainable for Australian agriculture. It is therefore unrealistic to expect Australian agricultural research workers to provide Australian farmers with technologies that would enable them to match European benchmarks.

Government involvement in extensively funding the R&D process implies that government will have a close interest in ensuring that the funds it provides are used appropriately. However, government objectives are not necessarily consistent with economic efficiency, and its demands for accountability may include bureaucratic and political dimensions that are inconsistent with an efficient R&D process.

Analytic techniques

Allocative efficiency with non-rival goods

Consider the market for a non-rival, homogeneous and perfectly divisible good whose marginal cost of production is MC (figure 5.3). There are three consumers of this good whose marginal valuations are MBa, MBb and MBc respectively. If the producer contracted with individual c to purchase the commodity, they could agree to produce and consume quantity c but, because the commodity is non-rival, individuals a and b could consume level c without paying for the commodity (i.e. they would be 'free-riders'). However, had the contract been with b, a higher level of output could have been contracted, but a and c could have been free-riders; and an even larger output could have been contracted between the producer and a, with b and c as free-riders.

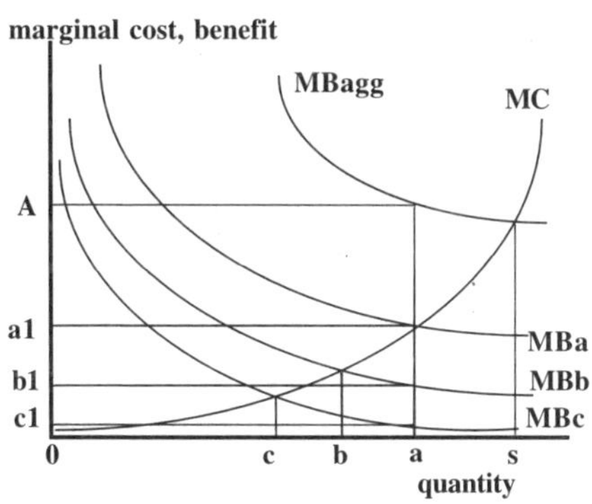

Source: adapted from Randall (1987, figure 9.3)

Figure 5.3 Allocative Efficiency with Non-rival Goods

Market equilibrium with *rival* goods is based on a demand curve parameterised on price, and aggregating across consumers' demanded quantities at those prices. With non-rival commodities, however, once a price is struck with one consumer, there is no incentive for any other consumer to pay for that level of output (although they may be willing to pay for additional output). The optimal non-rival commodities solution is obtained by noting that, at some level of output (say a), each of the consumers has a non-zero marginal valuation of the commodity (a1, b1, c1 respectively). If each consumer were approached secretly, each would be prepared to pay their respective valuation for quantity c. The aggregate marginal valuation of quantity c is therefore A. If this process were repeated for each level of output, an aggregate marginal benefit curve (MBagg) could be derived. The optimal level of output, at which marginal cost equals aggregate marginal benefit, is quantity s. While this analysis identifies the optimal *level* of output for a non-rival commodity, the practical achievement of efficiency requires an appropriate *institutional structure*—whether market or government—which yields an identical solution.

The translation of this analysis into the research context yields additional difficulties. Research *output* is a non-homogeneous good, comprising a complex array of information, products, processes, and mechanisms for disseminating information, products and processes to end-users. Research may also require the preservation of existing information, and its bequest to new generations (the education process). Identifying research beneficiaries' marginal benefit curves would be exceedingly difficult. Characterising research in terms of its outputs requires the even more difficult transformation of the conventional continuous analysis (figure 5.3) into discrete analysis.

Measuring benefits of new technology

If research output comes in discrete packets, a problem of evaluating the appropriate level of research funding is to measure the benefits of this output. The conventional way of measuring research benefits is to consider the effect of research on the production of commodities. This procedure simplifies the analysis because, for most agricultural industries, output is continuous and perfectly divisible and may be roughly assumed to be homogeneous; additionally, on the supply side at least, markets are generally atomistic. Research output is considered as being disembodied, and available to the user without cost. Edwards and Freebairn (1981, chapter 3) considered the problem for an agricultural commodity in a closed economy with supply S1 and demand D; equilibrium is at price c and output Q1 (figure 5.4). Suppose a new disembodied technology shifts the supply curve to S2 by reducing marginal costs of production. The new equilibrium is (x, Q2) and the gross benefits of the research are measured by the increase in total surplus from area qba to area qyz, equivalent to area abyz.

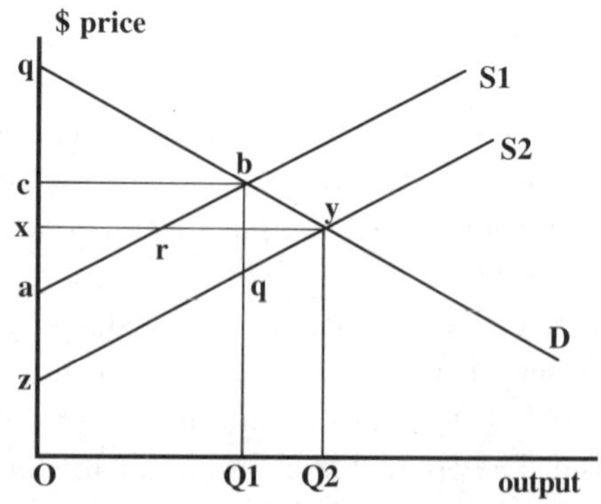

Figure 5.4 Benefits of Disembodied Research in a Closed Economy

Area abyz comprises the sum of areas abqz and byq. Area abqz (a trapezium) is equivalent to the area of a rectangle with height (a – z) (= distance bq) and base 0Q1; thus area = (a – z).0Q1.[7] The area of triangle byq is 0.5.(a – z).(Q2 – Q1). Gross research benefit—equivalent to area abyz—may be derived algebraically as:

$$\text{(5.i)} \qquad \text{area abyz} = (a - z).0Q1 + 0.5.(a - z).(Q2 - Q1) = 0.5.(a - z)(0Q1 + 0Q2)$$

Estimation of the gross value of the technology shift becomes increasingly difficult to estimate as these assumptions are relaxed (cf. Davis et al. 1987).

One further useful result can be derived from figure 5.4. At equilibrium b, consumer surplus is area qbc; while at equilibrium y, consumer surplus is area qyx. Thus the increase in consumer surplus with the supply curve shift is area cbyx:[8]

$$\text{(5.ii)} \qquad \text{area cbyx} = 0Q1.(c - x) + 0.5.(0Q2 - 0Q1).(c - x) = 0.5.(c - x).(0Q1 + 0Q2)$$

The change in producer surplus as a consequence of the technology-induced supply curve shift is defined as the difference between the gross research benefit and the change in consumer surplus, estimated algebraically as:

$$\text{(5.iii)} \qquad \text{change in producer surplus} = 0.5.(a - z)(0Q1 + 0Q2) - 0.5.(c - x).(0Q1 + 0Q2) = 0.5.(0Q1 + 0Q2).\{(a - z) - (c - x)\}$$

In this simple model, the annual gross benefit of research, and the benefits to consumers and producers, can be defined algebraically in terms of the prices and quantities of output at the initial and final equilibria, and the size of the induced supply curve shift (a – z). Alternatively, benefits could be determined from the supply curve shift, the initial equilibrium price and quantity, and supply and demand elasticities. The latter approach is especially useful in *ex ante* analysis.

The key assumption of the preceding analysis for Australia's agricultural industries is that of a closed economy, since the major Australian agricultural industries—wool, wheat, meat, sugar, manufactured dairy products—export most of their output. Davis et al. (1987) showed how the preceding model could be modified with exports. This analysis requires a familiarity with the concept of excess supply and excess demand. In panel (a) of figure 5.5 is depicted a conventional supply and demand diagram for a market with domestic demand and supply (D and S). Above price P0, supply exceeds demand, and the country is a potential exporter; these potential exports are 'excess supply' (ES, panel (b) of figure 5.5). Conversely, below P0, demand exceeds supply and the country is a potential importer with 'excess demand' (ED). With two potential trading countries, there is trade equilibrium if one country's excess supply equals the other's excess demand.

This framework was used by Davis et al. (1987) to derive the economic effects of R&D with trade (figure 5.6). In figure 5.6(a), an exporter has an initial supply curve

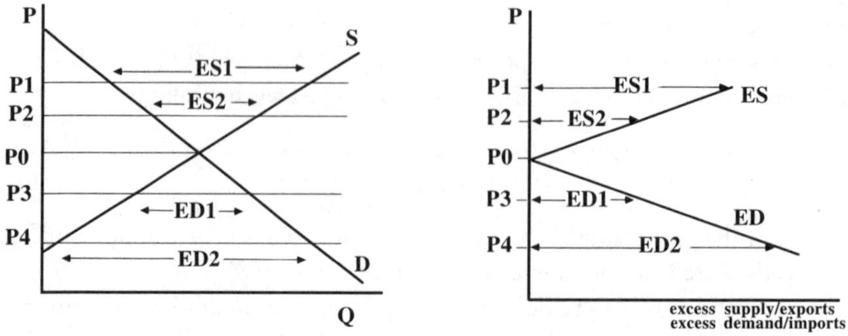

(a) supply and demand

(b) excess supply and demand

Figure 5.5 Derivation of Excess Supply and Excess Demand

Sa and domestic demand D; its excess supply curve (ES) for world prices above PE is shown in figure 5.6(b). The market conditions for a potential importer are shown in figure 5.6(c); its excess demand schedule for world prices below PI is ED (figure 5.6(b)). With the initial conditions, world trade equilibrium is established at price Pw and trade volume q2. Suppose technological change shifts the exporter's supply curve from S to S′ (figure 5.6(a)). The exporter's excess supply curve shifts from ES to ES′, and world price equilibrium shifts to the lower price Pw′ and higher trade volume q2. The trade effect in the importing country is two-fold: at the lower price Pw′, total consumption rises but, because price has fallen, domestic production falls. In the exporting country, consumption also rises but, because domestic demand is assumed less elastic than supply, the increased domestic consumption is outweighed by the increase in output; hence exports rise.

The welfare effects on the exporter can be derived by reference to figure 5.6(a). The initial total surplus (TS) for the exporting country (sum of consumer and producer surplus, CS + PS) is:

(5.iv)
$$TS = CS + PS = zhg + gim$$

After the technological shift, the sum of producer and producer surplus is:

(5.v)
$$TS' = zsn + njk$$

Thus the total change (Δ) can be expressed as:[9]

(5.vi)
$$\begin{aligned}\Delta &= (zsn + njk) - (zhg + gim)\\ &= ghsn + njtr\end{aligned}$$

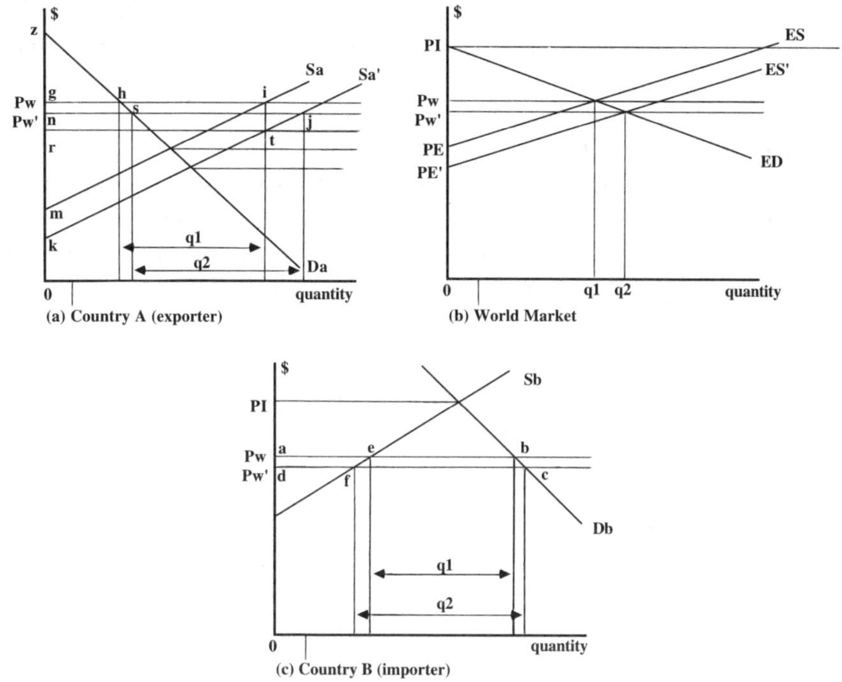

Source: Davis, Oram and Ryan (1987)

Figure 5.6 Gross Benefits of Technical Change with Trade

By analogy with the closed economy case, area ghsn may be determined algebraically using the world prices before and after the technological shift, and corresponding domestic consumption. Area njtr may be determined algebraically from output after the supply shift, change in total output, the size of the supply curve shift and the change in equilibrium price. If the initial equilibrium and relevant elasticities are known, the expected value of an expected research shift (k) could be estimated *ex ante* (cf. Davis et al. 1987, chapter 2).

This analysis may be made more realistic by relaxing assumptions. Davis et al. (1987) considered the consequences of the technology developed in the exporting country being applicable in the importing country, shifting its supply curve and giving rise to research 'spillovers'. Additionally, in the preceding analysis it was assumed that producers and consumers trade directly; Freebairn et al. (1982) extended the analysis to allow for farm input supply and post-farmgate production, and research that could reduce costs in those sectors as well as in the farm sector. Where improved technology benefits only part of a domestic industry, those not having access to it will lose from others' adoption if the demand curve is not infinitely elastic and output price falls (cf. Edwards & Freebairn 1981).

In the preceding analyses, the *single period* benefits of agricultural R&D have been estimated. However, R&D is an investment decision—once R&D funding is committed, it is expected that resulting new technologies will yield a supply curve shift in some or all future periods. The benefits of R&D are not simply the estimated benefits in a single period, but those in all relevant subsequent periods. Any investment decision would require discounting future benefits (cf. Davis et al. 1987, chapter 2). A more complete analysis of the returns to R&D would also need to consider how long it takes for a new technology to be adopted ('adoption lag') and the rate at which it is taken up ('adoption rate'), and how many producers finally adopt the new technology ('adoption level').

Estimating supply curve shifts

A key parameter required for the preceding studies is the size of the supply curve shift (k). In *ex post* analysis, supply curve shifts may be estimated econometrically. In *ex ante* analysis, it is necessary to link possible research projects with their expected effects on supply curves (cf. Davis et al. 1987). Methods for estimating supply curve shifts include:

- Estimate supply curve shifts directly—at an aggregate level, Binswanger (1978) estimated the rate of disembodied technological change in a cost function model; this analysis can be extended to simultaneous estimation of technological change in input and product markets using a variable profit function model (Lopez 1980) or ad hoc supply models.
- Estimate production function shifts directly—production functions may be estimated with conventional inputs and variables relating to R&D inputs or outputs, and the relationship between research inputs and farm output measures the shift in the production function due to research (cf. Norton & Davis 1981). Similar estimates may be obtained in a two-stage process using estimates of total factor productivity (e.g. Thirtle & Bottomley 1988).
- Yield shifts—experimental data might be used to estimate the shift due to new technologies (cf. Godden 1987).
- Synthesise supply curve shifts—supply curves at farm or aggregate level may be synthesised using mathematical programming; changes to input parameters resulting from new technologies will allow resulting supply shifts to be estimated.

Conclusion

Market failure is probable in the R&D process, particularly for new knowledge and disembodied technologies. Government may therefore play a central role in funding R&D to avoid such failures. Where government funds—and especially where it undertakes—R&D, there may also be significant distributional consequences of its decisions. A well-developed economic approach exists for estimating the benefits of disembodied technological change, but considerable difficulties may be experienced in determining appropriate government policy from empirical analysis.

ISSUES FOR DISCUSSION

1 Can government increase allocative efficiency by subsidising goods or services? Discuss with reference to agricultural R&D by the public sector in Australia.

2 In a 1992 Australia Day address the Chairman of CSIRO, Professor Adrienne Clarke, argued that taxation incentives and other allowances to encourage science and technology are necessary for Australian firms to reach international competitiveness. In the context of agricultural R&D, evaluate Professor Clarke's claim that such assistance would not be 'a return to protectionism'.

3 In commenting on recent changes to research funding in CSIRO, Professor Ian Lowe of Griffith University indicated that Australia's R&D funding should be increasingly directed to 'industries of the future' and implied that agriculture is not an 'industry of the future'. Discuss the appropriate funding of agricultural R&D in Australia in the context of this comment.

Endnotes

1 The Industry Commission (1994, volume 3, figure QB.2) reported the following approximate estimates of multifactor productivity (or total factor productivity) for the period 1975/76 to 1992/93: agriculture 2%; mining 0%; manufacturing 2%; electricity, gas and water 3%; construction 0%; wholesale and retail trade 0%; transport, storage and communication 3%; recreational; personal and other services –2%. Males et al. (1990, tables 1 and 2) reported the following estimates for agricultural industries: all broadacre 2.2%, crops 5.5%, mixed crops and livestock 2.4%, sheep 0.2%, beef 0.1%, sheep–beef 2.4%, dairy 1.9%.

2 Until major consolidations of agricultural RIRFs in the 1980s, the states also had relatively small research funds, and thus had some control over RIRF-funded research, even if only through appointments to their Boards or Committees.

3 In 'mission' research, the study begins with an empirical problem, and the underlying science is only investigated to the extent necessary to answer the problem.

4 The argument in this section has benefited from discussions relating to a project funded by the Rural Industries Research and Development Corporation within the Department of Agricultural Economics, University of Sydney.

5 Proving the old adage that there is nothing new under the sun, see McEwen's 1952 comment on 'bench marks' in Box 1 of the Introduction.

6 The chief executive of the Walt Disney organisation commented that his organisation, relying on creative people, could not be managed like an industrial company ('60 Minutes', Channel 9, 18 June 1995).

7 The area of a trapezium with base Q and vertical height k is equal to the area of a rectangle with base Q and height (k + A) *less* the area of two triangles of base Q and height A. Thus, trapezium area = Q.(k + A) – 2{0.5.Q.A} = Q.k.

8 The area of a rectangle with sides bc (= 0Q1) and (c – x) and a triangle with base (0Q2 – 0Q1) and height (c – x).

9 Δ = (zsn + njk) – (zhg + gim)
= ghsn + (njk – gim)
= ghsn + [(njtr + rtk) – (givr + rvm)]
= ghsn + [njtr + (rtk – rvm) – (givr)]
= ghsn + [njtr + mvtk – (givr)]
= ghsn + [njtr + (mitk – itv) – (gitr – itv)]
= ghsn + njtr since mitk = gitr (area of parallelogram)

6 Agricultural marketing: general issues

Since the Second World War, agricultural marketing has been a key area of government intervention. For constitutional reasons, agricultural marketing policy was implemented by both state and Commonwealth governments. There was a significant reduction in its importance in the 1980s and early 1990s as both Commonwealth and states retreated from intervention. This retreat reflects the general fervour for deregulation that reached its own apogee (or nadir) during the late 1980s. A principal aim of this chapter is to explain and evaluate changing attitudes towards government intervention in marketing.

Policy issue

Farmers have often felt that, while their production methods were efficient, they obtained lower returns from selling their produce than market conditions warranted. There was, therefore, something wrong with the marketing process which merited government intervention to redress the economic imbalance back in favour of farmers. The key issue is whether or not there has been, or is, an efficient marketing system for agricultural products in Australia. If there is not, output of and prices for agricultural commodities may be sub-optimal. Inefficient marketing systems may also have distributional effects since there may be consequent income transfers among market participants. Because a large proportion of the output of major agricultural industries is exported (e.g. wheat, wool, meat, sugar, dairy), the efficiency of the marketing process also involves export efficiency.

Prior to the Second World War, lack of good physical and electronic communications meant that farmers were largely dependent on local buyers. With few local buyers, there might have been insufficient competition to ensure farmers received the competitive price for their products. This possibility can be illustrated using a conventional monopsonistic model (figure 6.1). For a monopsonist who sells in a perfectly competitive market, the infinitely elastic demand curve Dm in figure 6.1 represents the price at which it can sell its product on the open market. While the supply curve S in figure 6.1 is the aggregated marginal cost curves of the (assumed) atomistic producers, to the monopsonist the supply curve represents the *average* cost curve if the monopsonist pays the same price to each firm from which it purchases.

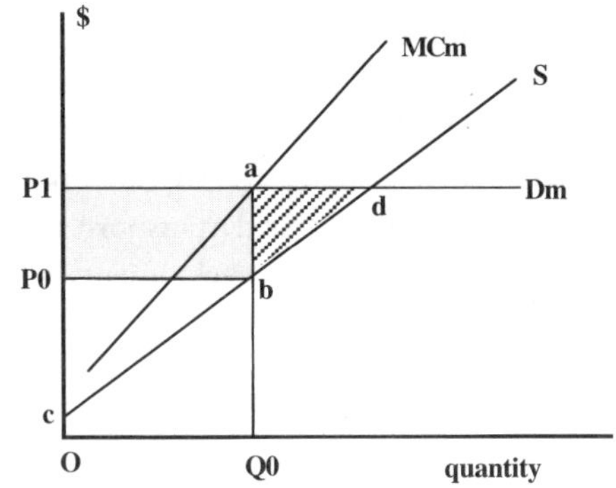

Figure 6.1 Monopsonistic Buyer Model

The monopsonist's marginal cost curve lies *above* the supply curve.[1] The familiar algorithm MC = MR indicates that the monopsonist would buy Q0 from farmers at P0, and would then re-sell this quantity at P1 in the competitive market. The monopsonist obtains a rent of P1abP0, which is its return from the communications barriers that prevented farmers selling directly on the end-use market. Area adb is the efficiency loss from poor communications. The total area P1dbP0 represents the gains to farmers from better communications; area adb represents social gains in this industry from better communications which break down monopsony buying. In contemporary terminology, local monopsony arose from and was nourished by information transactions costs. Subsequent developments in communications and transport technology have reduced the power of local monopsony.

If the local monopsonist had bargaining power as well as market power, it could also price discriminate in its purchases, only paying P0 for its *marginal* purchases, paying less than P0 for its infra-marginal procurement and paying bQ00c in total (figure 6.1). Poor communications, particularly in a geographically dispersed industry like agriculture, would aid the buyer in practising price discrimination. More complex models would be required to analyse situations where there were a few local buyers constituting an oligopsony. There was a widespread, and superficially justifiable, belief that, in the period before the Second World War, farmers were susceptible to local market imperfections, which could have resulted in both income transfers from farmers to local purchasers, and sub-optimal farm output. More general explanations of uncompetitive markets have referred to the need for 'countervailing power' to protect small sellers against larger buyers with market power (cf. James 1971, pp. 51–5; Walker & Roberts 1988, pp. 93–4; Campbell 1973).

Following the Second World War, Australia experienced rapid and sustained economic growth. One outcome of this growth was the development of a better country road network, and good mass communications. New electronic technologies included the telephone—especially with the advent of STD in the country in the 1960s—television and, since the early 1980s, international electronics communications based on cheap and readily-accessible computer technology, satellites and optical fibre cable. Good physical and electronic communications greatly increased the effective size of the market for individual farmers. Instead of being solely dependent on local buyers, farmers gained good access to information about the current state of markets and projections about likely market developments. Good transport systems offered them the opportunity to bypass local buyers. The same developments in transport and communications also reduced the costs of marketing firms. Lower transport and communications costs also reduced the dependence of raw material users on local buyers, and enabled them to buy directly from individual farmers.

Policy intention

While the explicit intervention in marketing occurred post-farmgate, the ultimate objective was to improve the efficiency and/or profitability of farm production.

Efficiency

The belief that the agricultural marketing system was inefficient was partially responsible for government intervening in agricultural marketing. The standard intervention was the 'statutory marketing authority' (SMA), which was granted control over the marketing process. Other transactions costs such as property rights, negotiation and revenue collection may have also reduced the efficiency of agricultural markets.

Other forms of inefficiency might arise from variations in external price and domestic production. Thus the overt intention of the Wool Reserve Price Scheme was to smooth wool prices through time. Wheat marketing arrangements were known as 'stabilisation' even though was no mechanism for transferring income between wheat crop pools to enable intertemporal stabilisation of either prices or incomes. The National Farmers' Federation (1981, p. 46)—during its transition from 'interventionism' to 'deregulationism'—argued that 'Governments have responded to perceived consumer preference for stable domestic prices rather than more volatile export prices'. In most cases where 'stabilisation' was the claimed intention of policy, however, it was unclear as to whether there actually was such a 'perceived consumer preference'. No rationale was proffered as to why consumers were concerned about unstable prices in products with marketing schemes (e.g. wheat, dairy, sugar and dried fruit) but not those without (e.g. meat, fresh fruit and vegetables). Consumer preferences for food price stability have probably changed as the share of food in household budgets has declined. Further, the consumer price of processed farm produce (e.g wheat) is highly buffered by processing costs, and thus quite large variations in grain prices translate into small changes in final product price. There were also clear difficulties in implementing stabilisation:

- Instability could arise from either unstable domestic production or international price; however, neither necessarily led to income instability, since there could be offsetting price and production effects.
- Instability often arose from unstable international prices in markets where Australia was such a small player that it had no economic or political power to influence prices and where previous attempts to stabilise prices had been spectacularly unsuccessful (e.g. International Wheat Agreement, International Sugar Agreement). The principal exception was in fine wool.
- Instability also arose from unstable production within Australia, for which there was no feasible preventative. Indeed, perceived preventatives such as irrigation could actually exacerbate the degree of production instability.

Distribution

Campbell and Fisher (1982, p. 66) argued that prices are not only the free market mechanism for allocating resources but are also 'the primary determinants of income'. The real objective of government intervention in marketing might, therefore, simply have been to improve farm incomes by raising prices—as long as there were no offsetting or detrimental quantity effects. The euphemism for distributional intentions in agricultural marketing was 'orderly marketing'. This entailed creating producer-controlled or producer-dominated marketing monopolies which could be 'used very loosely to endorse practically any (approved) measure taken by governments on behalf of farmers' (Watson & Parish 1982, p. 327). Sieper (1982, p. 7) also argued for a distributive interpretation of agricultural marketing policy where, 'Roughly speaking, the rule that appears to have been observed in Australia is that each protected industry "owns its own demand curve" the elasticity of which bounds the size of the available transfer'. An agricultural industry with a statutory marketing scheme could be viewed as a monopolist (figure 6.2). If the statutory marketing monopolist restricted production to Q0 (where marginal revenue and marginal cost are equated) and sold Q0 at price P0, the shaded area in figure 6.2 represents the industry's rent. In practice, however, the operation of marketing schemes was rarely a textbook monopoly.

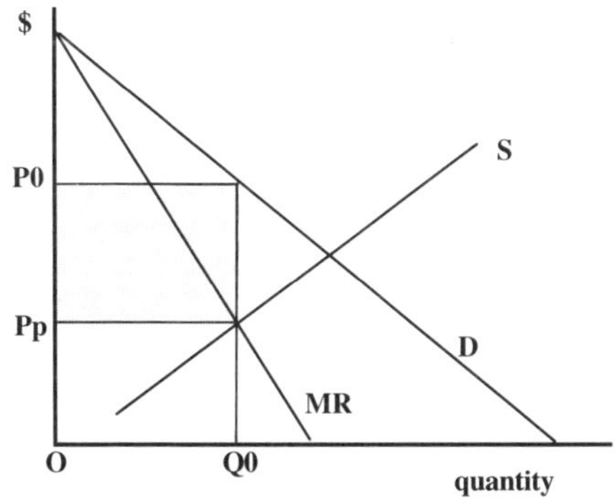

Figure 6.2 Agricultural Marketing Schemes as Statutory Monopolies

First, a monopolist must be able to restrict output to force up the price but, since nearly all agricultural industries had an atomistic structure, effective monopoly required supply control. Few Australian agricultural marketing schemes had effective mechanisms for restricting output—market milk, eggs, sugar and rice were the principal industries in which supply control was practised.

Second, effective monopoly requires that the demand curve be less than perfectly elastic. For most major Australian export commodities, however, it has proved hard to demonstrate that industries are not price takers (wool is the principal exception). Domestic demand curves for many food types may be inelastic in aggregate but, where there is substitutability between products—e.g. between types of meat such as beef, sheep, pork, poultry and fish—effective marketing control would have required a scheme covering all the products that were highly substitutable.

Overall, the monopoly power of most agricultural industries was muted and SMAs only forced higher prices on some domestic consumers.

The combination of highly elastic export demand and relatively inelastic domestic demand created opportunities for creative, partial monopoly schemes. In figure 6.3, product could be sold on the domestic market (with demand curve Dd) or the export market (demand Dx); optimal output is Q1. Ignoring exporting costs such as transport, if the same price rules in both markets (i.e. Px), then Q2 would be sold on the domestic market and (Q1 – Q2) on the export market. However, if a marketing authority had control over *domestic* sales, then total sales revenue would be maximised by supplying Q0 on the domestic market (where *domestic* marginal revenue equals the export price) and exporting the remaining (Q1 – Q0).[2] This scheme does not result in production inefficiency, since the optimal free market output Q1 is still produced. There is consumption inefficiency (welfare loss abc in figure 6.3) and a transfer from consumers to producers (P0acPx).

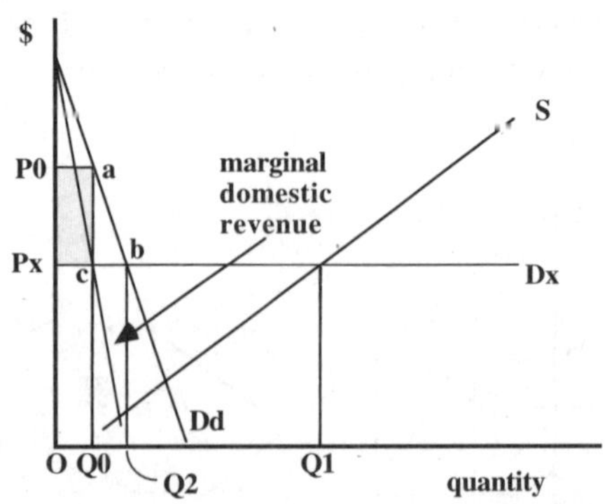

Figure 6.3 Agricultural Marketing Schemes as Partial Monopolies

Australia contributed two fascinating institutional responses to the practical implementation of marketing schemes. The first response was a *private enterprise* solution to maintaining high domestic prices while simultaneously ensuring sufficient exports to prevent these prices being undermined. This private enterprise response was the Commonwealth Dairy Produce Equalisation Committee Ltd, which operated from 1935 to 1977. While there was no overt government intervention in this scheme until the 1970s, covert government assistance was needed to restrict dairy imports (e.g. from New Zealand), which would have undermined the domestic price. A second, contrasting institutional response to marketing schemes was the innovative use of the free interstate trade provisions of the Australian Constitution (s.92) to attempt to quarantine interstate trade from marketing controls. Individuals like James and Ubergang, who disliked the compulsory nature of Commonwealth or state statutory marketing schemes, used s.92 to seek relief (cf. Chapter 3).

Evaluation

The Australian economy has matured since agricultural marketing schemes were first enacted and any valid original rationale for this intervention has been superseded. There has been an evolution away from particularly interventionist schemes; some schemes have disappeared (e.g. wool's reserve price scheme, and statutory egg marketing in New South Wales) or been substantially modified (e.g. wheat marketing where domestic marketing has been deregulated).

Area of agricultural sector

Government intervened primarily in the agricultural marketing sector by imposing controls on flows of product, on product prices, or controls on the timing of payments. There were some on-farm interventions, especially where supply control was practised (e.g. delivery quotas on wholemilk, input quotas on hens for egg-laying, land quotas for sugar cane production, water controls in rice production, licensing of opium poppy production). However, only a small number of marketing schemes had extensive on-farm intervention—indeed, many schemes failed because there were no controls on farm production.

While marketing intervention largely occurred off-farm, the effect of intervention permeated the entire agricultural sector. For example, the averaging of marketing costs prior to the 1988 wheat marketing arrangements reduced the costs of producers most distant from markets relative to the costs of closer producers. When New South Wales market milk authorities prevented distant dairy producers from supplying lucrative metropolitan markets, distant producers were restricted to the lower priced manufactured product markets even though improvements in bulk transport and refrigeration enabled them to compete with producers closer to major markets. The form of marketing arrangements thus affected the location of industries.

Policy orientation

The Commonwealth is constitutionally empowered to regulate export and interstate trade through s.51(i) and (iii). Except where the corporations power (s.51(xx)) permits, the Commonwealth is unable to regulate intrastate trade, and states may regulate intrastate marketing subject to ss.90–92. States also attempted to control product and food quality and safety as *de facto* marketing schemes to protect their domestic producers. Statutory marketing arrangements traditionally comprised:

- Commonwealth schemes, where exports dominate—e.g. wool;[3]
- state schemes, where within-state consumption dominates—e.g. market milk;
- state schemes, where a state dominates the Australian production of a commodity —e.g. sugar (Queensland), rice (NSW);
- cooperative arrangements, where neither exports nor domestic consumption dominate—e.g. wheat, manufactured dairy products.

Governments also became interested in improving marketing processes and, again, there is no clear constitutional guidance as to where this power lies. Both the Commonwealth and the states have established schemes to encourage 'innovative' marketing.

Policy mechanism

Lloyd (1982, p. 353) commented that the diversity of Australian agricultural marketing arrangements (cf. Box 6.1) contrasted with the relatively uniform systems in Europe and the USA. In Australia, regulatory methods were preferred to fiscal measures, resulting in most farm support being funded by consumers rather than taxpayers.

Box 6.1 Mechanisms for regulatory intervention

- Price discrimination—between export and domestic markets (e.g. manufactured dairy products); between different domestic markets by state marketing schemes (e.g. market milk), by end use (e.g. human consumption versus industrial wheat); by product type (e.g. shell eggs versus egg pulp).
- Import protection—a form of agricultural market intervention by itself, or as an essential accompaniment to other intervention. Marketing schemes where domestic prices were raised above potential import prices required some form of import protection to prevent imports lowering the domestic price. Import protection may take three basic forms: import embargoes (e.g. wheat until 1983, sugar until 1988), high import duties, or other countries' voluntary restraints (e.g. New Zealand's voluntary dairy restraints prior to Closer Economic Relations). Import protection could also be used to prevent influx of disease, although it is sometimes difficult to determine whether or

not disease prevention—important though it is in a relatively isolated and disease-free environment like Australia's—is also for supply control.

- Export subsidies—essential elements of some schemes, which maintained domestic prices above import prices, and where domestic production exceeded consumption (e.g. manufactured dairy products, wheat). Export subsidies could become export taxes if export prices exceeded domestic prices (e.g. around 1950 for the wheat). Export subsidies were required to induce sufficient producers to export rather than market domestically so that the desired domestic price was not undercut. Without the export subsidy, quotas would have been required to control supply onto the domestic market.
- Production controls—production could be directly controlled by licences or quotas on key inputs, such as land in sugar production, hens in egg production, or water in rice production.
- Market share quotas—sales of milk for human wholemilk consumption have been controlled by delivery quotas such as in New South Wales; emergency delivery quotas were also used in wheat production in 1969–70.
- Output subsidies—usually shunned because of their visibility and cost to the Treasury; their use was usually restricted to emergencies (e.g. the wool deficiency payment scheme of 1970–71) or they were rolled into 'stabilisation' schemes such as the wheat stabilisation or 'potholing' arrangements of the 1980s.
- Price setting—prices could be set explicitly as in the case of market milk at the state level, or implicitly by the operation of a two-price scheme.
- Vesting and compulsory acquisition—to enable an SMA to acquire sufficient product for export so that domestic supply was controlled to support the desired domestic price; in practice, all of a product was normally acquired.
- Export controls—to encourage or mandate product description schemes, supporting export monopolists (e.g. wheat, sugar), supporting sharing in export markets where there is regulated access for total Australian exports (e.g. US beef market), regulating access to export markets when there are implicit or explicit agreements on minimum international prices.
- Pooling—farmers could be paid the average return for a commodity delivered to a common pool (e.g. wheat); as marketing became increasingly flexible, premiums and discounts could be introduced for quality, and differential payments for other marketing characteristics such as distance from markets. A 'pool' with premiums or discounts for all commodity characteristics is essentially an individual price scheme.

- Buffer funds—could be used to smooth prices over time by 'taxing' producers when prices were above some arbitrarily determined level, and rebating from the fund when prices fell below an arbitrarily determined level.
- Buffer stocks—used in the wool industry, with an authority purchasing wool when prices fell, and reselling when prices rose (cf. Chapter 12).

Sources: Lloyd (1982, 366–79), Mauldon (1990)

Marketing schemes were implemented by a combination of legislation, regulations and administrative decisions, and limited by Constitutional provisions and the operation of other legislation. Legislation was required to create marketing arrangements and administering authorities at state or federal level. Significant administrative discretion was vested in marketing authorities, particularly following 'corporatisation' of Commonwealth SMAs in the 1980s (Commonwealth of Australia 1986, pp. 65–9; Industry Commission 1991b). For example, the Australian Wool Corporation set overall indicator prices and intervention prices for each type of wool in the Reserve Price Scheme; the Australian Wheat Board, in cooperation with state governments and bulk handling authorities, determined the varieties which would be received into particular silos under the pre-1988 wheat marketing arrangements.

The 'active' forms of marketing intervention listed in Box 6.1 were supported by 'passive' intervention which permitted exemptions from restrictive trade practices under the Federal *Trade Practices Act 1974*. Agricultural marketing previously obtained six forms of exemption to the Act's provisions, but exemptions had expired by the early 1990s (Hilmer et al. 1993, pp. 93, 139–44). Many agricultural producers were exempt from the Trade Practices Act because of Constitutional limitations to the Commonwealth's corporations power. Voluntary conduct under statutory marketing arrangements may be specifically approved by Commonwealth or state legislation; mandatory marketing legislation—such as the vesting or compulsory acquisition of product—did not contravene the Trade Practices Act and therefore did not require trade practices exemption. Hilmer et al. (1993, pp. 143–4) proposed the elimination of most agricultural marketing exemptions from trade practices except for certain export contracts. In April 1995, the Council of Australian Governments agreed to support the Hilmer reforms—including proposals to terminate the shield of the Crown exemptions of state and territory entities, and to include intrastate unincorporated businesses under competition legislation.[4]

Campbell and Fisher (1982, p. 67) argued that:

> Many influential economists believe that it is inadvisable to manipulate agricultural prices in order to achieve income or welfare goals. They argue that prices should be allowed to

> perform their primary function of allocating resources in accordance with changes in supply and demand, and that income goals should be achieved by supplementary payments of one kind or another.

Apart from committing the logical fallacy of appealing to authority, the Campbell and Fisher argument mirrors the slow moves to disentangling efficiency and distributional aspects of Australian agricultural policy. Similar arguments occurred in Western Europe and the USA where 'decoupling' was used to describe the separation of efficiency and distributional policy. In principle, however, efficiency and distribution cannot be totally severed.

'Decoupling' does not solve the economic problem of marketing policy. If there is market failure in the marketing chain, the extent of this market failure must be assessed to determine whether or not intervention to address it is warranted. Given the past extensive government intervention in marketing, recent emphasis has been on eliminating *government* failure rather than market failure.

Institutional structure

The important institutional players in statutory agricultural marketing were traditionally the SMAs—both the Boards and their employees—and the farmers who were the intended beneficiaries of SMA activities and key members of the SMA boards until the mid-1980s. The SMAs and their boards were important because they administered 'orderly marketing' and because they were key players in implementing policy changes. For example, in IAC/IC inquiries into agricultural marketing arrangements, the SMAs —and the policy framework in which they operated—were the focus of the inquiry. The SMAs were also the repositories of quantitative and other data whose analysis was essential to evaluating the efficiency of existing and alternative policy mechanisms. Once the IAC/IC had reported, the SMAs also participated in non-public evaluations of the 'feasibility' of the IAC/IC's recommendations—e.g. through Agricultural Council. Finally, the SMAs provided governments with the opportunity to play in the public choice market, both directly through appointments to SMA boards and indirectly through managing the policy framework in which the SMAs operated (e.g. Watson and Parish 1982, p. 327).

Producers' interests in the activities of SMAs were concentrated and direct, and National/Country Party politicians at least accorded them primacy in developing the policy framework for marketing agricultural products (cf. Chapter 3). Conversely, consumer interests were diffuse and indirect, and it was difficult for consumers to discover the effects of marketing schemes on the price they paid for products, or the quality of these products. Even after the establishment of the IAC and its searching inquiries into marketing of wheat and dairy products in the 1970s and 1980s, consumers—and even consumers' organisations—appear to have been largely ignorant of the effect of agricultural marketing policy on Australian food markets.

Policy performance evaluation

Three levels of policy performance evaluation can be distinguished. At the professional level, most notable Australian agricultural economists cut their professional teeth on evaluating agricultural marketing schemes, especially dairy, wheat or wool.[5] These professional analyses fed into formal policy evaluation processes which, particularly after 1972, became increasingly public.

Irregular analysis of agricultural marketing policy was included in more general national agricultural inquiries (e.g. Harris et al. 1974, Balderstone et al. 1982); agricultural marketing policy also arose in the economy-wide competition policy report (Hilmer et al. 1993). Between 1974 and 1991, the IAC/IC undertook three inquiries into the wheat industry, four inquiries into dairying, about twenty inquiries into various aspects of Australian fruit-growing industries, two inquiries into sugar, and inquiries into other industries' marketing arrangements (cf. Industry Commission 1991b, Appendix C). However, despite the Commonwealth–state marketing scheme for eggs, there was never an IAC/IC inquiry into this industry; similarly there was no IAC/IC inquiry into wool marketing despite the Reserve Price Scheme. While the dairy industry was investigated by the IAC/IC, the principal focus was on manufactured dairy products (butter, cheese, skim milk powder and casein), which were the subject of Commonwealth marketing schemes, and not on market milk and related fresh products which were dominated by state intervention.

In parallel with the IAC's 1988 inquiry into wheat marketing, the Commonwealth and states jointly commissioned a Royal Commission into Grain Storage, Handling and Transport (1988). The Commonwealth also commissioned a special inquiry into SMAs (Commonwealth of Australia 1990). Following the collapse of the wool reserve price scheme, the Commonwealth commissioned several reports into wool marketing (cf. Chapter 7). The declining importance of agricultural marketing policy is illustrated by two Commonwealth policy statements during the 1980s: the marketing of agricultural products warranted a separate chapter in the White Paper on Economic and Rural Policy (Commonwealth of Australia 1986); by 1988, agricultural marketing had simply become one section of the government's policy on 'The External Environment, Marketing and New Opportunities' (Kerin & Cook 1988). States also inquired into agricultural marketing arrangements (cf. Industry Commission 1991b, pp. 24–8), and the regulated prices of some products in some states were regularly 'reviewed'.

Finally, funding for agricultural marketing schemes was sometimes explicitly or implicitly related to annual budgets. For example, before the mid-1970s, when the Australian Wheat Board was increasingly required to borrow from the private sector, the first advance to farmers was a contingent liability on the Federal Budget. There was some annual Treasury scrutiny of the funding for agricultural marketing arrangements.

Economic issues

Some possible market failures are illustrated in table 6.1. In the domestic market, it is improbable that there are major contemporary failures arising from serious transactions costs problems in property rights. Because most agricultural products are physical, non-rivalry problems are rare, and well-developed property rights of a developed economy provide good excludability and a sophisticated—albeit expensive—legal system to resolve disputes over revenue collection. Information transactions costs in the domestic market have been falling as improved communications technologies have lowered the costs of market information systems. The development of food description systems and product labelling has largely eliminated one potential source of market failure. An area of increased interest in information concerns food safety: how efficient is the market for information about food safety, and how do consumers obtain adequate information for the safe storage and use of food? (cf. Chapter 8).

Table 6.1 Possible Contemporary Failures in Markets for Australian Agricultural Products

	Market	
	Domestic	*Export*
Market failure		
Property rights/exclusion	..	restructuring economies
Information	food safety	new markets
Negotiation	..	new markets product description
Revenue collection	..	restructuring economies
Incomplete markets for risk		restructuring economies
Market imperfections		
Economies of scale/natural	industrial processing	international trading
Monopoly	retailing	industrial processing
Barriers to entry	..	technology
Anti-competitive behaviour	trade practices regulation	international agreements (multi- and bi-lateral)
Government intervention	packaging and labelling imports	import barriers quarantine export subsidies

Australia's geographical dispersion, excellent internal transport and communications, and export capacity in most major products provide a substantial defence for domestic consumers against incomplete risk markets. Only towards the end of the eastern states' drought of the early 1990s, for example, was the import of grain seriously considered—and this primarily to support the export-oriented lot-fed beef industry.

Australia potentially suffers problems of market imperfections because of its small population size and dispersed markets. Efficient technologies for processing agricultural products for the domestic market may be large-scale, creating barriers to entry and opportunities for anti-competitive behaviour. Similarly, increasing retail concentration, including supermarkets and other food distribution, provides the opportunity for inefficiencies to emerge. But while the 'competitive fringe' of small independent butchers and greengrocers has been increasingly pressured by supermarket expansion in this area, bread manufacturers and supermarket retailing of bread have been increasingly pressured by hot bread shops. Government intervention through direct involvement in marketing activities would probably not reduce such problems, since government trading firms themselves may implement similar, or other, forms of anti-competitive behaviour. Increasingly, Australian governments are relying on trade practices legislation to reduce the impact of anti-competitive behaviour in private markets.

The importance of Australian agricultural exports means that competitiveness in international markets affects domestic agricultural and resource policy. For example, transactions costs problems in foreign markets—because of undeveloped social infrastructure in developing countries or fractured infrastructures in former centrally planned economies—may create a (temporary) role for Australian governments in assisting Australian firms to enter these markets (cf. table 6.1). Such assistance may include assisting foreign governments to establish infrastructures compatible with developed market economies, or assisting Australian firms to survive initial difficulties in operating in these markets. Assistance in infrastructure development may involve developing institutions to better specify property rights, develop negotiation or revenue collection systems, or assist in the development of public information systems. Where Australian firms enter new markets, or develop markets for new products, government may have a role in helping to overcome transactions costs, especially where other Australian firms may benefit from developing these markets.

Australia is a small player in international markets, and Australian firms are likely to be small compared to international firms operating in similar industries. There may therefore be a substantial role for the Australian Government in seeking reasonably competitive international markets in which Australian firms have a chance of competing. Many such problems arise from government failure in the form of other countries' protection for their domestic industries. The Australian Government's principal activity in this area was to seek a reasonable outcome from the Uruguay round of the General Agreement on Tariffs and Trade. From the 1970s, the Australian Government also cooperated with the New Zealand Government in reducing restrictions on trans-Tasman trade, which improved export prospects for some Australian agricultural industries, and increased competition on the domestic market for others. Australia has also taken a leading role in developing APEC, the forum for Asia–Pacific Economic Cooperation, which has sought to avoid increasing protection and wherever possible reduce trade barriers in the Asia–Pacific region.

Analytic techniques

Derived demand

Because marketing policy affects the farm production sector, a simple model of this linkage is desirable. Demand for primary products is derived from the final demand for a manufactured product, and the market for the services used in the manufacturing process (figure 6.4). In the upper-right quadrant the demand for final product is presented. The ratio by which the raw material is transformed into the final product is shown in the bottom-right quadrant—if, for example, one tonne of wheat is transformed into 0.8 tonnes of bread (allowing for non-flour by-products, unavoidable wastage, and the addition of non-flour ingredients), then the slope of the transformation ratio in this quadrant would be 0.8.

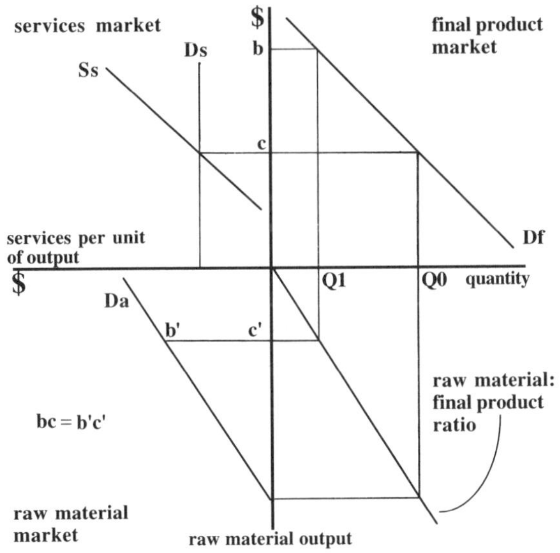

Figure 6.4 Linking Raw Material and Final Product Markets

The 'services' market for transforming raw material into final product is presented in the upper-left quadrant. This services market comprises all marketing processes: assembly of the raw material, storage if required, transport to processing plant(s), processing itself, wholesaling and retailing (including export, if required). The upper-left quadrant simplifies all these processes into a single services market. The horizontal axis of the upper-left quadrant indicates the quantity of services incorporated in *each unit* of the final product. The supply curve indicates the marginal cost of adding a unit of marketing services to the final product. The demand curve indicates the relationship between the volume of services included in the final product, and the price consumers would be prepared to pay for these services.[6] Given the services market depicted in the upper-left quadrant, the *minimum* price at which the commodity will be sold is c, the price of the equilibrium package of services in the final product, and Q0 would be sold. If Q0 were sold, raw material would have to be acquired at zero price for the sale of the final product to be profitable to its producers.

Knowing that quantity Q0 would be sold at price c, with corresponding raw material price zero, locates one point on the derived demand curve for the raw material—i.e. the point at which this derived demand crosses the quantity (i.e. vertical) axis in the bottom-left quadrant. Another quantity–price pair for the final product (e.g. quantity Q1 and price b) allows the derived demand curve for the raw material to be determined, assuming a linear derived demand relationship. At Q1, the final product price b is divided between the services market (price c, assuming the equilibrium level of services is invariant to the quantity of final product sold) and raw material suppliers (b–c). The transformation relationship in the bottom-right quadrant of figure 6.4 enables point b′ on the derived demand curve to be determined (bc = b′c′).

This model permits analysis of intervention in the market services sector. Suppose marketing intervention resulted in increased costs: i.e. the market services supply curve (Ss) moved upwards. The equilibrium level of marketing services added to the final product would decline (unless demand for services is perfectly inelastic) and the minimum price at which the final product is sold would rise. The derived demand curve for the raw material would contract towards the origin in the bottom-left quadrant, and both farm output and farm price would fall. Conversely, if policy intervention decreased the cost of marketing services, the derived demand curve for the raw material would expand. This was the logic behind the Royal Commission into Grain Storage, Handling and Transport's recommendations to increase efficiency in the storage and handling of grain to improve the profitability of grain production on farm (cf. Chapter 7).

Quotas

Simple partial equilibrium models may also be used to examine marketing issues such as market quotas. A simple model of a competitive industry in a closed economy with conventional supply (S) and demand (D) is presented in figure 6.5. With no intervention in the market, equilibrium occurs at quantity Q0 and price P0. An effective quota Qq would result in a market price Pq, and would create an income transfer of Pqabd from consumers to producers, and a welfare loss of acb.

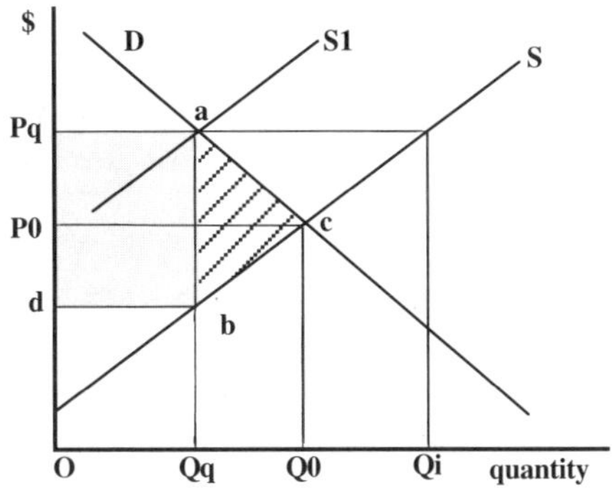

Figure 6.5 Market Quota, Competitive Industry in a Closed Economy

The use of quotas gives rise to two problems, in addition to the direct welfare losses. The first is that an economic incentive exists to operate outside the statutory cartel. If the quota supports a high price—e.g. Pq in figure 6.5—the marginal return for an additional unit of quota is Pq. When all producers act as if Pq is the marginal return, there is an incentive for output to be as high as Qi unless monitoring and enforcement is intense. Because marketing schemes with quotas are a political creation, actual or potential production exceeding the quota limit will generate pressure from beneficiaries of economic rents to invest in monitoring and enforcement to protect the rents. Thus, for example, New South Wales dairy industry authorities in the 1970s mounted a 24-hour guard with powerful lighting to prevent consumers purchasing milk from a local dairy. In the 1980s, the New South Wales egg industry expended considerable resources pursuing egg producers operating outside 'orderly marketing' arrangements. There is often a tendency to feel sympathy for those opposing statutory marketing arrangements: however, as indicated by figure 6.5, there are major benefits for those ignoring statutory cartel arrangements . . . as long as not too many other producers do the same.

The second problem emanates from the economic rents—area Pqabd in figure 6.5. If these rents are channelled directly into consumption or savings, no additional problems arise. However, these rents may be imputed into land values if particular areas must be farmed in order to qualify for participation in rent distribution, or rents may be imputed into licence values if tradeable quotas are issued. In either case rents become imputed into production costs. Alternatively, rents may permit investment in 'gold-plated' capital equipment when zinc-plated items would do, or become appropriated as specialised labour payments. Additional regulation may mandate the use of gold- rather than zinc-plated production processes. Average production costs will rise, dragging up marginal costs and contracting the industry supply curve. The industry supply curve could conceivably rise as high as S1 (figure 6.5), dissipating all economic rents as payments to factors of production. Should marketing arrangements subsequently be terminated, extensive restructuring would be required to shift the industry's supply curve back to S. If the original marketing intervention led to the mandating of 'high quality' standards in the industry, it might be extremely difficult to change both producer and consumer perceptions—or government perceptions—that these standards are actually dispensable.

Price equalisation

Price equalisation schemes are similar to a simple quota scheme, except that there is a secondary market in which product 'surplus' to the high priced market can be sold. To induce manufacturers to participate and farmers to support the scheme, returns across high and low priced markets are 'equalised'. The scheme is illustrated in figure 6.6. Domestic demand is Dd, export demand is assumed to be infinitely elastic (this assumption can be readily relaxed), and supply is S1. In a free market, Qf would be produced, of which Qc would be consumed domestically, and (Qf–Qc) exported at the ruling export price Px. In an equalisation scheme, domestic price is raised to Pd and maintained by controls on imports and by levying manufacturers (Pd–Px) on domestic sales. Total revenue from the levy—viz. area PdabPx—is averaged over *all* units sold, yielding the curve Pe as the average price for production above Qd. If no producer is sufficiently large, curve Pe also represents marginal returns. Thus, with equalisation, equilibrium occurs at output Qe; Qd is sold on the domestic market at price Pd, and (Qe–Qd) is sold on the export market at price Px.

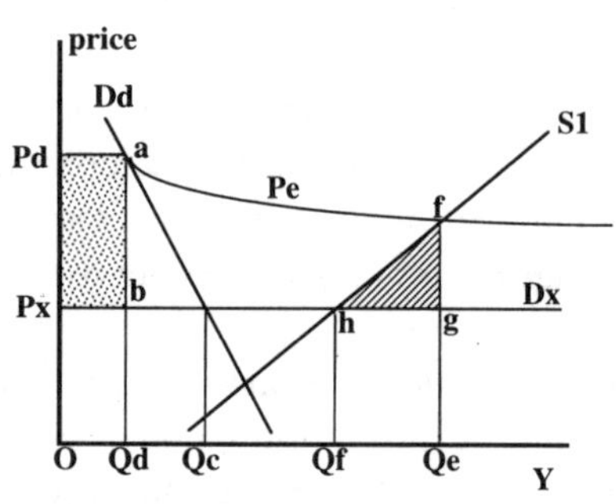

Figure 6.6 Equalisation Scheme

Early analyses of equalisation schemes focused on the output expansion induced by the scheme (output level Qe instead of Qf) with associated allocative production inefficiencies measured by area fgh.[7] However, the major economic effect of the scheme potentially occurs through the transfer PdabPx. If this transfer induces economic effects—such as over-intensive use of inputs, which shifts the supply curve upwards—then, like a quota scheme, substantial inefficiencies will be caused by what appears to be a simple transfer arrangement.

Equalisation schemes also affect income variability. Suppose domestic returns (D = P.Q, where P and Q are domestic price and consumption respectively) are non-stochastic, whereas export returns are a random variable (X = Px.Qx, Px and Qx are export price and exports respectively), and d is the fixed share of domestic consumption in total production. Then average returns R:

(6.i) $$R = d.D + (1 - d).X$$

and the variance of average returns is var(R):

(6.ii) $$\text{var}(R) = (1 - d)^2.\text{var}(X)$$

since var(D) = 0. The larger the share of non-stochastic domestic sales, the smaller the variability in total returns. If returns are not equalised and some producers sell on domestic markets while others are exporters, then some producers will have no variability in their returns, whereas exporters will bear the full variability of the market where returns are stochastic.

Buffer-stock

In a buffer-stock scheme, an authority purchases a commodity when prices are low and re-sells it when they are high. The underlying principle is that prices are cyclical and it is possible to hold stocks for a sufficiently short time that the costs of stock holding are small, but sufficiently long for prices to rise. The problem is illustrated in figure 6.7. Production is assumed to be fixed at level Q, while demand oscillates between D1 and D2. With no intervention—and assuming private buyers do not undertake a storage/speculative function—prices oscillate between P1 and P2 because of the oscillation in demand.

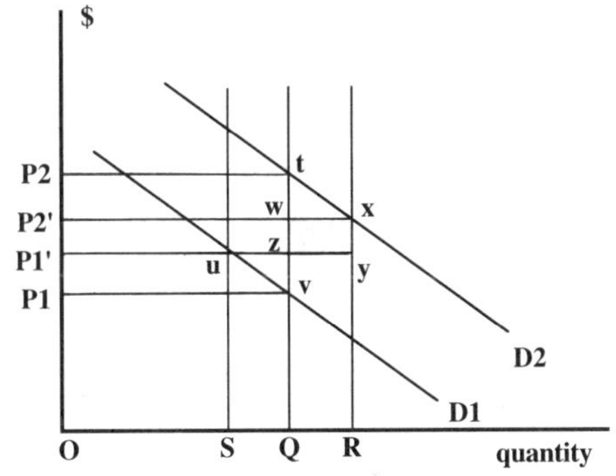

Figure 6.7 Buffer Stock Scheme

Now suppose an authority operates in the market: when demand and price are low, the authority purchases Q–S, and stores this output until demand rises to D2; because of the intervention, price rises to P1′. When demand rises to D2, the authority sells

R–Q ≡ Q–S; instead of price being P2, the extra sales reduce price to P2′. By purchasing Q–S at P1′ and selling this at P2′, the storage authority makes a *visible* profit equal to area wxyz in figure 6.7. However, since such schemes are supposed to benefit producers (and even the community as a whole), their allocative effects must be considered.

The total benefits (TB) of the scheme as illustrated in figure 6.7 are:

(6.iii) TB = visible profit + increased income when prices are low – decreased income when prices are high – stockholding costs
= wxyz + P1′zvP1 – P2twP2′ – SHC

where SHC is the stockholding cost (including physical and financial costs of holding stock and the operating costs of the intervention authority). In Figure 6.7, the analysis is simplified by the demand shift being assumed to be parallel, thus (P2–P2′) = (P1′–P1), and therefore P1′zvP1 = P2twP2′. This particular problem reduces to a simple comparison of the visible profit wxyz and the stockholding cost SHC. If the visible profit exceeds the stockholding cost, then the operation is profitable.

In practice, evaluation of a buffer stock scheme will be more complex for the following reasons:

1. The analysis assumes that the authority has perfect knowledge of the future—but if perfect knowledge were available, it would not require a single authority to perform the stockholding function. If the scheme were profitable for the authority, it would be profitable for individual producers unless there were economies of scale in storage. However, the opportunity cost of physical storage may be low for individual producers—e.g. in the case of wool, a woolshed corner would serve—and so economies of scale in financing an authority's storage must be sufficiently large to offset the producer's low physical storage costs. (In the case of wool, the authority's private borrowing costs were lower than growers' because the former were guaranteed by the Federal Government.)
2. Imperfect knowledge of the future—if there are sufficiently large economies of scale in acquiring adequate information to run a storage scheme, an authority may obtain information required to run a stockholding scheme more efficiently than producers.
3. The authority must purchase the commodity prior to its sale—thus the value of commodity sales out of the buffer stock in subsequent periods must be discounted to validly compare stock investment and returns.
4. Demand curve shifts may not be parallel as assumed in figure 6.7, and are likely to be more complex than a simple oscillation between 'low' and 'high' values—more complete analysis of a buffer stock scheme would require a simulation of the expected future values of commodity sales.

This analysis suggests it would be desirable to model variability—and policies to deal with variability—in a more systematic way.

Stabilisation model

Turnovsky (1978) presented a formal model of variability, in which conventional non-stochastic demand and supply curves are modified by incorporating an error term (Box 6.2, Equations B6.1a,b). The demand and supply equations have quantities as the dependent variable, rather than the more common graphic representation which has price as the dependent variable (i.e. price on the vertical axis of a market diagram). Equilibrium price and quantity (equations B6.2a,b) are found by simultaneously solving equations B6.1a,b; the parameters of this solution are the parameters of the demand and supply curves, and the random error terms u_t, v_t. Suppose a buffer stock authority holds price at P_S where expected demand equals expected supply (equation B6.3). The authority must continually trade some amount $x_t = u_t - v_t$ (equation B6.4), the difference between actual demand and expected demand at the expected price P_S, and actual supply and expected supply at the expected price P_S (figure 6.8). Investigating the welfare effects of price stabilisation requires definition of base case welfare, and this is taken to be the sum of producer and consumer surplus at the unstabilised equilibrium given by equations B6.2a,b (because of the form of the supply and demand equations (cf. figure 6.8), producer surplus is defined as the area *under* the supply curve). The difference between producer surplus with and without stabilisation is defined by equation B6.7, and an algebraic form given in equation B6.8; similarly the change in consumer surplus is defined by equation B6.9 (Turnovsky presented an algebraic form of this equation) and the change in total welfare is defined by equation B6.10. The *expected* changes in producers', consumers' and total welfare are derived from using the expectations operator on equations B6.8–B6.10, and the resulting expressions are given in equations B6.11a–11c (Box 6.2). These results may be interpreted as follows:

1 producers lose (gain) from price stabilisation if price instability only comes from random shifts in demand i.e. via u_t (supply, i.e. v_t) [equation B6.11a];
2 consumers lose (gain) from price stabilisation if price instability only comes from random shifts in supply i.e. via v_t (demand, i.e. u_t) [equation B6.11b];
3 where both supply and demand are random, individual group's gains are indeterminate, and depend on the relative sizes of σ_u^2 and σ_v^2, and upon supply and demand slopes;
4 total gains from stabilisation are always positive (equation B6.11c), with gainers in principle able to compensate losers (i.e. there is a potential Pareto improvement):
   - total gains/losses are strictly proportional to variances (variances only occur in numerators of equations B6.11);
   - if either demand or supply is infinitely elastic ($a \to \infty$ or $b \to \infty$), then gains tend to zero;
5 results depend crucially on linearity of model and additivity of disturbances:
   - non-linearity leads to minor modifications of results; results are, however, highly sensitive to non-additive disturbances.

Box 6.2 Simple Model of Stabilisation

Supply and demand are represented as:

Demand: $D(P_t) = A - a.P_t + u_t \qquad a \geq 0,\ u_t \sim (0,\sigma_u^2)$ (B6.1a)

Supply: $S(P_t) = B + b.P_t + v_t \qquad b \geq 0,\ v_t \sim (0,\sigma_v^2)$ (B6.1b)

with u_t and v_t distributed independently, and with zero covariance between u_t and v_t:

$$E(u_t,v_t) = 0$$

1. *unstabilised equilibrium*:

price: $P_t = (u_t - v_t + A - B)/(a + b)$ (B6.2a)

quantity: $D(P_t) = S(P_t)$
$= (B.a + b.A)/(a + b) + (a.v_t + b.u_t)/(a + b)$ (B6.2b)

2. *stabilised equilibrium*:

- suppose a *buffer stock* authority operates to hold price at P_s where expected demand equals expected supply:
  $E[D(P_t)] = E[S(P_t)]$ (B6.3)
  this implies that the authority must continually trade some amount x_t:
  $x_t = v_t - u_t \qquad \text{where } E(x_t) = 0$ (B6.4)
  i.e. in any period, the authority will trade (buy or sell) the difference between output and consumption *at the price P_s*, and the *expected* value of the trade is zero (i.e. the expected buffer stock is self-liquidating).

- taking expected values of equations (B6.1a) and (B6.1b) [i.e. following (B6.3)], and solving:

  stabilised price: $P_s = (A - B)/(a + b) = E(P_t)$ (B6.5)
  consumption: $D(P_s) = (B.a + b.A)/(a + b) + u_t$ (B6.6a)
  production: $S(P_s) = (B.a + b.A)/(a + b) + v_t$ (B6.6b)

  equation (B6.5) equals the expected value of (B6.2a) – self-liquidating buffer stock implies that price stabilised at *arithmetic* mean.

3. *producers surplus* without price stabilisation is (P_t = unstabilised price):

$$\pi_t = \int_{P(0)}^{P_t} S(P')dP'$$

where P(0) is supply price at zero output (for finite integral, requires finite y-axis intercept)

4. *producers surplus* with price stabilisation at price P_s is:

$$\pi_s = \int_{P(0)}^{P_s} S(P')dP'$$

and the *difference* between producers surplus without/with stabilisation is GP (from equations [B6.6c] and [B6.6d]):

$$GP = \pi_s - \pi_t = \int_{P_t}^{P_s} S(P')dP' \qquad \text{(B6.7)}$$

substituting (B6.1b), equation (B6.7) integrates to:

$$\begin{aligned} GP &= 0.5(P_s - P_t).[S(P_t) + S(P_s)] \\ &= -0.5[(u_t - v_t)/(a + b)].[2(B.a + b.A)/(a + b) \\ &\quad + [(2a + b).v_t + b.u_t]/(a + b)] \end{aligned} \qquad \text{(B6.8)}$$

5. *change in consumers' welfare* with stabilisation is G_C:

$$G_C = 0.5(P_t - P_s).[D(P_t) + D(P_s)] \qquad \text{(B6.9)}$$

note that, if (B6.7) is an increase (decrease) in *producer* surplus, then (B6.9) is a decrease (increase) in *consumer* surplus

$GP > 0$ implies $P_s > P_t$ (from (B6.8)); but if $P_s > P_t$ then $G_C < 0$ (from (B6.9)), and vice versa

6. *total welfare* is the *sum* of producers and consumers surpluses, then total welfare change is G:

$$G = GP + G_C = 0.5(P_s - P_t).[S(P_s) - D(P_s)] \quad \text{since } S(P_t) = D(P_t) \qquad \text{(B6.10)}$$

equations (B6.7), (B6.9) derivable by simple trigonometry from the appropriate figure.

7. *expected gains from stabilisation* are (using E(.) on equation (B6.8) and corresponding substitutions for (B6.9) and (B6.10)):

$$E(GP) = [-b.\sigma_u^2 + (2a + b).\sigma_v^2]/[2(a + b)^2] \quad \text{(B6.11a)}$$
$$E(G_C) = [-a.\sigma_v^2 + (2b + a).\sigma_u^2]/[2(a + b)^2] \quad \text{(B6.11b)}$$
$$E(G) = (\sigma_u^2 + \sigma_v^2)/[2(a + b)] > 0 \quad \text{(B6.11c)}$$

8. *Inferences*
   - i Producers lose (gain) from price stabilisation if price instability *only* comes from random shifts in demand i.e. via u_t (supply, i.e. v_t) [equation (B6.11a)]
   - ii Consumers lose (gain) from price stabilisation if price instability *only* comes from random shifts in supply i.e. via v_t (demand, i.e. u_t) [equation (B6.11b)]
   - iii Where *both* supply and demand are random, individual group's gains are *indeterminate*, and depend on the *relative* sizes of σ_u^2 and σ_v^2, and upon supply and demand *slopes*
   - iv *Total* gains from stabilisation are *always* positive (B6.11c), with gainers *in principle* able to compensate losers (*potential* Pareto improvement)
     - *total* gains/losses are strictly proportional to variances [variances only occur in *numerators* of equations (B6.11)]
     - if *either* demand *or* supply infinitely elastic ($a \to \infty$ or $b \to \infty$; remember that supply and demand curves are *inverses* of usual representation, where 'infinite elasticity' means *zero* slopes), then gains tend to zero (infinite denominators in (B6.11c) implies $E(G) \to 0$)
   - v *Results depend crucially* on linearity of model and additivity of disturbances
     - *non-linearity* leads to *minor modifications* of results; results *highly sensitive* to ***non-additive disturbances***

Source: Turnovsky (1978)

This model is one of the simplest forms in which the consequences of non-deterministic supply and demand might be investigated. More realistic models of markets with non-deterministic supply and demand require quantitative modelling to evaluate the consequences of intervening in such markets and, in particular, the sensitivity in such models to errors in estimated parameters.

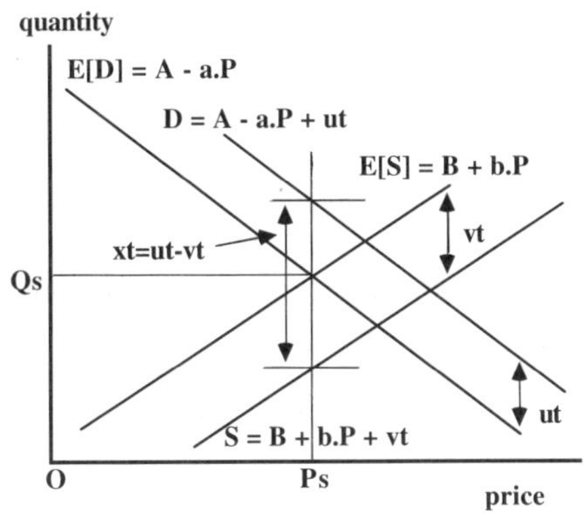

Figure 6.8 Buffer Stock Authority's Purchases in Turnovsky's Model

ISSUES FOR DISCUSSION

1 Is stabilisation of the prices of Australian agricultural products 'a good thing'?
   i How could the net benefits of price stabilisation for Australian agricultural products be examined? what are the key parameters and variables affecting this net benefit? what affects the distribution of these benefits?
   ii Why might a price stabilisation scheme which had positive net economic benefits actually fail?
2 Government intervention in the marketing of Australian agricultural products has evolved over the past seven decades. Compare the importance of 'economic efficiency' and 'public choice' explanations of this evolution of government intervention.
3 Agricultural production and marketing are unstable activities. Discuss whether or not this instability justifies government intervention to mitigate its effects.
4 Despite a review recommending deregulation of the New South Wales rice industry, the State Government stated in April 1996 that it was 'not going to subject [rice growers] to an ideologically driven experiment in deregulation' on the grounds that the rice industry was 'pretty efficient' and 'it's good for consumers' because the price of rice 'has fallen 0.5 per cent each year since 1984'. Thirty per cent of Australia's consumed rice is imported, and 90 per cent of Australian production is exported. Comment.

Endnotes

1 If the industry supply curve (S) is:

$$P = c + d.Q$$

where P is product price corresponding to output Q, the monopsonist's total cost of purchasing product Q is:

$$TC = Q.(c + d.Q) = c.Q + d.Q^2$$

and the monopsonist's marginal cost (MCm) is:

$$MCm = c + 2.d.Q$$

2 If total output Q1 is given, and the monopolist domestic seller has to choose the amount to sell domestically (Q0) where Px is the export price, and the domestic demand schedule is P = a – b.Q0, then the monopolist's optimal value of Q0 is determined by using calculus to find the maximum value of TR = P.Q0 + (Q1 – Q0) with respect to Q0.

3 The Commonwealth has also entered international arrangements for price stabilisation, price raising or market sharing for particular agricultural commodities—e.g. International Wheat Agreement, International Sugar Agreement.

4 During the 1996 Federal election, the National Party indicated concerns about the thorough implementation of a national competition policy, indicating a desire to exempt SMAs and some small business (e.g. newsagents and pharmacists).

5 Probably the most elegant Australian agricultural marketing policy analysis was the initially anonymous 'Milking the Australian Economy' (Lewis 1967) which, even after three decades, retains vividness and vitality—and a depressing relevance to contemporary dairy marketing policy.

6 If the 'services' sector is simply 'processing', a particular product in the top right-hand quadrant corresponds to a *vertical* demand curve for services in the top left-hand quadrant. If 'services' are simply storage, handling and transport, and demand is aggregate demand—e.g. excess demand—final product demand may have a sloping services demand curve.

7 Production inefficiencies could be avoided by restricting production to Qf using a quota scheme. Quota has a value above zero if it is the mechanism for distributing equalisation rents (area PdabPx in figure 6.6).

7 Agricultural marketing policy: case studies—wheat, wool, sugar

The form and effects of agricultural marketing policy in Australia differ among the major agricultural industries, and these forms and effects also change over time. In this chapter key policy issues are illustrated by focusing on three industries around 1990. Domestic marketing in the wheat and related grains industries was deregulated in the late 1980s, following an Industries Assistance Commission (IAC) inquiry into wheat marketing (section 7.1). The Reserve Price Scheme for marketing Australia's wool clip collapsed in the late 1980s under the weight of a huge stockpile, and wool marketing was deregulated (section 7.2). The IAC recommended deregulating sugar marketing, and this was partially accomplished (section 7.3).[1]

7.1 DEREGULATING WHEAT

The question 'How best to market the Australian wheat crop?' invites another: 'Best for whom?' In a perfectly competitive economy, 'best' for individual producers or consumers—taking into account the legitimate interests of other producers and consumers—is 'best' for the economy.

Policy issue

Wheat marketing policy since the Second World War has attempted to 'balance'—or, at least, reconcile—interests within wheat farming, within the wheat industry (including downstream activities), and between the wheat industry and other interests (Whitwell & Sydenham 1991). By the mid-1980s, the issue had become whether this marketing policy structure was still relevant to the wheat industry, and the Australian economy generally. Lloyd (1982, p. 364) had noted a more parsimonious government attitude towards the Australian wheat and other industries and, since its first wheat report in 1978, the IAC had been urging the deregulation of wheat marketing arrangements.

Policy intention

The last of the wheat 'stabilisation' acts required the Australian Wheat Board to 'perform its functions with the objects of securing, developing and maintaining markets for wheat and maximising the return to growers from the marketing of wheat' (IAC 1988a, p. 30 quoting the Commonwealth's *Wheat Marketing Act 1984*). The Wheat Board claimed that it could exert market power in international markets, and therefore the revenue of wheat producers could be maximised by having a single export marketer.

Although the wheat marketing schemes were titled 'stabilisation' schemes, intertemporal stabilisation was not possible because the revenues from each year's crop ('pool') were kept separate from each other. 'Stabilisation' took three possible forms: unstable export prices for wheat were equalised with more stable domestic prices; returns were stabilised using external funds, principally from taxpayers via underwriting; and, where the Australian Wheat Board developed export markets for *Australian* wheat, sales continuity 'stabilised' export quantities even if not prices. Wheat 'stabilisation' had distributional effects including discrimination between consumers and farmers, since domestic prices could be higher or lower than export prices. Price discrimination was possible between different domestic users—e.g. under the 1983–88 scheme, the same domestic wheat consignment attracted different prices for human food, industrial, or livestock feed uses. Price discrimination was also practised between farmers supplying the same grade of wheat, depending on how marketing costs were assessed.

There was intensive intervention at the state level for storage, handling and transport arrangements for the wheat crop. Each state had a bulk handling authority (BHA), which handled wheat as well as other grains. The BHAs were either statutory authorities (New South Wales, Victoria and Queensland)[2] with substantial grower influence over BHA activity, or farmer cooperatives (South Australia and Western Australia). Effectively, BHA financing was publicly underwritten because there were exclusive agreements between the Australian Wheat Board and BHAs, and the borrowings for BHAs were underwritten by state governments. In Queensland, Victoria and parts of Western Australia, legislation prohibited road transport of wheat to port terminals, while in some states it was inhibited by charges or the lack of road receival facilities. Intrastate wheat transport was thus largely undertaken by the railways (IAC 1988a, p. 41).

Area of sector

Wheat marketing policy was principally concerned with selling the wheat crop. However, since wheat prices were affected by selling arrangements or intervention in storage, handling and transport, wheat marketing policy was a central determinant of allocative efficiency. If the Wheat Board extracted monopoly rents on export markets, returns to growers would be higher. If the domestic market price was higher (lower) than export prices—once allowances were made for different costs of supplying these markets—average returns to growers would be higher (lower). If BHAs were technically or allocatively inefficient through lack of effective competition, the costs of the system would be higher—and growers' price would be lower. The impact of the marketing system reached back to the farm level, not only through price but also via restraints on growers as to the varieties they could deliver to their local silos. When the system ran out of control, as in the over-production of the late 1960s, delivery quotas were imposed on farmers.

If the Wheat Board interpreted its legislated objective of maximising growers' returns literally and in the short run, monopoly exploitation of the domestic market could squeeze the profitability of domestic wheat users. Of particular interest were wheat-using industries with export potential—particularly lot-feeding and gluten.

Policy orientation

From 1987–88 to 1990–91, approximately 85 per cent of all Australian wheat was exported (ABARE 1993, table 98). Control of export marketing, the Commonwealth's domain if it legislates (cf. s.51(1) of the Constitution), dominates growers' returns. However, for constitutional reasons (cf. section 3.2), until the late 1980s wheat 'stabilisation' schemes were cooperative arrangements among the Commonwealth and states. Under interpretations of the 'free interstate trade' provision of the Constitution (s.92), pre-eminent until the late 1980s, wheat marketing arrangements in eastern Australia were always vulnerable to farmers' selling their wheat across borders and claiming the protection of s.92 (cf. Chapter 3). Part of the bargain between Commonwealth and states over legislative competence in wheat marketing was to recognise state storage, handling and transport monopoly arrangements (IAC 1988a, pp. 41–2). Highly variable export prices, and the relative financial strength of the Commonwealth, also facilitated the emergence of underwriting arrangements by which taxpayers contributed to minimising falls in farmers' wheat prices (with one major public contribution in 1986–87).

Policy mechanism

The principal legislative framework for peacetime wheat marketing was the Commonwealth's wheat stabilisation acts from 1948, which created the Australian Wheat Board as the statutory marketer for the Australian wheat crop. Complementary state wheat marketing acts enabled the Wheat Board to engage in intrastate trade. Legislation establishing the Wheat Board endowed it with substantial administrative powers since it was empowered to make trading arrangements for acquiring and disposing of the crop. Under schemes prior to 1988, some administrative decisions were jointly taken by several agencies: for example, the Wheat Board, BHAs and state agriculture departments decided the silo locations at which particular types of wheat would be accepted. Grain transport involved decisions between the Board and BHAs, and between BHAs and state railways, where they were used. The 1983 wheat marketing arrangements are summarised in Box 7.1, and the 1989 Wheat Marketing Act in Box 7.2.

Box 7.1 Wheat Marketing under the 1983 Act

- Australian Wheat Board (AWB) has complete power over export markets, could use private traders
- AWB operated through 'authorised receivers' in each state (BHAs, except for State Wheat Board in Queensland)
- AWB major trader of domestic wheat for human consumption and industrial uses
  - price depended on end-use
  - grower-to-buyer arrangements for human/industrial/stockfeed uses: direct bargaining between growers & buyers, possibility of direct delivery, reduced BHA charge, possibility of discounts/premiums relative to AWB pool prices, growers received payment through pool
  - stockfeed permits issued by AWB except in Queensland, permit price covered statutory levies and BHA charge; by end of scheme, most stockfeed wheat sold under permit, not through AWB
- wheat classes: Prime Hard, Hard, Australian Standard White (ASW), Soft and Durum; segregations: General Purpose No. 1 (lightweight), No. 2 (weather-damaged), Feed
- varietal control: 'best' marketable varieties, State Wheat Advisory Committees decided which varieties could be delivered at which silos without discount (provided they satisfied 'receival standards'—protein and physical properties)
- underwriting
  - GMP (guaranteed minimum price) for ASW, usually calculated as 95 per cent of the average of the estimated gross pool return for ASW in current season, and the estimated gross pool returns for the two lowest of the previous three seasons (less certain costs); PGMP (preliminary estimate of GMP) declared prior to 1 October and GMP by 1 March—if GMP < 0.9*PGMP then GMP = PGMP; GMPs for other categories determined from ASW by margins
    - Commonwealth Government payments to AWB when net pool return for a category < GMP for that category
    - gross payments were underwritten (i.e. including AWB costs)
    - underwriting arrangements first triggered for 1986–87 pool

- pooling: central feature since 1948
  - uniform price to growers for wheat in same pool, increasingly differentiated for varietal discounts/dockages, category premiums (Prime Hard etc), protein (within Prime Hard & Hard)
  - AWB costs averaged across all growers; BHA, wharfage & some other handling generally pooled within states
  - payments: first advance = 0.9*PGMP – (AWB costs + some handling deductions) (interim first advance within 3 weeks of delivery, final first advance)
  - 3 deferred payments systems
  - deferred delivery system

Source: Industries Assistance Commission (1988a)

Box 7.2 Key Features of 1989 Wheat Marketing Act

- deregulation of the domestic wheat market
- introduction of the Wheat Industry Fund to provide a capital base for the Australian Wheat Board (AWB) (funded by a 2 per cent levy on farmgate value of all wheat sales, later continued to 30 June 1999)
- AWB permitted to trade in grains other than wheat
- Guaranteed Minimum Price replaced by Commonwealth Government underwriting of AWB borrowings, with a sunset clause to 30 June 1994 (later extended to 30 June 1999)
- AWB objective to maximise net farm returns to producers —> minimise storage, handling and transport costs, costs to be passed back to individual producers wherever possible
- AWB permitted to have more than one grain receiver in a state
- AWB Board changed from primarily growers to commercial expertise
- no sunset provision on life of AWB

Source: Ryan (1993)

Institutional structure

Interest groups in the wheat industry include wheat growers and their lobby groups (state farmer organisations, Grains Council of Australia, National Farmers' Federation), those engaged in storing, handling and transporting grain (both organisations and employees), domestic stockfeed industries, domestic grain processors, overseas users, and domestic consumers. Because of the high level of wheat exports and low relative value of grain in final product, domestic consumers are irrelevant as an interest group.

While the Grains Council represented the farm voice for the wheat industry, there is not—and has not been—a single voice in the industry. Individual firms will always benefit by operating outside a cartel. Even within the cartel, arrangements favoured some growers at the expense of others (e.g. pooling of BHA or Wheat Board costs). Growers whose crops had low marketing costs had strong incentives to have the regulatory structure changed to disaggregate such costs.

In states where it was mandatory for grain to be transported by rail, it was in the interests of the railways—and their owners (the state governments), employees and taxpayers—to maintain arrangements that increased the railways' revenues. Conversely, it was in the interests of the road haulage industry to lobby for the deregulation of grain haulage. The symbiotic relationships among BHAs, railways and state governments also contributed to opportunities for political decisions, such as the location of the wheat loader in Port Kembla (NSW), and the forgiving of a substantial portion of New South Wales Grain Handling Authority's debt when it was privatised.

Until the late 1980s, domestic processors (other than the lot feed industry) strongly supported retaining highly regulated marketing arrangements. There apparently were significant benefits from the relationship between a monopolist and a small number of large buyers. Even where selling arrangements resulted in higher prices for some wheat types, there were advantages in administrative arrangements such as the allocation of wheat to domestic processors. Exemption from the anti-competitive provisions of the Trade Practices Act enabled domestic wheat processors to negotiate while knowing the conditions under which their domestic rivals were buying grain. Thus all purchasers could be guaranteed common purchasing arrangements, reducing their costs of competition.

Finally, agricultural economists generally supported deregulation of the wheat industry—and indeed all agricultural industries—to increase allocative efficiency. Ironically, Australian agricultural economists with expertise in wheat transport became consultants to the Royal Commission into Grain Storage, Handling and Transport, but their contracts apparently prevented them from commenting publicly during the debate about deregulation of the post-farm sector.[3]

Policy performance evaluation

The Industries Assistance Commission inquired into wheat marketing arrangements in 1978, 1983 and 1988. The Commonwealth and state governments jointly commissioned the Royal Commission into Grain Storage, Handling and Transport, a rare example of a joint Royal Commission. There were several state inquiries into grain marketing, such as a 1981 report into the Queensland grain distribution system, subsequently reviewed by the Queensland Planning Committee in 1986; the 1981 Carmichael inquiry into the New South Wales grain handling system; and CANAC Consultants Ltd's report into the Victorian grain handling system, which was also considered by Lloyd (1986). There were also inquiries into the efficiency of related infrastructure such as railways, and ports and road charging.

Economic issues

In reviewing the pre-1989 marketing arrangements, the Royal Commission into Grain Storage, Handling and Transport (1988, p. xxv) concluded that:

> Economic efficiency, or making the best use of the community's resources, is not achieved for a number of reasons. Principal amongst these is that most grain is channelled along a restricted number of predetermined paths with minimal scope for using lower cost paths. The outcome is that the cost of moving the total grain crop is higher than it would otherwise be.

Even if past intervention in wheat marketing had been economically justifiable, whether or not continued intervention could still be justified in principle depended on the existence of contemporary market failures. Possible effects of deregulation on information, exclusion, negotiation and revenue collection transactions costs are summarised in Box 7.3. These changes suggest the absence of transactions costs sufficiently serious to have justified retaining a complex regulatory environment for wheat marketing. There were still grounds for considerable contention as to whether or not the deregulation of export marketing was appropriate in view of possible negotiation transactions costs—i.e. the imbalance in market power between individual wheat growers and large grain traders.

One major qualification to this conclusion related to the effect of transport deregulation. It is not clear that deregulating grain haulage would necessarily increase allocative efficiency, even if it were to the advantage of the wheat industry. Since roads are publicly provided at relatively low cost to their users, whereas rail is a private good, the increased carriage of grain by road would not necessarily lead to any corresponding increase in road costs being borne by the beneficiaries of road grain haulage. However, governments faced this problem with respect to the efficiency of the transport system generally; it was not appropriately addressed solely through consideration of grain transport.

Box 7.3 Potential Changes to Transactions Costs with Wheat Market Deregulation

Market segment	Transaction cost	Comment
Grain marketing	Information	• information flows a consequence of market structure: public authorities and rural media continue to disseminate; AWB as (sole) export seller, disseminate information to attract desirable stocks for export; private firms disseminate to acquire stocks • information increasingly cheaper to acquire/store/retrieve/disseminate • risk: futures markets (overseas); limited capacity to establish prices before planting but potential for growth of contracts
	Exclusion	• exclusion difficult for both public and private information: some incentive not to reveal private information to recover its costs of acquisition; incentive to form information 'clubs' to average costs of private information
	Negotiation: market power	• farm enterprise substitutability limits domestic market power (unless grain monopolies or agricultural product monopolies) • export market power remains (AWB); limited substitution among grains in use increases domestic buyers' power in short run
	Revenue collection	• private buyers not guaranteed like AWB
Storage, handling	Information	• competitive fringe in storage, plus ending of sole receiver status of BHAs, create competitive pressures on BHAs to contain storage/handling cost increases because of knowledge of costs of storage/handling alternatives
	Exclusion	• easy at receival • impossible after receival unless lots so large that individual segregation possible
	Negotiation	• existing BHA monopolies in each state likely to remain dominant with deregulation of wheat grain market

		• corporatisation, if involves state subsidy for 'social' component of storage/handling, permit greater focusing on profits—but not necessarily economic optimisation if monopoly
		• privatisation, also permit greater focusing on profits—but not necessarily economic optimisation if monopoly
	Revenue collection	• private handlers not guaranteed like BHAs
Transport	Information	• mixed modes for grain transport allow greater identification of least cost operational means of shifting grain
	Exclusion	• poor excludability on roads: lower private costs in this mode, although social costs may be higher ('externalities' of road transport); general road transport problem, not just wheat transport
	Negotiation	• relative market power of AWB/private grain firms cf. road hauliers
	Revenue collection	• private hauliers not guaranteed like railways

Analytic techniques

Wheat marketing was a price equalisation scheme (figure 7.1 cf. figure 6.6). Suppose domestic demand is Dd, supply is S1, and export demand Dx is infinitely elastic at price Px. If the domestic price is held at price Pd1 (above the export price), and returns are averaged over domestic and export markets, Pe1 is the average price curve and, because wheat production has atomistic producers, Pc1 is also the marginal revenue curve. Equilibrium output is Qe, and since Qd is used on the domestic market, (Qe – Qd) is exported. Efficiency costs are the hatched triangle, with transfers from consumers to producers represented by the hatched rectangle. If supply to the domestic market were deregulated, domestic price would fall to Px since producers would be indifferent as to which market they supply, and domestic use would rise from Qd to Qc; production would fall to Qf, and exports would fall from (Qe – Qd) to (Qf – Qc). Conversely, if domestic price were held at price Pd2 (below the export price), the equalised price curve is Pe2: Qg is equilibrium production, of which Qh is consumed domestically and (Qg–Qh) is exported. Efficiency costs are the dotted triangle, with transfers from producers to consumers represented by the dotted rectangle.

Deregulation of domestic selling in this case raises domestic prices to Px, with total production rising to Qf, domestic use falling to Qc, and exports rising to (Qf–Qc) from (Qg–Qh).

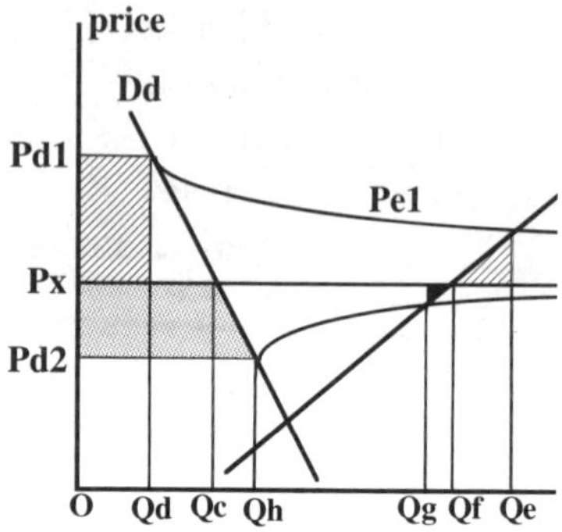

Figure 7.1 Idealised Representation of Australian Wheat Market

Deregulation of the domestic wheat market was accompanied by a search for efficiencies in the storage, handling and transport of wheat. The existing state systems remained in place, although some were corporatised or privatised, but the Wheat Board was liberated from the necessity of using them. The Wheat Board was able to seek the least cost system of moving any domestic grain it marketed, and could also seek least cost systems of moving its export grain to the seaboard. Export demand is given by line De in the right-hand side of figure 7.2. The services represented by storage, handling and transport are assumed to have an inelastic demand (Ds in the left-hand side of figure 7.2). The initial supply curve for storage, handling and transport services is Ss1 in the left-hand side of figure 7.2. The quantity of services supplied *per unit of exports* is Qs, at price Ps1 *per unit of services*. The corresponding derived demand curve for wheat is Dw1 (= De – Ps1) in the right-hand side of figure 7.2, and Qw1 wheat is grown. If deregulation of storage, handling and transport services achieves its objectives of lowering the cost of these services, the supply curve for these services falls—e.g. to Ss2. The quantity of services supplied per unit of exports remains unchanged at Qs, but the equilibrium price of these services falls to Ps2. The corresponding derived demand curve for wheat rises from Dw1 to Dw2 (= De – Ps2) and, with an unchanged wheat supply curve, the equilibrium output of wheat rises to Qw2.

Subsequent developments

An attempt to privatise the Australian Wheat Board was rejected by growers in 1992. Amendments to wheat marketing regulations in 1992, including extension of the Wheat Industry Fund levy until 1999, strengthened the power of the Board to function as a trading entity and maintained its export selling monopoly (Ryan 1993).

Flugge (1996) commented on a continuing debate about the appropriate structure of the Board, identifying a preference for greater commercial flexibility while retaining the export monopoly.

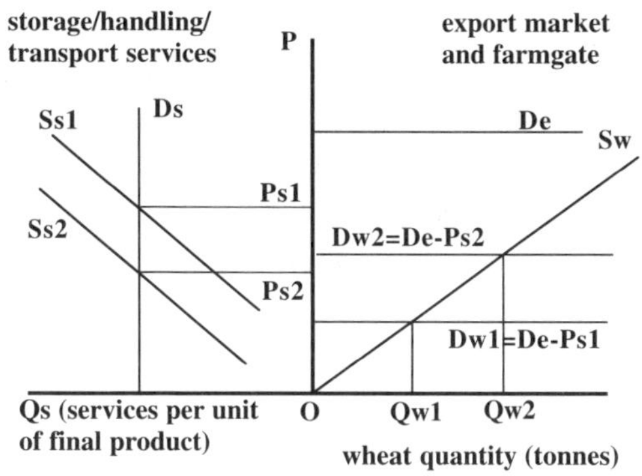

Figure 7.2 Effect of Increasing Storage/Handling/Transport Efficiency

7.2 SHRINKING WOOL

Since the early 1950s wool boom, the real price of wool has declined secularly, with occasional deep troughs and high peaks (Garnaut et al. 1993, figure 2). Australian wool production began rising from the early 1980s and this increase intensified because wool prices rose relative to cattle and wheat prices, and good seasonal conditions increased the wool cut (Garnaut et al. 1993, figures 9, 12, 13, p. 52). With strong demand for wool in 1987–88 and 1988–89—particularly stemming from China and the then Soviet Union becoming major wool buyers—real wool prices rose sharply, as did the price of wool relative to cotton and artificial fibres (Garnaut et al. 1993, figure 5). The market indicator for wool was raised to 870 cents per kg clean in 1988–89 following a strong rise in the indicator in 1987–88; the market indicator was held at 870 cents in 1989–90.

Exports of wool to China declined sharply from 1988–89, as did exports to the former Soviet Union and Eastern Europe from 1990–91 (Garnaut et al. 1993, pp. 34–5). Wool prices started falling in late 1988–89 and continued to decline in 1989–90 while the indicator was held steady at 870 cents per kg clean. Hence the Australian Wool Corporation's purchases under the Reserve Price Scheme rose rapidly in 1989–90; by year's end, stocks were more than twice as high as the previous highest level. The market indicator was lowered to 700 cents per kg clean in June 1990; the Reserve Price Scheme was suspended in February 1991 and abandoned in July 1991 (Garnaut et al. 1993, p. 33). The wool stockpile was 0.824 million tonnes when the Reserve Price Scheme was suspended.

Policy issue

Most agricultural products experience fluctuating prices, arising from the interaction of variable supply and demand. Supply is variable primarily because of the unpredictable, and largely uncontrollable, impact of weather on farm production. Agricultural supply may also be variable in mixed farming because farmers switch enterprises in response to changes in relative profitability. Demand is variable partly because of changes in consumers' preferences. For an agricultural exporter, demand may be variable because—with a given world demand for a commodity—supply variability in other countries will affect excess demand. Since Australia is the principal supplier of fine wool, domestic supply conditions and world demand are the major determinants of wool prices.

The efficiency issue concerning variable prices in the wool industry is whether or not resource allocation can be improved by public intervention to stabilise prices. A related distributional issue is whether or not Australian wool prices can be stabilised upwards—i.e. whether the Australian wool industry possesses sufficient market power to raise wool prices absolutely, or relative to the prices of competing fibres (e.g. cotton, artificial fibres).

If the distribution of wool prices were known—e.g. random fluctuations about a known mean or trend—opportunities would exist to reduce the variability of, or even stabilise, prices. Whether growers' or community income would be increased by stabilisation would depend on the characteristics of the supply and demand shifts (cf. Chapter 6). The desirability of stabilisation would also depend on growers' attitudes to risk, whether or not the costs of stabilisation were borne by growers or others, and whether growers actually wished to stabilise prices or some related variable such as income. Opportunities to improve welfare by modifying variable prices do not depend solely on a central stabilisation authority. Individual growers may seek to reduce their price uncertainty by entering into forward contracts with purchasers, by operating in futures markets,[4] or by their private stockholding activities.

Despite wool growers voting against a grower-financed wool reserve price scheme in 1965, the dramatic fall in wool prices from 1969 to 1971 led to the establishment of a scheme to smooth price fluctuations and eventually to a buffer stock scheme (the Reserve Price Scheme) to stabilise wool prices (cf. Piggott 1990, pp. 294–7; Vines et al. 1991, pp. 15–16). This scheme survived several sizeable stockpiles and their dissipation, but failed to survive the massive stockpile built up by 1990. The primary focus of this section is the failure of the Reserve Price Scheme and the effects of this failure.

Policy intention

The explicit intention of the Reserve Price Scheme changed over time. Originally the intervention was 'to protect growers against "disastrously low" prices and to provide them with some protection against short-term falls in the market ("potholes")' (Vines et al. 1991, p. 16). By the early 1980s, the Australian Wool Corporation reported the objective as being 'to provide greater price stability and predictability in Australian currency terms for Australian wool while also aiming to achieve the maximum sustainable price for growers in the long term' (Vines et al. 1991, p. 16). This objective was modified by 1990 to 'provide greater stability of wool prices at maximum commercial levels, subject to the costs of stockholding' (Vines et al. 1991, p. 16).

Area of sector

The principal explicit focus of the Reserve Price Scheme was the stabilisation of market prices for wool. This would also stabilise wool prices at farmgate unless stockholding costs increased faster than wool prices as the stockpile increased. If the real objective was to stabilise wool prices upwards, farm incomes would also be stabilised upwards, unless there were effects on production or marketing costs that offset the price effects. Changes in the relative profitability of the wool enterprise could—and did—affect wool supply.

If the Reserve Price Scheme stabilised or raised wool prices, wool users could also be affected. The Australian Wool Corporation consistently argued through the crisis of 1990–91 that wool users valued the price stability conferred by the Reserve Price Scheme. Ignoring any effect of the scheme on the price *level*, this argument is likely to be correct because growers, not processors, were bearing the costs of stabilisation, including storage costs. If users had been required to bear or share price stabilisation costs, it is unclear whether they would have continued to support the scheme or would have preferred market-based stabilisation measures such as forward contracts, futures markets or private stockholding.

Policy orientation

Since the wool industry is primarily export oriented, national mechanisms underwrite policy through the Commonwealth's export powers (s.51(i) of the Constitution); the Commonwealth's sole power over excises (s.90), enabling it to raise levies on wool production to fund the Reserve Price Scheme, R&D, and promotion; the size of the industry relative to the economies of the individual states; and the engagement of the Australian wool industry at a national level with the International Wool Secretariat (Australia provided some 76 per cent of the IWS's funding). The Secretariat's activities important to Australia were principally the demand-creating roles of promotion, and support to solve technical problems of using wool (Garnaut et al. 1993, chapter 7; see also Vines et al. 1991, section 6.6).[5]

Policy mechanism

Under the 1972 Wool Industry Act, the Commonwealth Minister was responsible for key decisions regarding market support for wool, and the Australian Wool Corporation carried out these decisions. The 1987 Wool Marketing Act devolved power for the 'management and day-to-day running of the wool industry' to the Australian Wool Corporation and the Wool Council of Australia (Vines et al. 1991, pp. 15–16). Under the 1987 Act, the Minister retained powers to intervene in 'exceptional circumstances'—these powers were used, for example, to forcibly reduce the minimum reserve price from 870 cents to 700 cents per kg in May 1990 (Vines et al. 1991, pp. 29–30).

Institutional structure

Organisations in the wool industry during the 1990–91 wool crisis formed an interlocking web of decision-making and influence, comprising:

- Australian Wool Corporation—facilitating wool marketing (principally administering the Reserve Price Scheme), promoting wool use, funding and administering wool R&D, and managing wool stores;
- Interim Wool Industry Policy Council—'to assist in the formulation of decisions regarding the industry's future and to ensure that the views of the major industry sectors are taken into account';
- Wool Council of Australia—'recognised in the relevant Commonwealth legislation is the peak body representing Australian wool growers'; the Corporation and the Industry Policy Council were required to report to the Wool Council, and the Wool Council was intimately involved in the selection of members to the Corporation's board and the Wool Research and Development Council;
- Wool Research and Development Council—investigated industry research needs, recommended to the Corporation on the administration of R&D funds, and supervised research activities; and
- Australian Wool Surveillance Authority—supervised wool testing services in Australia (which were largely undertaken by a previously statutory company, the Australian Wool Testing Authority) (Vines et al. 1991, pp. 26–9).

Following the demise of the Reserve Price Scheme, the Vines Committee recommended a new set of institutions (Vines et al. 1991, section 4.3), although a somewhat different set of new institutions was actually created (cf. Garnaut et al. 1993, chapters 6, 8, pp. 79, 82–3):

- Australian Wool Corporation—having the Corporation's original functions other than the Reserve Price Scheme and liquidation of the stockpile;
- Australian Wool Industry Council—providing advice to the Minister on matters affecting the industry, proposing overall strategic directions, and constituting an industry forum;
- Wool Industry Advisory Committee—comprising growers, brokers, exporters, private treaty and processor representatives, operating under the auspices of the Corporation, and managing the auction system;
- Wool Research and Development Corporation—undertaking R&D activities for the wool industry; and
- Australian Wool Realisation Commission—liquidating the wool stockpile and other assets and debts of the Reserve Price Scheme.

The Garnaut Committee recommended a further set of phased changes to these institutions: abolishing the Council, privatising some functions of the Corporation, corporatising and then privatising the Commission, and merging the Corporation's residual functions with those of the Wool Research and Development Corporation (cf. Garnaut et al. 1993, chapter 1, especially figure 1).

Policy performance evaluation

The collapse of the Reserve Price Scheme provoked a flurry of investigations into the future of the wool industry. Investigations included official government inquiries, such as Vines et al. (1991) and Garnaut et al. (1993), and excursions into value adding (e.g. Lawson et al. 1993). The Australian Bureau of Agricultural and Resource Economics undertook several studies into related issues (e.g. ABARE 1990, 1992a,b). Studies were commissioned by non-government bodies (e.g. Stoeckel et al. 1990). Professional economists had been engaged early in the debate on a buffer stock scheme (cf. Piggott 1990) and continued to be engaged in debates about the future of wool after the collapse of the Reserve Price Scheme (e.g. Bardsley 1991, Chisholm et al. 1994). Neither the Industry Commission nor its predecessor evaluated the Reserve Price Scheme in particular or wool marketing in general because, superficially at least, the Reserve Price Scheme was grower-funded.

Economic issues

Efficiency arguments about the Reserve Price Scheme and its collapse concern whether or not there were sufficiently large market failures that policy could overcome, and rectification of which would raise Australian wool-growers' incomes. Market failures might arise from transactions costs, particularly in information. A key question is: 'Was market information so poor when the Reserve Price Scheme was implemented that growers actually benefited from its operation?' Since information technologies improved markedly over the Reserve Price Scheme's lifetime, information available by its demise was probably sufficient for growers, buyers, processors and consumers to manage price risk and uncertainty without statutory intervention. A secondary question is: 'Are there aspects of wool's characteristics either not currently known or not sufficiently widely known that limit the efficient use of wool as a fibre and which would not otherwise, or not efficiently, be addressed by the wool trade?' Such informational failures might be addressed by promotion and research activities undertaken, for example, by the International Wool Secretariat. Market imperfections, for instance extensive economies of scale in the manufacture of competitive artificial fibres, might limit incentives for fibre users to investigate the properties of a single fibre such as wool.

Other transactions cost issues involved property rights—for example, alternative ways of describing a commodity such as wool may create potential efficiency gains from encouraging the use of superior product description technologies (e.g. sale by description), especially where such technologies could reduce the costs of handling and transporting the commodity. However, these issues were addressed separately from the Reserve Price Scheme. With negotiation transactions costs, market intervention might overcome market failures or imperfections arising from the unequal bargaining power of large processors compared to small domestic wool producers. But a stabilisation scheme would not overcome unequal bargaining power, unless there were major information asymmetries.

While these efficiency arguments might suggest a prima facie case supporting intervention, it is an empirical matter as to whether or not intervention mechanisms could be satisfactorily designed and implemented so that their benefits exceeded their costs. The Reserve Price Scheme was a form of insurance scheme where growers paid premiums (the wool tax) to accumulate financial reserves to buy and hold wool when prices were low. Any insurance scheme can bankrupt, especially in its early years, or when there is a major change to the form of the risk being covered. With the Reserve Price Scheme there was a major change as the Australian financial sector, including exchange rates, was deregulated during the 1980s. While exchange rates were regulated and stable, stabilisation only required forecasting international demand. However, the elimination of exchange rate controls meant that price stabilisation in Australian dollars also required forecasting of future exchange rates. Successful stabilisation with such uncertainty may not have been possible for a central agency because the informational demands may simply have been too high. Some events leading to the collapse of the Reserve Price Scheme—such as major restructuring of societies and economies in Eastern Europe, and political and economic instability in China—were probably unforecastable, even though these markets had become central to the viability of Australian wool price stabilisation.

The successful upwards stabilisation of Australian wool prices had several prerequisites. First, overseas buyers needed to value stable prices to the extent that they were prepared to pay sufficiently higher average prices to offset the total costs of running the Reserve Price Scheme. Second, prices could have been stabilised upwards if supply control could have been exerted either always, or at times of downwards pressure on prices, so that average prices were higher than otherwise. Except for providing marketing information about possible future price movements that may have affected growers' intentions, the Wool Corporation never had supply control instruments that could have supported upwards price stabilisation. Third, upwards stabilisation might have resulted from the Corporation's participation in demand-creating promotional activities. Such activities were, however, logically and practically distinct from the Reserve Price Scheme.

Finally, even if the principal objective of the Reserve Price Scheme had been simply to stabilise prices, it is unlikely that the politicians, bureaucrats and interest groups who were intimately involved in its creation and succour could have permanently resisted the temptation to use the scheme for distributional reasons unconnected with price stabilisation.

Analytic techniques

The price of wool relative to competing enterprises such as cattle and wheat, and the ratios of wool to cattle and sheep output, rose throughout the 1980s (Garnaut et al. 1993, figures 12 and 13). These observations are consistent with an expansion of demand from D0 to D1, which increases both equilibrium price (from P0 to P1) and equilibrium output (from Q0 to Q1) (figure 7.3). A run of favourable seasonal conditions after the drought broke in the early 1980s also contributed to increased sheep numbers and wool cut per head (cf. Garnaut et al. 1993, p. 52).

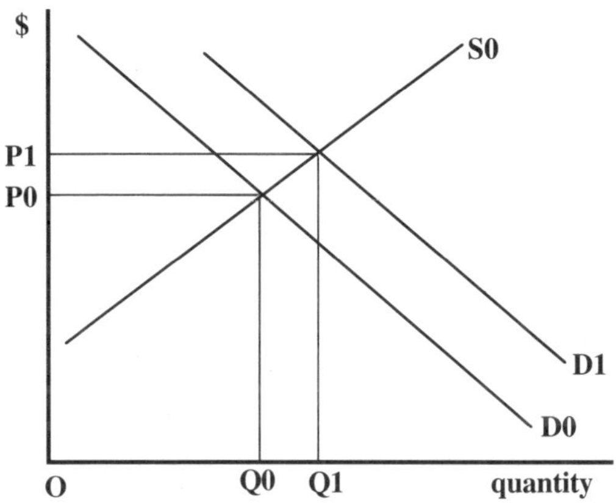

Figure 7.3 Demand and Production Shifts with Rising Wool Prices

The analysis of buffer stock schemes in Chapter 6 proceeded on the basis that the variability of supply and demand was known. However, even if normal variability were known, unpredictable abnormal shifts could place such intolerable pressure on a buffer stock scheme that it might bankrupt. Suppose there is no supply variability and 'normal' demand variability is between D1 and D2 and, within this range, the buffer stock scheme is just solvent if prices are held in the band P1′ to P2′ (figure 7.4 cf. figure 6.7). Occasionally demand falls to D3—an analogy would be the withdrawal of the former Soviet Union, Eastern Europe and China from wool-buying in the late 1980s and early 1990s. When demand falls to D3, large purchases—(Q – T) rather than the 'normal' (Q – S)—and very large expenditure (corresponding to area qzrs) are required to prevent the price from falling below P1′. If the scheme is only just solvent with demand confined to the range D1 to D2, it must become insolvent if demand falls to D3. The problem would be compounded if demand fell to D3 for more than one production period, as it did after 1989. This illustration does not prove that the wool Reserve Price Scheme was fatally flawed—it merely shows the kind of analytical framework that would be required for an empirical analysis of the conditions under which the viability of such a scheme could be compromised.

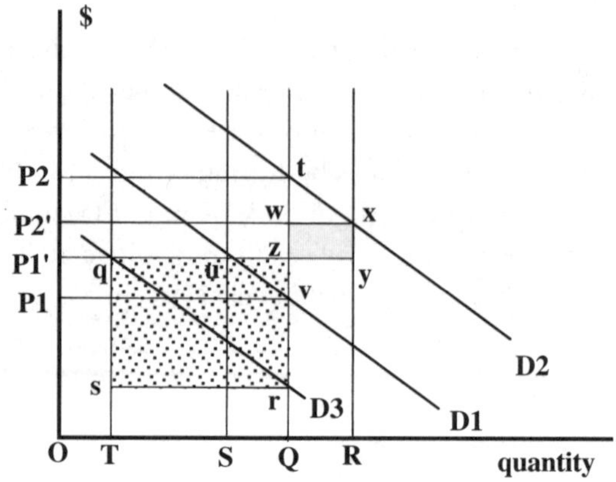

Figure 7.4 Buffer Stock Scheme

When the wool crisis struck in 1990, the Australian Bureau of Agricultural and Resource Economics undertook a study of the impact of various policy options on wool profitability. ABARE's modelling showed that the lower the wool tax rate, the less favourable the present value of future wool production (ABARE 1990, table 4 and figure 8). This modelling also showed that the lower the floor price, the greater the profitability of wool production. Sensitivity analysis was also undertaken in relation to changes in wool demand and exchange rates, and drought and interest rates, and the wool tax rate (ABARE 1990, tables 6–9). The incidence of costs over time was also analysed, showing that the lower floor price being considered at the time (700 cents per kg clean compared to the existing floor of 870 cents per kg) inflicted the same degree of financial pain sooner than the higher floor price, but recovery also commenced sooner (ABARE 1990, figure 10). Bardsley (1991, figure 8) showed that a rapid fall in the subsequent floor price of 700 cents per kg clean to about 450 cents per kg clean resulted in recovery to a long-run equilibrium price of about 560 cents per kg clean within about four years.

Subsequent developments

In 1994–95, approximately 3.1 million bales (557 kilotonnes of greasy wool) remained unsold in the stockpile, corresponding to 77 per cent of that year's wool production. Additionally there were high brokers' stocks (Groesch et al. 1996). Wool International, the stockpile liquidator, was committed to selling 187 000 bales per quarter up to mid-1997 (Richardson 1995).

7.3 DISSOLVING SUGAR

Queensland produces about 95 per cent of Australia's total sugar production. Eighty per cent of Queensland's sugar is exported and approximately 75 per cent of Australian sugar (predominantly as raw sugar). The major domestic uses of sugar are in beverage and confectionery manufacture, and in retail trade. Major sugar substitutes are sweeteners made from carbohydrates such as wheat and maize starch, and synthetics such as aspartame (Industry Commission 1992a, chapter 3). Three Queensland groups control 71 per cent of sugar milling capacity, and eight other Queensland mills have 23 per cent of milling capacity; the remaining 6 per cent of capacity is in New South Wales. Although CSR Ltd dominated domestic sales of refined sugar until 1989, the New South Wales refinery now supplies about 25 per cent compared to CSR's 69 per cent; the remainder is supplied by Millaquin. About 80 per cent of farms have 'assignments' (like quotas) less than 90 hectares. Sugar cane is highly perishable, and needs crushing within about 16 hours of cutting; it can, however, be stored while growing in the field although sugar cane content changes in the mature cane.

Edwards (1993, p. 1) commented 'For all my working life I have been engaged in the examination of agricultural policy in Australia. I have found it harder to fathom what goes on in the sugar industry than to come to grips with any other industry'. Edwards noted that part of the reason for this difficulty is that 'there has been less information available on the sugar industry than on other rural industries'. But the lack of information is probably only the consequence of the complexity—and value—of the policy arrangements which pervaded nearly every nook and cranny of the sugar industry. Until the early 1990s, the sugar industry was the epitome of cradle-to-grave socialism, where the product was nationalised on production and its distribution tightly controlled by the state. The consequence of this intervention was, as the Industry Commission (1992a) noted and the poet Les Murray (1994) celebrated, the dominance of small family farms in the Queensland sugar industry:

> and the small family farm incurs deep hate
> because work there is owned and adult.
> Covet it, destroy its supports, cheat it,
> but slavery comes back when we defeat it.

The following discussion focuses primarily on marketing arrangements at the time of the Industry Commission's 1992 sugar inquiry, and subsequent proposals for reform.

Policy issue

As with wheat and wool, the principal issue concerned the best marketing system for this important agricultural export industry. The form of marketing intervention, and the marketing tentacles that reached back to the farm level, also had major distributional consequences of profound social significance in the sugar-growing areas.[6]

Policy intention

The sugar industry has one of the longest histories of regulation of all Australian agricultural industries. The Industry Commission (1992a, pp. 33–6) briefly surveyed its regulatory development.[7] Such a rich regulatory history drew most of the key players into the regulatory web, and there was thus no simple policy intention in sugar marketing. The key justifications for intervention in sugar marketing were, at the production end, to protect growers from the buying power of the mill (ironically, mills were also concerned about the power of organised grower groups); to ensure orderly expansion, within the capacity of industry infrastructure, which did not threaten established economic interests; to enable the equitable coordination of cane harvesting and delivery to mills; to protect growers, potential growers and millers from non-viable investment decisions; to protect mills against the threat of closure arising from the transfer of cane supply between mills; and to preserve and increase average industry returns (Industry Commission 1992a, p. 41).

At the marketing end, the industry justified intervention through a management and services package, which provided management and control of raw sugar quality; long-term sales contracts; logistical control, particularly through the bulk sugar terminals; economies of size; and product promotion and market development for identifiably Queensland sugar (Industry Commission 1992a, p. 71).

Industry also nominated other advantages, including price premiums, countervailing market power, maintenance of quality standards, stabilisation of returns, and provision of assistance (Industry Commission 1992a, p. 76).

Area of sector

Regulation of sugar marketing ranged across the entire industry. Land that could be used for cane production was regulated by the assignment system; and sugar produced on non-assigned land received a price of $1 per tonne (Industry Commission 1992, p. 185). Sugar produced on assigned cane land was divided into two classes: sugar associated with a farm 'peak' was paid a 12 per cent higher price (Industry Commission 1992a, pp. 37–9).

Production controls also affected cane milling. Marketing regulations affected cane and milling suppliers by influencing the size of the margin between milling and final consumers, and thus the derived demand for cane and raw sugar.

Policy orientation

Since the sugar industry is predominantly export oriented, the Commonwealth's export powers derived under s.51(i) of the Constitution would be expected to give it leverage to control the industry. However, states are not excluded from fields involving exporting unless their activities conflict with legitimately exercised powers of the Commonwealth (cf. Chapter 3). Since sugar is essentially a Queensland industry, and the New South Wales portion of the industry is entirely domestically focused and small, the Queensland Government would be the effective determinant of Australian sugar industry policy if the Commonwealth chose to vacate the field. Except for the sugar import tariff, which was introduced in 1989 when the import embargo was lifted, and anti-dumping mechanisms, the regulatory command of the Australian sugar industry was effectively Queensland's.

Policy mechanism

The central policy mechanism in the sugar industry was Queensland's *Sugar Industry Act 1991*, which implemented the regulatory structure outlined above (cf. Industry Commission 1992a, Appendices C and D). This Act also managed R&D for cane growing in the Bureau of Sugar Experiment Stations[8] and Productivity Boards, which performed an extension role in the industry (Industry Commission 1992a, Appendix E). The relationship between growers and mills was regulated by local awards that regulated cane harvesting and delivery to mills; transport, handling and crushing of cane by mills; and cane payments (Industry Commission 1992a, pp. 39–40). The Queensland Sugar Corporation was responsible for marketing all raw sugar produced in that state and it compulsorily acquired all raw sugar produced in Queensland. The Corporation sold raw sugar to refineries mainly for domestic consumption, and CSR Ltd was the sole export marketing agent for the Corporation in all international markets except New Zealand.

Institutional structure

Apart from the statutory organisations noted above, the institutional structure of the sugar industry is nicely illustrated by those who submitted to the Industry Commission's 1992 inquiry (cf. Box 7.4a).[9] Many of these institutions also made submissions to the Sugar Industry Task Force, established by the Commonwealth government to review the Industry Commission's 1992 report (Box 7.4b).[10] All Task Force members were intimately involved in the sugar industry, and the one sugar user member came from an industry with limited import competition or export prospects. It is easy to sympathise with Edwards' (1993, p. 22) conclusion that it would have been surprising 'if the interaction of interest groups represented on the Task Force led to the view that the maximum net benefits to Australia was an important criterion.'

Policy performance evaluation

The Industry Commission noted four reports on the sugar industry (in 1985, 1989 [2 reports], 1990) following its 1983 report. In the decade prior to the Commission's 1992 report, ABARE's published research appears to have been primarily involved with the international sugar trade rather than the domestic industry. The Commission's 1992 report was followed by the report of the Sugar Industry Task Force (1993).

Economic issues

Interventions in the sugar industry appear to have had distributional rather than efficiency justifications. The assignment system, particularly since the freezing of farm peaks in 1982, discriminated in favour of 'old' producers. Assignments were justified by the industry for protecting asset values (Industry Commission 1992a, pp. 44–5) but—whether or not assigned land had a higher value than unassigned land—it is difficult to see why the purpose of government intervention should have been to protect the asset values of those producers who had assignment, and thus reduce the profitability of those firms who could not enter the industry because they could not obtain assignment.

Box 7.4 Sugar Industry Interest Groups

(a) Sugar industry participants in Industry Commission's 1992 inquiry:

- farmers—Australian Cane Farmers Association, Canegrowers (and 10 of its regional groups), Davco, NSW Canegrowers Association
- millers and refiners—Australian Sugar Milling Council, Bundaberg Sugar Company, CSR Ltd Sugar Mills Group, CSR Ltd Refined Sugars Group, Mackay Sugar Cooperative, Manildra Harwood Sugars, Maryborough Sugar Factory, Mossman Central Mill, NSW Milling Cooperative, Proserpine Cooperative Milling Association, South Johnstone Mill, Tully Sugar
- users—Australian Soft Drink Association, Coca-Cola Amatil Ltd, Confectionery Manufacturers of Australia, Sugar Users Group (joint submission by Australian Soft Drink Association and Confectionery Manufacturers of Australia), Tooheys Ltd
- government—Commonwealth (Department of Foreign Affairs and Trade, Australian Bureau of Agricultural and Resource Economics, Sugar Research and Development Corporation, CSIRO, Great Barrier Reef Marine Park Authority), Queensland (Queensland Government, Queensland Sugar Corporation, Department of Primary Industries, Bureau of Sugar Experiment Stations), NSW (Treasury)
- lobbyists—Centre for International Economics (study for CSR)
- other—Australian Raw Sugar Industry (joint submission by Queensland Sugar Board, Canegrowers and Australian Sugar Milling Council).

(b) Membership of the Sugar Industry Task Force 1992.

- a Member of the House of Representatives who represented a Queensland electorate with a major sugar industry, and who was also a sugar grower;
- the Chairman of Canegrowers;
- the President of the NSW Cane Growers Council;
- the Chairman of a sugar milling company and deputy chairman of the Australian Sugar Milling Council;
- the National President of the Australian Workers' Union (this union covers 75 per cent of all unionists employed across all sectors of the sugar industry; Sugar Industry Task Force 1993, p. 35);
- Deputy Managing Director of CSR Ltd.
- Director, Group Supply, Coca-Cola Amatil.

As in any market, there are clearly possibilities for market failure. However, no great information, property rights or revenue collection transactions costs appear to have led to significant market failures. Negotiation transactions costs appear to have been superficially possible, but closer examination suggests they are unlikely. While growers are small relative to the mills they supply, this may partly stem from previous administration of the assignment system, which appears to have restricted the growth of farm firms. The New South Wales mills, and ten of the twenty-five Queensland mills, were owned by growers or cooperatives, and it is unlikely that they would have systematically exploited growers. Even where mills were incorporated (seven owned by CSR, six by Bundaberg Sugar, and two others), there were strong incentives for the mill, and growers as a group, to establish mutually agreeable supply conditions both within a season and over time to ensure the viability of both mill and growers. Contractual arrangements between millers and growers to allocate the benefit of high-sugar-content cane during the harvesting season were possible without land assignment. Contracts independent of land assignment could similarly have been used to develop supply arrangements that were mutually beneficial to growers and millers.

On the marketing side, some of the justifications for intervention—e.g. the Sugar Corporation's marketing services package and maintenance of minimum quality standards—were directed at objectives that would have been achieved with large firms such as CSR, who would operate in the absence of statutory intervention. Others—such as price premiums, especially those domestic premiums supposedly supported by the sugar tariff—were primarily distributional, although they had potential allocative effects such as restricting the growth of sugar-using industries in Australia. The countervailing market power argument confounded the large number of small growers with the small number of domestic refiners. However, most of Australia's sugar was exported raw, and so the bogey of a large refiner dominating a large number of small producers was improbable. What was more important was the relationship between growers and millers which, as previously argued, was characterised by sufficient symbiosis for bilateral monopoly to result in a satisfactory outcome for both parties. Further, since the major price premiums were supported by the tariff, and there was strong competition on the domestic market, trade practices legislation would have resulted in allocatively efficient domestic prices if the statutory intervention had been eliminated.

Finally, the provision of assistance is a social welfare matter better handled by the normal assistance mechanisms—such as the Rural Adjustment Scheme, Farm Household Support, or standard social welfare measures—rather than being channelled through efficiency distorting industry arrangements.

Analytic techniques

Inefficient production

The assignment system restricted the quantity of sugar that could be produced (Qa in figure 7.5) (cf. Industry Commission 1992a, pp. 47–50).

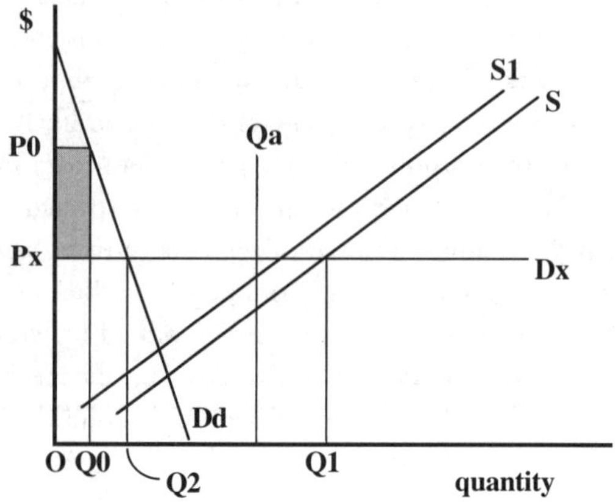

Figure 7.5 The Australian Sugar Industry

Because the production constraint operated through a land restriction on cane production, the constraint is not strictly a vertical line. The problem is illustrated in figure 7.6: ignoring the capital cost of land, taking the price of the land input as being the cost of technically optimal cultivation, and assuming the firm has a budget constraint C0, the optimal production occurs at point b where the line representing the relative price of inputs (Z) is tangential to the highest achievable isoquant Y1. Suppose, however, that the firm has an area of cane assignment represented by L1; then the firm's optimisation problem is to maximise:

(7.i) $$\Pi = Y - p.F - (C0 - pz.L1)$$

where Π is net returns, Y is the firm's output, p is the price of fertiliser and chemicals (normalised with respect to the price of cane; i.e. the cane price = 1) and F is the quantity of these inputs applied, and pz is the normalised price of cultivating area L1. The first order condition for profit maximisation with respect to fertiliser and chemicals is:

(7.ii) $$\partial Y/\partial F = p$$

whereas, without the land constraint, the firm's first order condition for profit maximisation with respect to fertiliser and chemicals is:

(7.iii) $$(\partial Y/\partial L)/(\partial Y/\partial F) = pz/p$$

or

(7.iv) $$(\partial Y/\partial F) = \partial Y/\partial L.p/pz$$

and, unless $(\partial Y/\partial L(L0))/pz = 1$, the land constrained and land unconstrained first order conditions are entirely different. The unconstrained first order conditions of equations (7.iii–7.iv) must result in an output level (Y1) higher than any other constrained solution (e.g. Y0) (cf. figure 7.6). Land-constrained optimisation forces the firm to use additional resources (= pz.[L0 – L1]) in applying fertiliser and chemicals. The observed supply curve will be the technically inefficient curve S1 rather than the efficient supply curve S (cf. figure 7.5, and cf. Edwards 1993, pp. 6–7). Because the administration of assignment required sugar to be repeatedly grown on the same assigned land, the crop was more susceptible to pests and diseases, thus more control inputs were required, and the production system was even less efficient than if greater flexibility allowed preferred rotations. The maintenance of land quotas in cane farming, and limited transfer rights, resulted in a largely small-scale industry even though substantial economies of size appeared to be obtainable from modest increases in farm size (Industry Commission 1992a, pp. 148–9). Other inefficiencies were also induced by the assignment system (cf. Industry Commission 1992a, chapters 5 and 11).

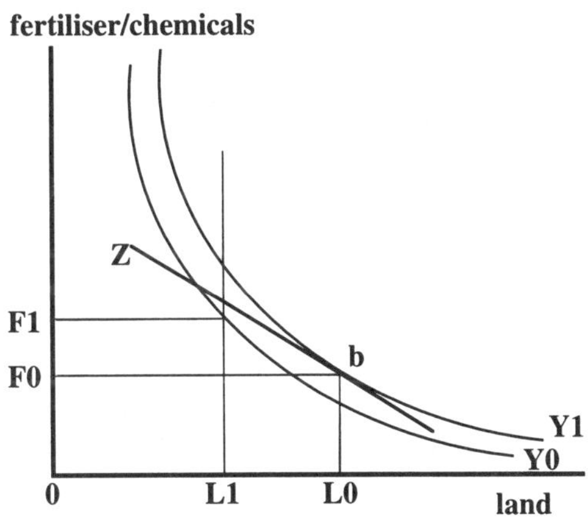

Figure 7.6 Firm Producing with a Land Area Constraint

If externalities resulted from runoff contaminated with fertilisers and chemicals, allocative inefficiencies caused by sugar production would have been greater because of the land constraint (cf. Edwards 1993, pp. 6–7).

Sugar grown on assigned land received two different prices. Sugar grown within mill (or farm) peak received a price 12 per cent higher than sugar grown outside mill peak, and 'peaks' had been frozen since 1982 (Industry Commission 1992a, p. xi). Thus, for example, new growers who had received new assignment since 1982 would have received only the lower price for their cane. The possible effects of differential pricing can be seen in figure 7.7, which contrasts the extreme cases of grower A all of whose cane was 'peak', and a new grower B who had no 'peak' cane. For the technical reasons outlined above, the new grower would have had lower production costs (long-run average cost curve LRACb and marginal cost curve LRMCb) than the older grower (long-run average cost curve LRACa and marginal cost curve LRMCa). At price P1 for pool 1 sugar, grower A produces Qa and, at price P2 for pool 2 sugar, grower B produces Qb sugar. The dotted rectangle in figure 7.7 represents the additional total cost of producing cane using the technology of a grower on old assigned land compared to the costs of producing this quantity of cane on new assigned land.

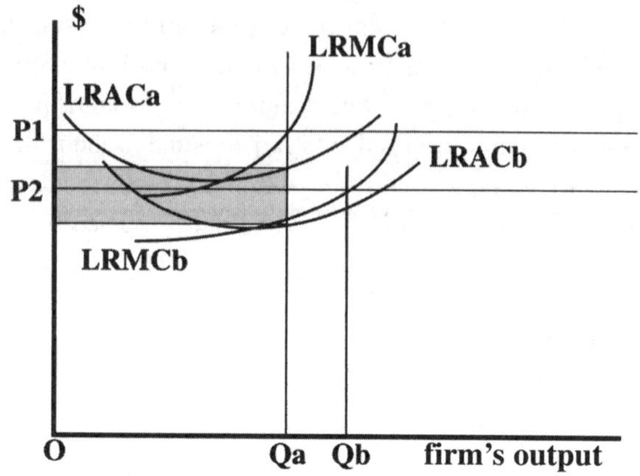

Figure 7.7 Effects of Differential Cane Pricing

Marketing

The previous embargo on sugar imports was lifted in 1989, and replaced by a tariff of \$115 per tonne on raw and refined sugar, subsequently reduced to \$55 per tonne. Because the sugar industry was essentially export oriented, the Industry Commission (1992a, pp. 96–7) argued that the tariff only resulted in domestic sugar prices above import parity because of limited competition in the domestic sugar market. Thus, for example, in figure 7.5, without the dominant influence of the Corporation and CSR, domestic sugar refiners would have competed the price down from P0 to somewhere above Px (where the margin above Px reflected import parity). The pricing regime for domestically utilised sugar possibly prevented a refined sugar export market emerging, although there are clearly other factors which would also affect the decisions of potential export refiners (cf. Industry Commission 1992a, pp. 135–8).

Subsequent developments

The cane-growing area in Queensland increased 30 per cent between 1989 and 1996, and that state's raw sugar production increased 37 per cent. The availability of assignment was claimed not to be constraining cane production. By 1996 domestic capacity for refined sugar production exceeded domestic consumption, there were four independent domestic refiners, and prices were lower than in 1989 (Rutledge 1996). A review of the structure of the industry in Queensland under National Competition Policy Principles was conducted during 1996 (cf. Chapter 14).

ISSUES FOR DISCUSSION

1 The wool industry's reserve price scheme collapsed in 1990. Discuss whether this collapse was likely to have been associated with an increase or decrease in economic welfare.

2 On 10 March 1993, the New South Wales Minister for Agriculture said:

'The Labor Party was the architect of the collapse of the wool industry, and should be held responsible for the destruction caused by its policies.'

Evaluate this argument in economic efficiency and public choice terms.

Endnotes

1 This chapter purposely avoids the dairy industry, the Australian rural industry with the most complex regulatory arrangements, primarily because there is so much otherwise available over such a long period of time. The novice policy economist could test their grasp of the subject by doing their own analysis of dairy policy.

2 Some have been subsequently privatised.

3 While (Australian) agricultural economists continue to address policy issues in their research, and debate these issues extensively among themselves at their conferences and in their journals, they appear to have become increasingly unwilling to participate in public debates on policy issues. This situation may partly reflect the bureaucratisation of the profession (in particular, the numerical dominance of the Australian Bureau of Agricultural and Resource Economics) and increasing demands on academics' time. I am indebted to Bob Batterham for several discussions on this issue.

4 Australia had a greasy wool futures market until it was made redundant by the Reserve Price Scheme.

5 Garnaut et al. (1993, p. 121) commented that the IWS had 'propped up' activities of the International Wool Textile Organisation, mainly comprising European textile manufacturers, for many years.

6 cf. Les Murray's poem quoted in the text; the mythological tentacles of the sugar industry reached as far south as Melbourne, setting of Ray Lawler's *The Summer of the Seventeenth Doll* and scene of its triumphal first staging.

7 The Commission's historical review carefully skirted the issue of the Queensland sugar industry being initially dependent on 'indentured' labour 'recruited' from the Pacific Islands whose repatriation was a precondition for Federation of the Australian colonies into a 'White Australia'. 'Import protection was granted in 1901 as a condition of Queensland joining the Commonwealth' (Industry Commission 1992a, p. 33) because sugar could not be grown profitably at the wage rates expected by the replacement white labour force without tariff protection and a bounty. Insufficient protection was apparently offered because half Australia's domestic consumption was still imported by 1914 despite output almost doubling (Shaw 1990, p. 11).

8 Other R&D was undertaken by the Sugar Research Institute, funded by all Australian mills. The Sugar Research and Development Corporation is a Commonwealth statutory authority which collects levies from growers and millers, and matches these funds with taxpayers' funds to support sugar-related research (cf. Industry Commission 1992a, Appendix E).

9 At a time of re-focused interest in compulsory unionism, it is interesting that apparently every cane grower in Queensland had to belong to Canegrowers.

10 Edwards (1993, p. 22) commented on the Commonwealth Government's 1993 'sugar industry package' which was in response to the report of the Sugar Industry Task Force (1993) that 'The report of the Sugar Industry Task Force was not available to interested persons for critical examination before the Government announced its decision. What the Task Force recommended, and how it justified those recommendations, is still not known.' This was not the only Industry Commission report where secret negotiations followed Commission reports. As a public servant, the author was personally involved in non-public negotiations under the auspices of the Standing Committee on Agriculture following the Industries Assistance Commission's reports into dairying (1976) and wheat (1988). Particularly in the case of the former, discussions had little to do with net benefits to Australia. (On leaving Australia prior to the conclusion of the dairy negotiations, the author's supervisor commented that he was very confident that major reforms would result from these discussions. On returning to Australia five years later, his first task involved the IAC's 1983 dairy inquiry—and nothing of substance had changed in his absence.) The Sugar Industry Task Force's Report was dated 'February 1993' and the Federal Government's sugar policy announcement (Crean 1993) was dated 2 February 1993!

8 Marketing issues: value adding and food safety

The dominant government response to perceived problems in agricultural marketing up to the 1980s was constructing marketing schemes to improve returns to primary producers. In the 1980s, new markets for agricultural products emerged as a new policy theme and one old theme, food safety, re-emerged.

8.1 VALUE ADDING

Policy issue

The contemporary focus on value adding as a strategic objective for firms, industries or government is an economic nonsense.[1] The term 'value adding' has been appropriated from the national accounts literature, where it literally means the returns to the fixed factors of land, labour, capital and enterprise (cf. Mauldon 1991). If firms are profitable, they add value; the key issue must therefore be *additional* value added. Since additional value adding inevitably means investment, its evaluation requires comparison between the additional value added available from developing a particular activity, and the opportunity cost of the factors that would be drawn into the relevant industry. Only if the opportunity cost of these factors were zero would expanded value adding opportunities in one industry necessarily be economically advantageous. In the contemporary political context, 'value adding' typically means additional domestic processing; but greater value might be obtained in other ways. For example, McMaster and McKay (1991, pp. 35–6) showed that research and development adds value by developing wheat varieties better suited to overseas markets or by improving production processes in other countries to better enable them to use Australian wheats. Improved processing technology that reduces the processing costs associated with using Australian raw materials increases value adding in Australia, even if the processing occurs overseas.

Australian firms previously undertook value-adding activities which have subsequently moved offshore. When transport costs were high, it was efficient to wash wool in Australia prior to export to reduce the costs of transporting dirt. Once rail supplanted other forms of wool transport within Australia and efficient industrial processes were developed to scour wool, wool washing in Australia was replaced by overseas wool scouring. Similarly, Australia exported flour and other processed wheat products until bulk handling and transport replaced bagged wheat. It then became more efficient to transport grain for overseas processing rather than to transport processed wheat. The form of processed exports also changes with technology. The early export of 'meat' from Australia was in the form of fat rendered from carcases. The development of canning technology created opportunities for canned meat exports, subsequently supplanted by frozen meat, itself subsequently partially replaced by chilled meat. Rather than exporting carcase meat, boxed meat allows additional services to be added domestically. But value can also be added domestically by exporting live animals rather than meat, if the exported value of the live animal is higher than the value of meat.

What is the appropriate role of government relative to firms' marketing strategies? If free markets result in optimal—or, at least, satisfactory—allocation of resources, intervention by government in firms' marketing decisions requires some evidence of market failure. The contemporary value-adding catch-cry can also be viewed as a philosophy or political program in which it seems self-evident that more processing of raw materials should be done at home. Such programs have more in common with autarchy or mercantilism than with rational economic behaviour.

Policy intention

A charitable interpretation of contemporary emphasis on value adding is that it represents an attempt to improve Australia's national income by expanding the processing of domestically produced raw materials. However, such expansion would only be likely to increase national income if there were barriers to efficient domestic processing. A less charitable interpretation of value adding in a public choice framework is that firms who see technical potential for additional processing are simply seeking government subsidies, and that government is prepared to provide these subsidies for political gain.

Area of sector

Almost any area of the agricultural and resources sector—and, indeed, non-resource processing industries including the services sector—are candidates for value adding and potential recipients of government largesse. Farmers who grow new products, processors who transform new or existing products in novel ways, and exporters who find new markets are clearly value adding—as are all the service industries who support these activities, such as institutions supplying finance and escort agencies tendering to the needs of purchasers in potential new markets.

Policy orientation

The principal focus of value-adding rhetoric has been on enhancing domestic processing of raw materials, including agricultural raw materials, for export. The main emphasis has been on determining opportunities in new foreign markets, and assisting firms to profit from these opportunities. The primary responsibility has been the Commonwealth's because of its export powers under s.51 of the Constitution. However, the Commonwealth's export powers are not exclusive and thus the states, as long as their activities are not inconsistent with those of the Commonwealth, also undertake activities to promote value adding. But taking advantage of new export opportunities often also requires a domestic focus. This domestic focus might be quite general (e.g. reducing bureaucratic barriers to exporting), or firm-specific (e.g. assisting firms to acquire or develop new technologies or processes, or hire those with new skills, to sustain export capabilities).

Policy mechanism

Programs supporting value adding are principally legislative or administrative. Legislation may be either generic (where initiatives from any industry may be supported —e.g. export enhancement schemes) or possibly even industry-specific (cf. Box 3.3). Value adding may also be supported administratively, or through research, information collection/evaluation, lobbying, or finance for these activities. New programs may be developed within existing bureaucratic infrastructure (e.g. departments of Industry Technology and Commerce, Foreign Affairs and Trade, Primary Industries and Energy at the federal level, or New South Wales Department of State Development at the state level), or specific bureaucratic organisations may be developed to deliver these services (e.g. AUSTRADE at the federal level, NSW Agsell within New South Wales Agriculture at the state level).

Institutional structure

The policy drive for value adding provides one of those happy coincidences where beneficiaries are concentrated, and losers are diffuse and difficult to identify and therefore unable to affect the policy outcome. Potential beneficiaries are the firms who can obtain additional financial assistance to support expansion of their private activities. Such firms are likely to be supported by their current or potential workforce, who have the opportunity for additional or more secure employment. The bureaucracy benefits by having additional programs to administer. Politicians benefit by having recipient firms which support their policy, programs to announce, and by having successful beneficiaries to regularly announce and promote. The winners are easily identified, and it is an obvious 'motherhood' that such programs employ 'more' workers.

Policy performance evaluation

Policy analysis of value adding has been extensively undertaken by those who are potential beneficiaries of government programs. Many agricultural and agriculture-related inquiries by the IAC/IC recommended abolishing government intervention, which would reduce processing costs and encourage value adding (cf. Industry Commission 1991b, Appendix C). Analysis by economists has tended to be less enthusiastic about prospects for value adding (e.g. Cashin 1988, Lloyd 1988, 1990, ACIL Australia P/L 1989, pp. 52–3) and some geographers have also reacted cautiously (e.g. Sorensen 1993, pp. 283–4).

Economic issues

The focus on value adding has principally been one of emphasising the opportunities for individual firms or industries to develop their export potential. Barriers to efficient domestic processing might be overcome by reductions in transactions costs.

If there were inadequate property rights in either raw materials or transformed products, then rectifying these deficiencies might lead to additional domestic processing and/or additional raw material production, which would also lead to more domestic processing. Property rights deficiencies might result from inadequate intellectual property rights, such as patents, or inadequate raw material property rights. Even if property rights were demonstrably defective, their improvement would not necessarily lead to greater domestic processing. More rapid technological advance from strengthening property rights might reduce costs more in overseas processing countries than in Australia, and improved property rights might lead to processing moving offshore rather than being concentrated within Australia. Domestic innovation would be more likely to advantage domestic processors but, in an increasingly global market, perverse outcomes cannot be ignored. Improved property rights for raw material production might increase domestic production and thus increase opportunities for domestic processing. For example improved definition of property rights for open-access natural resources such as ocean fisheries might lead to greater domestic exploitation of fish stocks. Whether or not these opportunities were taken up, however, would depend on the profitability of doing so. Even with increased domestic exploitation of natural resources, processing might well be most efficiently undertaken overseas. It is an empirical matter as to whether or not improved property rights would favour additional value-adding activities in Australia.

Particularly for firms or industries first entering exporting, or for new export markets, generic information about markets is non-rivalrous and largely non-excludable. The cultural knowledge that is required to do business in a differently structured society may be expensive to develop, but cheap to transmit to others. Even discovering the potential profitability of possible new markets is expensive, and this information is non-rivalrous and largely non-excludable. Maintaining proprietary ownership of this knowledge may be difficult because it is unlikely to be subject to intellectual property rights, and may easily 'escape' through employees being hired by rival firms. Thus, the initial knowledge base required to enter a foreign market might be a candidate for some form of government intervention. Such intervention might take the form of government providing information—analogous to government providing statistical services—or as direct government subsidy of firms entering new markets, to compensate for set-up costs.

Additionally, firms entering new markets may need to adopt different methods of bargaining suited to the different cultural context. These firms would also need to develop strategies for operating within a different legal system. These new bargaining methods and strategies will be non-rivalrous and partially non-excludable. Again, there might be some justification for government intervention.

Operating in new markets may raise the cost of ensuring payments for services rendered and commodities provided. For example, importing firms in countries with inadequate access to hard currency may prefer to barter rather than pay via financial instruments. Such risks are exacerbated when the political systems of new markets are unstable. If firms are very risk-averse and appropriate insurance is hard or expensive to obtain, and society itself is less risk-averse, there may be a role for government to assume some of the risk of operating in new markets.

Since market failure constitutes a necessary but not sufficient condition for government intervention, it may be preferable to leave firms to bear the costs of developing new territorial markets or markets in new products, rather than commit government expenditure to assisting them. This will occur, for example, when firms can keep marketing and negotiation information secret and so appropriate sufficient quasi rents from market development to make market development worthwhile, even if this information subsequently leaks to rival firms.

More workers are not necessarily employed by value-adding programs, nor do their benefits necessarily exceed their costs. If programs promoting and financing additional value adding have general equilibrium effects, it may be difficult to identify detrimentally affected firms and industries whose contraction enables the firms favoured by these programs to expand. Disadvantaged firms are unlikely to know that their reduced profitability is caused by increased taxes to finance value-adding programs or increased demands for inputs also financed by these programs. Especially where firms detrimentally affected by value adding are small and geographically dispersed, such as in agriculture, their power to limit any detrimental effects of these programs will be at best minuscule.

Analytic techniques

The potential effect of value-adding on a firm can be seen in figure 8.1. Three different technologies are shown for the production of a 'traditional' and a 'value-added' product. Production possibility curves PP1 and PP2 are production options before the firm has invested in productive capacity. In PP1 it is relatively easy to substitute between the 'traditional' and 'value-added' enterprises; in PP2, substitutability is more difficult. With production possibility curve C, substitution is relatively easy for a small additional output of the 'value added' product, but further substitution is extremely difficult.

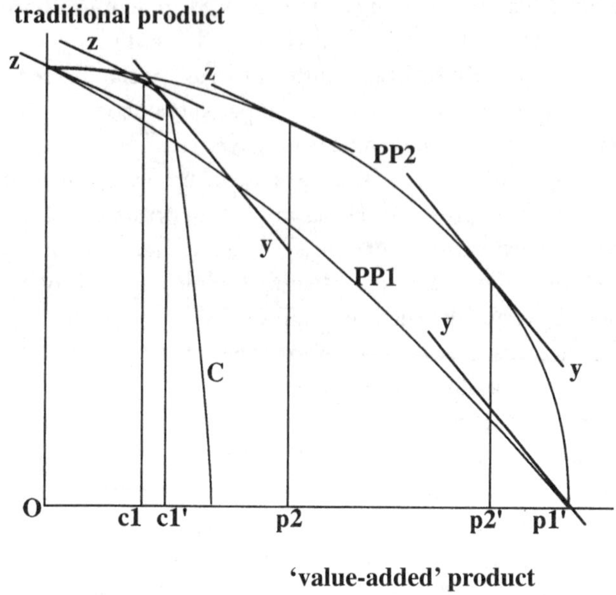

Figure 8.1 Value Adding—the firm's choices

If the traditional product is relatively valuable—represented by relative price lines z—output is biased against the 'value-added' product irrespective of the available technology: 0, c1 and p2 of the 'value-added' product are produced for production possibility curves PP1, C and PP2 respectively. However, if relative prices change to y (the 'value-added' product becomes more valuable), production of the 'value-added' product changes to p1′, c1′ and p2′ respectively (figure 8.1). With technologies where substitution is relatively easy—e.g. PP1—small changes in relative prices change output from all 'traditional' product to all 'value-added' product. Conversely, where a complete shift from one enterprise to another is either difficult or very expensive—e.g. technologies C—large shifts in relative prices may only result in a small shift from the 'traditional' to the 'value-added' product.

Changes in the technologies of 'traditional' and 'value-added' products also affect optimal 'value added' output (cf. the production possibility curve shift in figure 8.2). If technological change is faster in the 'value-added' enterprise then, even if relative prices (z) remain unchanged, optimal output of the 'value-added' product will increase faster than optimal output of the 'traditional' product (v1 to v2 cf. t1 to t2). However, there will be opportunities for technological change in the 'traditional' product, and it is an empirical matter as to whether this potential—and thus the potential for greater growth—is better in 'traditional' or 'value-added' industries. In practice, the change in optimal output of traditional and value added commodities is the result of a complex interaction between the relative price and substitutability possibilities of figure 8.1, and the future technological opportunities of figure 8.2. The desirability of further value adding has four key technical aspects:

1 how much the *relative prices* of 'traditional' and 'value-added' products are likely to change in the future;
2 the *substitution possibilities* between 'traditional' and 'value-added' enterprises within both firms and the aggregate economy;
3 the future possibilities for *technological change* in 'traditional' and 'value-added' enterprises; and
4 whether or not substitution possibilities are only available with substantial *investment*, thus requiring a dynamic analysis.

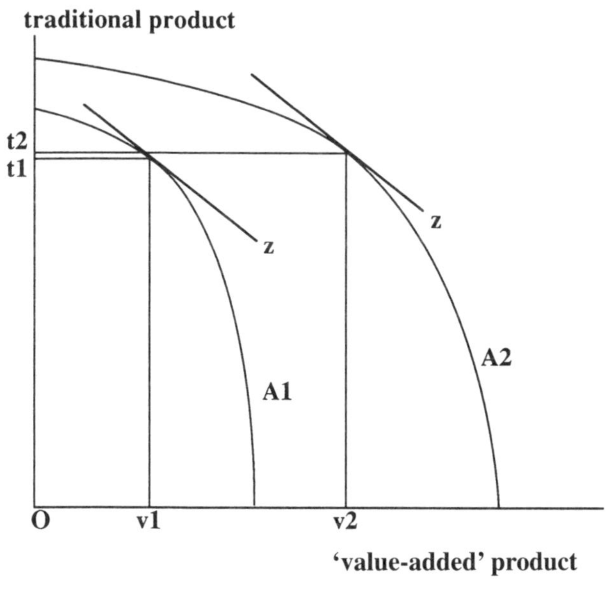

Figure 8.2 Value Adding with Technological Change—the firm's choices

If value-added production by Australian resource industries such as farming is, or becomes, more profitable than traditional products, and these new value-added products require processing, the question arises as to where this processing should be carried out. The same question also arises with respect to traditional products—should they be processed in Australia to provide more employment opportunities for Australian workers? The issues involved can be illustrated in figure 8.3 (cf. figure 6.4). First, suppose that relative costs for the Australian and foreign processing industries are as shown in figure 8.3—i.e. for technological, labour, capital, marketing or other cost reasons it is cheaper to process in the foreign country (supply curve Sf) than in Australia (supply curve Sa). Australian processing includes the cost of exporting and marketing the Australian-produced final product to the export market, and overseas processing cost includes the cost of transporting the Australian raw material to the processing country and marketing the final product. Assume the foreign demand for services is infinitely inelastic (Dfs) and the demand for the final product is Df. If the raw material is processed in Australia, demand for raw material is Da. Conversely, with overseas processing, raw material demand is Dfr and considerably more raw material will be produced in Australia than if processing were also undertaken in Australia (Rf compared to Ra). With overseas processing, producer surplus in the raw material industry would be larger than with domestic processing by an amount abcd. Conversely, however, producer surplus in the domestic processing sector would be reduced by the area xyz if product was processed overseas. It would be an empirical matter as to whether the gain in the domestic processing sector from Australian processing (area xyz) exceeded the loss in the raw material sector from domestic processing (area abcd).

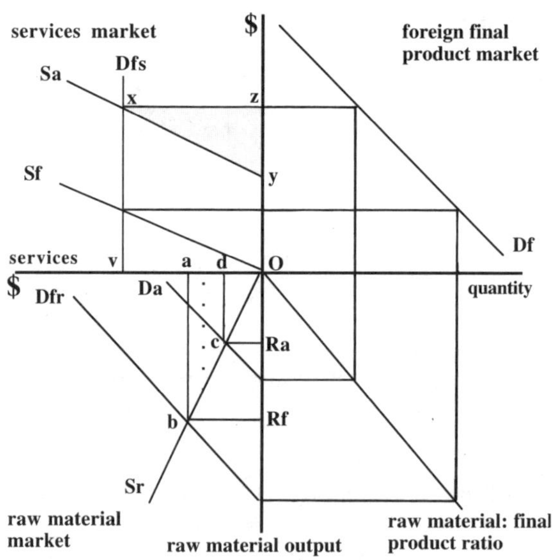

Figure 8.3 Optimal Processing Location: domestic or overseas processing

Second, suppose that initially there was no domestic processing and Rf raw material was exported. Subsequently, government policy promotes value adding, which can only be obtained with processing costs represented by Sa, and domestic raw material output falls to Ra. Suppose that producer surplus from domestic processing (area xyz) exceeds the reduction in raw material producer surplus (area abcd). Value adding appears to be efficient. But a partial equilibrium assumption underlies this conclusion —that the net change in resource cost (Ovxy – RfbcRa) is the total cost of expanding domestic processing. If, however, such expansion results in reduced producer surplus in industries which contract to provide additional resources for domestic processing, the total cost of value adding includes these producer surplus losses. Because it will be difficult to identify empirically those industries detrimentally affected by value adding, computable general equilibrium modelling would be required to estimate the economic effects of value adding policies.

If Australian processing is cheaper than overseas processing, it would be better for both raw material and processing industries if the processing were undertaken in Australia. But processing would not necessarily occur in the most efficient location. If existing processing firms have fixed capital or a large investment in a specialised labour force, the switch to superior processing locations will not occur immediately this switch becomes statically efficient. Where there are large adjustment costs of investment in physical or human capital, firms may rationally decide to depreciate their existing investment prior to switching location, or the switch may be made gradually rather than immediately. Foreign governments may also intervene in the relocation decision: should they wish to retain processing in their own countries, they may subsidise firms to stay rather than create (temporary) unemployment, irrespective of global efficiency. Where raw material processing firms are major polluters, changing government attitudes to the social costs of pollution may also influence location decisions.

Value adding has been viewed primarily in a partial equilibrium context. Where there has been an interest beyond the industry, it has usually only had a partial macroeconomic focus—i.e. improved export performance measured by balance of trade and balance of payments deficits. However, value adding inherently has *general equilibrium* dimensions, which will generally be difficult to discern and measure, particularly when the focus of expanding value-adding activities is on an individual firm. A danger of the present focus on value-adding industries is that it could become an alternative form of industry protection. Since the principal beneficiaries of these policies are likely to be processing industries, agriculture and other resource industries could be disadvantaged by an inappropriate value-adding focus in the same way they were disadvantaged by tariffs.

8.2 FOOD SAFETY[2]

Agricultural food may be contaminated in a variety of ways. Contaminants include micro-organisms, naturally occurring toxins, heavy metals, pesticide residues, herbicides, fungicides and veterinary chemicals, and food additives (National Food Authority 1992). Some individuals have intolerances to particular foods or components of food (e.g. diabetes, coeliac disease, and lactose intolerance). Food safety issues also include the nutritional balance of individual foodstuffs and the diet as a whole.

Contamination of foodstuffs may commence with agricultural inputs, and may occur at any point in the agricultural sector between farm and household consumption (cf. figure 8.4). Pre-existing contamination may be exacerbated by subsequent handling, storage, transport, processing or preparation.[3] Contaminated food is less valuable to consumers than uncontaminated food. Maintaining the uncontaminated state of food requires greater care and more inputs in the production of the raw material, and greater inputs during processing. Thus, ensuring food safety is a value-adding activity.

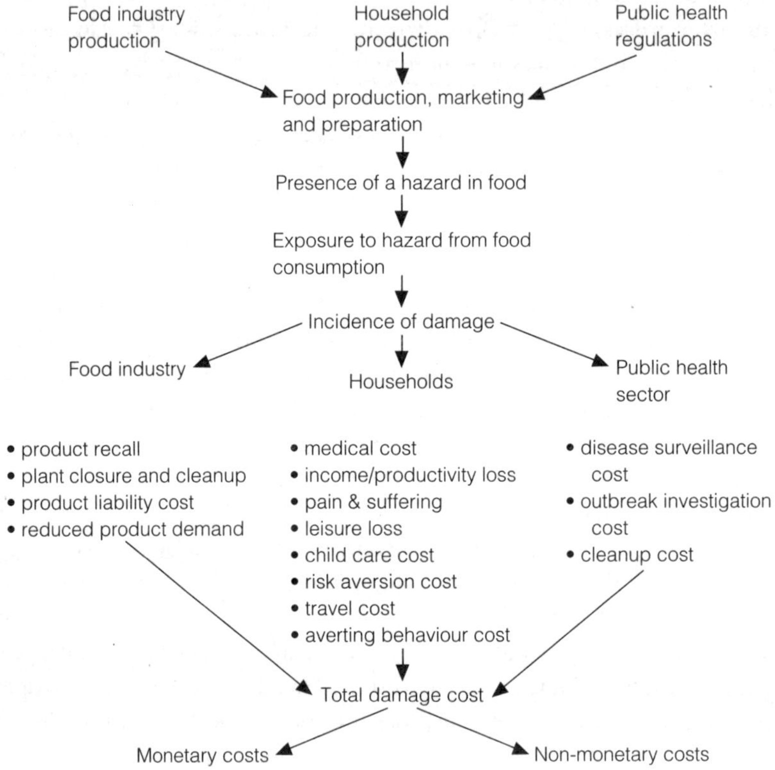

Source: Roberts and Pinner (1990)

Figure 8.4 Food Borne Disease in the Production Process

Food safety may be divided into 'risk' and 'hazard'. *Risk* is the probability of contamination occurring. *Hazard* is the severity of outcome of the contamination. Risks are not an unalterable function of the environment, but depend on procedures used in handling, storing, transporting, processing or preparing foodstuffs. Hazards partly depend on general levels of community health (healthy individuals are less susceptible to food contamination) and medical care (rapidly available medical care reduces the likelihood of severe outcomes). A simple categorisation of risks and hazards into 'low' and 'high' is shown in table 8.1. Food safety issues of low risk and low hazard (the asterisked cell) are unlikely to be of major concern *ceteris paribus*. While organisms like salmonella are common in foodstuffs they rarely cause major injury (high risk, low hazard). Conversely, while botulism is rare, its consequences are frequently fatal (low risk, high hazard). Estimates of risks and hazards of major food-borne diseases in the USA (table 8.2) are only related to hospital-based costs, and probably understate both the cost of hazard and also the differences among hazards (cf. Australian data on occurrence, table 8.3).

Table 8.1 Food Safety—risks and hazards

Risk (probability of occurrence)	*Hazard (severity of contamination)*	
	low	high
low	*	e.g. botulism
high	e.g. salmonella	e.g. heart disease

Table 8.2 Food-borne Diseases in USA—Incidence 1987–90 and Estimated Patient Costs[a]

Health effect (causative organism)	Incidence (1987–90)	Hospital costs ($/patient)[b]	Total costs ($/case)[c]
Cholera (*Vibrio cholera*)	337	4 259	5 100
Typhoid (*Salmonella typhi*)	4 778	5 015	29 700
Other salmonella (*Salmonella non-typhi*)	61 633	5 152	200
Shigellosis (*Shigella*)	21 376	3 160	1 200
Other food poisoning			
Staphylococcus (mainly *S. aureus*)	1 593		600
Botulism (*Clostridium botulinum*)	1 581		19 900
unspecified	19 894		
Amebiasis (*Entamoeba hisolytica*)	6 902	5 702	
Other protozoal intestinal diseases (mainly *Giardia lamblia*)	24 499	5 564	
Other intestinal (unspecified pathogens)	557 529	5 152	
Ill-defined intestinal infections	125 726	4 534	
Listeriosis (*Listeria monocytogenes*)	4 356	9 549	135 000
Viral hepatitis A [without coma]	49 078	6 114	
Taeniasis (tapeworms *Taenia solium* or *T. saginata*)	4 532	5 496	
Trichinosis (roundworm *Trichinella spiralis*)	525	4 328	
Unspecified gastroenteritis and colitis	2 122 757	3 709	
Mycotoxic effects		2 198	
mushrooms	1 184		
other [fish, berries, other plants]	1 607		
	Incidence (1987)[c]		
Bacillus cereus	5 000		200
Group A *Streptococcus*	500 000		350
Yersinia	3 250		450
Campylobacter jejuni and *C. coli*	2 100 000		700
Escherichia coli—enteric	50 000		1 200
Brucella	20		2 700
Clostridium perfringens	10 000		5 100

Source: Steahr & Roberts (1993) and Roberts & Foegeding (1991, table 6.9)

Note: a—in Australia, the hydatid-causing organism *Echinoccus granulosus* would need to be added to the 'worm' categories
b—hospital cost related to length of stay only
c—Roberts & Foegeding (1991, table 6.9)—1987 only; includes 'productivity' losses related to death and reduced future earnings.

Table 8.3 Comparison of Common Food-borne Diseases in Australia and USA (cases per 100 000)

	Australia (1992)	USA (1987)
Campylobacter	54.2	50.0
Salmonella	27.4	18
Hepatitis A	12.0	
Shigella	4.1	
Yersinia	3.4	
Listeria	0.3	0.7

Source: National Food Authority (1994, tables 4–5)

The preceding discussion focuses primarily on *acute* food hazards—i.e. those that are immediately life-threatening, or would be without reasonably prompt health care. Non-microbial acute food hazards also include over-indulgence in alcoholic beverages resulting in death from accidents, particularly in motor vehicles. Conversely, many food-borne hazards have *chronic* rather than acute effects. Nutritional imbalances may cause chronic effects such as poor growth or development, or increase susceptibility to disease (table 8.1). Sustained high alcohol intake may cause liver malfunction and brain damage. Many of these chronic problems are both high risk and high hazard. Some chemical contaminants also have chronic rather than acute effects.

Policy issue

The policy issue is simply the best level of food safety. A superficial response to this is that food ought to be absolutely safe. However, if food safety is a normal economic good, the marginal cost of food safety rises and/or the marginal utility of food safety falls. If these two characteristics hold, there will be an optimal level of food safety less than absolute safety. In Australia at least, food safety has been separated into the safety of individual food items, and the safety of the production processes.

Policy intention

In 1908, a conference of state premiers resolved that 'uniform legislation for the standardisation of manufactured food is desirable', and Prime Minister Alfred Deakin 'pledged to introduce uniform food laws in Australia' (National Food Authority 1994, p. 3). A Model Food Act for all Australia (NH&MRC 1985) was endorsed by health ministers and implemented in states and territories in the 1980s; Model Food Hygiene Regulations were developed in the early 1980s, but never implemented. The *National Food Authority Act 1991* gave effect to the National Food Standards Agreement of 1991 and established the National Food Authority (NFA). These arrangements were designed to:

- consolidate responsibility for domestic food standards development with a minimum number of decision-making levels;
- ensure uniformity between jurisdictions;
- establish objectives for the development of food standards;

- promote the coordination of domestic and international food standards;
- ensure an open and publicly accountable arrangement to allow input by all interested parties and provide for public hearings where appropriate; and
- retain the involvement of the states and territories in standards development and administration (NFA 1993, p. 1).

The NFA recognised that absolute food safety is unobtainable, and that 'Food safety must be viewed in terms of relative risk or degree of hazard' to be achieved by the Authority making a 'considered judgement' on the 'nature of a hazard and the extent of the risk posed by a food' (NFA 1993, p. 10). The NFA's principal objectives are to protect public health and safety; provide adequate information relating to food to enable consumers to make informed choices and to prevent fraud and deception; promote fair trading in food; promote trade and commerce in the food industry; and promote consistency between domestic and international food standards where these are at variance, providing it does not lower the Australian standard (NFA 1993, p. 9).

Because the NFA and its governing Act were established by agreement among the Commonwealth, states and territories, these policy intentions reflect those of all parties to this Agreement.

The NFA and the National Food Standards Code implement national standards for food composition and safety; labelling, advertising and packaging; permitted levels of contaminants and residues; and composition of some staple foods (NFA 1994, p. 4). However, these standards are achieved through processes regulated by separate food hygiene and cleanliness regulations; these are implemented by the states and not yet integrated nationally. The provisions of process regulation govern handling, storing, preparing and processing food for sale for human consumption; the cleanliness of utensils and articles used in connection with food production; and the cleanliness and hygiene of food handling staff (NFA 1994, table 1).

Area of sector

Food safety involves all areas of the food chain, although attitudes to and appropriate action over contaminants vary. For example, where contaminants can be graded out—e.g. stones or 'foreign' seeds in grain crops—it may be efficient to ignore contamination at the farm level and solve such problems by product cleaning. Conversely, where chemical residues are important, it may be difficult to ensure properties are innocent of contamination and infeasible to test them all for non-contamination because of the large number of relatively small dispersed firms. It may be more efficient to test product for contamination later in the food chain, and trace problems back to their source. Trace-back testing runs risks of adverse public sentiment and may thus only be efficient where contaminated product can easily be isolated from uncontaminated product (e.g. individual carcases) during processing, where processing dramatically dilutes product (e.g. through mixing individual deliveries of grain or milk), or where processing includes a de-contamination activity (e.g. pasteurisation of milk).

Policy orientation and policy mechanism

At the international level, Codex Alimentarius is a joint activity of the World Health Organization and the Food and Agriculture Organization. Codex aims to develop international food safety standards, with principal reference to trade in agricultural and food products. Its committees are responsible for food labelling, food additives and contaminants, food hygiene, pesticide residues, veterinary residues, food analysis and sampling, inspection and certification of imports and exports (anon. n.d.).

Domestic food safety is primarily the responsibility of the NFA, which has implemented a national system of food standards, and is developing standards for food hygiene processes. Other institutions involved in food safety include the Trade Practices Act which has oversight of claims about product safety, product liability and industry concentration; and the Australian Quarantine and Inspection Service (AQIS), which supervises safety aspects of the export of food and agricultural products through the Export Control Act, and similarly for food imports through the Imported Food Control Act.

Each of the states and territories has a food safety act. In New South Wales, for example, the *Food Act 1989* reflects the Model Food Act, and the *Food (Standards) Regulation 1989* implements the national food standards code. The New South Wales *Food (General) Regulation 1989* implements the National Food Standards Code, and other activities such as the inspection of premises (for other states, see Box 8.1). Through the National Food Standards Agreement of 30 July 1991, the Commonwealth, states and territories agreed that properly ratified changes to national food standards would, other than in 'exceptional circumstances', be automatically incorporated in state and territory food law. Ratification of changes requires that proposed changes be recommended by the NFA, and agreed by majority decision of the National Food Standards Council (a ministerial council). In New South Wales, for example, changes to the national food standards code as published in the *Commonwealth Gazette* are automatically incorporated in the *Food (Standards) Regulation 1989* without the normal requirement of publication in that state's own *Gazette*.

Other state legislation also affects food safety. In New South Wales, for example, the state's *Fair Trading Act 1987* regulates the supply, advertising, description, safe design and construction of goods, and provides for consumer protection and fair trading, and the *Trade Measurement Act 1987* controls weights and measures, which are of central importance in food sales. Food safety regulation in the dairy industry is undertaken by the state's Dairy Corporation which has moved to an 'auditing' model of food quality guarantee in which processors are responsible for assuring food quality and recording their activities. Records of food quality activities are then audited by the Dairy Corporation. The state's Meat Industry Authority registers meat premises and vehicles used in the meat trade.

Box 8.1 Australian Food Safety Legislation and Regulations

Australian Capital Territory	*Food Act 1992*
	Public Health (Sale of Food and Drugs) Regulations
	Public Health (Meat) Regulations
	Public Health (Eating Houses) Regulations
	Public Health (Dairy) Regulations
	Public Health (Fees) Regulations
	Public Health (Miscellaneous Amendments) Regulations
New South Wales	*Food Act 1989*
	Food (General) Regulation 1992
Northern Territory	*Food Act 1986*
	Law Reform (Miscellaneous Provisions) Act 1992
	Statute Law Revision Act 1987
	Food Standards Regulations 1988
	Public Health (Shops, Eating Houses, Boarding Houses, Hostels and Hotels) Regulations
	Food (Interim Provisions) Regulations 1986
Queensland	*Food Act 1981*
	Food Hygiene Regulations 1989
South Australia	*Food Act 1985*
	Meat Hygiene Act 1990
	Food Hygiene Regulations 1990
	Regulations (various) under the Food Act
Tasmania	*Public Health Act 1962*
	Public Health (Food Hygiene) Regulations 1977
	Public Health (Food Standards) Regulations 1986
Victoria	*Food Act 1984*
	Health Act 1958
	*Cleanliness (Foods, Drugs and Substances) Regulations 1984**
	*Health (Eating House and Food Premises) Regulations 1984**
	*Health (Food Appliances—Cadmium Plating) Regulations 1984**
	Health (Miscellaneous) Regulations 1985

Western Australia	*Health Act 1911* *Health (Food Hygiene) Regulations 1993* *Health (Meat Inspection and Branding) Regulations 1950* *Health (Adoption of Food Standards Code) Regulations 1992* *Health (Game Meat) Regulations 1992* *Meat Transport Regulations 1969*

Source: National Food Authority (1994, pp. 57–8)
Note: * lapsed due to sunset clause

Activities relating to food safety are also undertaken at the local level. In New South Wales, local government food inspectors are appointed under the *Food Act 1989* and are empowered to inspect premises, especially with respect to compliance with building ordinances.

An international trend in food safety regulation is to eliminate prescriptive requirements governing processes by which food safety should be achieved, and replace them with identification of the food safety objectives that must be achieved. These changes are currently being considered in Australia (NFA 1994). This alternative structure involves developing national food hygiene legislation incorporating the principles to which individual food businesses must adhere; developing and implementing food safety plans business; supplying appropriate training for food handlers; and developing appropriate documentation, including codes of practice and guidelines (NFA 1994, p. 19).

Institutional structure

There was a complex process for considering changes to the Food Standards Code until the late 1980s (IAC 1989a, figure 1). The *National Food Authority Act 1991* streamlined the interaction between Commonwealth and states by establishing the National Food Standards Agreement, which allows for the generally automatic implementation of Commonwealth decisions on the National Food Code. This streamlining was achieved via the creation of the NFA, and an advisory process consisting of the National Food Standards Council and the National Food Authority Advisory Committee. Many other bodies are also involved.

Several bureaucracies are concerned with chemical residue monitoring alone. The NFA undertakes the Australian Market Basket survey, which attempts to estimate the total dietary intake of particular contaminants. The National Residue Survey is undertaken by the Bureau of Resource Sciences to monitor residues of agricultural and veterinary chemicals in raw agricultural produce. The NFA has the power to regulate the maximum levels of contaminants and residues in food. Maximum Residue Limits (MRLs) for agricultural and veterinary chemicals are set by the Chemical Safety Unit of the Department of Health, Housing, Local Government and Community Services. The Chemical Safety Unit takes advice from the National Registration Authority for Agricultural and Veterinary Chemicals (Hooper 1995).

While the formation of the NFA has entailed some bureaucratic restructuring, the interest groups with a stake in food safety regulation have not disappeared. All the various groups identified in IAC (1989a, figure 1) still have interests to maintain, although the consultative and lobbying processes by which their interests are voiced may have been modified (cf. NFA 1993, chapter 10; interest groups involved in developing food standards policy are listed in NFA 1994).

Policy performance evaluation

There is little apparent evaluation of food safety issues in Australian economics literature. Agricultural policy statements note the importance of food standards regulation and its efficiency, and the intention of government to move towards uniform national standards (e.g. Commonwealth of Australia 1986, p. 55; Kerin & Cook 1988, pp. 48–52). The IAC's 1989 report into the Food Processing and Beverage Industries examined the impact of assistance policy and other regulations.

Economic issues[4]

The significance of economic issues arising from food safety is easily overlooked because regulations to mitigate their effects have been a growing part of everyday life since the mid-nineteenth century (cf. Walker & Roberts 1988, especially chapters 4 and 12). Many of these interventions have become so ingrained in the psyches of society and individuals that their original economic justification has long been forgotten. Changes to technologies and economic conditions, however, may change the optimal form and level of intervention for food safety. Current food safety practices may arise from problems with which consumers—and even regulators—in developed countries have little practical familiarity.[5] Changes to existing food safety infrastructure may therefore involve characteristics that may not be well understood.

Information transactions costs

Consumers are generally poorly placed to evaluate the safety characteristics of food purchases. Sight and smell are only rough and incomplete indicators of food safety. The complex instruments required to detect most microbial and chemical contamination, or the balance of nutrients in foods, result in significant economies of scale in food safety testing. Individual consumers will rarely, if ever, be able to justify investment in food testing equipment. Even groups of consumers—e.g. commodity and service testing 'clubs' such as the Australian Consumers Association—are rarely if ever able to test food at the batch level, which would be necessary to provide adequate information for informed decision-making by individual consumers about particular purchases. Two issues compound the testing problem. First, even food pronounced as safe after appropriate testing may be subsequently contaminated. Pronouncements on food safety are thus only conditional. Second, there may be significant delays between the ingestion of contaminated food and the onset of deleterious effects. These delays may prevent consumers 'learning by doing' through repeat purchases—or, at least, learning by making mistakes—about food quality. Consumers may only learn to associate food quality with hazards after the rapid onset of unpleasant and perhaps even irreversible effects; or there may be fatal, personal experiments which prevent 'learning by doing'.

Some consumers may be better informed than others because of better formal and informal education, which enables them to better access and better interpret food safety information. Higher incomes may allow some groups of consumers to belong to 'clubs' that provide food safety information *inter alia*.

A corollary of information transactions costs is that there will be significant non-rivalry in food safety information. Where particular goods, and especially particular batches of goods, are widely sold, information about the quality of these goods will be non-rival. Multiple testing of goods by individual consumers or consumer clubs for the same quality characteristics would be inefficient. A further corollary is asymmetry of information about food quality: sellers will generally know much more than consumers about the quality attributes of food.

Given information transactions costs, it is not surprising that governments have intervened extensively in the food safety market via information generation and dissemination, and by directly regulating production, distribution and retailing. But publicly provided information will compete with private information in the form of advertising. Especially with food safety issues concerning chronic effects such as nutrient imbalances, public health messages about links between fatty food intake, exercise and heart disease, for example, may be swamped by private advertising for 'fast' foods.

Negotiation transactions costs

Because of information asymmetries, equal bargaining between producers and consumers will be difficult unless consumers can obtain additional information. Sellers may recognise the existence of asymmetries, and seek their own remedies, such as establishing and maintaining reputation through product guarantees, brand naming, or the establishment of owned, franchised or licensed chains. These private sector responses will provide indirect information about product quality to consumers, and also encourage sellers to market products whose quality is commensurate with the reputation they wish to maintain. There is no guarantee, however, that reputation-preserving behaviour produces the optimal level of product quality.

Where there has been a breakdown in food safety, consumer redress will often pit an individual consumer against a large company well endowed with legal advice, better informed about food safety requirements, and with better access to public regulators of food safety. Without class actions for breaches of food safety, many incidents of liability may go uncompensated. Failure of compensation will have both distributional (and perhaps efficiency) effects on the affected consumer(s), and allocative effects via the incentive demonstrated by avoided liability. If producers learn that high quality legal advice, superior information about food safety requirements, and superior access to public regulators of food safety is a profitable substitute for food safety itself, there will be an under-provision of food safety and an over-investment in avoidance strategies.

Property rights transactions costs

Especially where contagious diseases are concerned, it is difficult to ensure that the disseminator of a disease is held responsible for its effects. Government intervention to minimise contagious disease has involved regulating noxious wastes, especially to prevent cross-contamination of wastes and food products; regulating contagious diseases by requiring notification of public health authorities; controlling disease entry into Australia via controls on people and products; and educating the public about communicable diseases, especially as concerns personal hygiene.

Non-deterministic problem

While the level of risk relating to food contaminants can be modified by the procedures adopted in food production, processing, distribution, storage and preparation, the particular risk associated with any given sample of food cannot be known with certainty. Thus food contamination, and its converse 'food safety', is a random variable. The risk of any particular food item being contaminated will therefore not be simply determined by production and storage processes, but will be probabilistic; similarly, the hazard of food contamination will also be a random variable.

If the risk of food contamination cannot be entirely eliminated, there will be complex trade-offs between the costs of obtaining greater food safety and the benefits of safer food. The 'optimal' trade-off between risk and costs and benefits will be complicated by deviations between observed risk and hazard, individual and social perceptions of the risks and hazards involved, and by the reactions to food safety breakdowns after they occur.[6]

For example, consumers appear to be highly concerned about the risk of chemical contamination of foodstuffs even though, in many cases, this contamination occurs at the limits of detection and the hazard is relatively low. Consumers also appear to be concerned about food safety in the producer–retailer chain, but less concerned about food safety in the home during storage and preparation. Consumers also appear to be more concerned about acute risks, even though the risk of mortality is low compared to chronic risks, such as heart disease, where both the risk and hazard are high. The latter, of course, may be explained by the long lag between exposure and hazard in the presence of positive discount rates, the lack of connection between any particular ingestion event and a particular hazard, and the swamping of public health messages by product advertising.

Where food contamination is a stochastic variable, the information asymmetry problem will be exacerbated because producers know more than consumers about their production processes, and the relationship between their activities and the risk of contamination.

Potential problems with market solutions to food safety from the consumer's perspective would lead rational consumers to demand regulatory intervention to ensure the maintenance of satisfactory levels of food safety. The potentially catastrophic personal effects of food safety breakdowns would lead rational consumers to demand a 'safe minimum standard' or 'precautionary principle', with minimum levels of food safety set very high for acute and hazardous risks. Continual development of technologies to detect food contamination will be a driving force for continual improvements in food safety requirements.

Public intervention to provide food safety might take several forms. The original English form of intervention was to vest the craft guilds with the responsibility of protecting their interests by self-regulation (O'Keefe 1968, pp. 1–11). During the nineteenth century in both Britain and Australia, direct regulation of food hygiene and food standards grew rapidly to combat serious public health problems before medical prevention or cure were available for food-borne diseases. Reliance on direct government regulation, implemented by the specification and enforcement of food safety rules, lasted until the 1980s. During the 1980s, two movements conjoined to produce an alternative response to food safety. Slow economic growth in Western democracies and the desire to reduce government intervention in the economy led to government interest in reducing all forms of direct regulatory intervention, including in food safety. Emphasis on the individual producer's responsibility for product quality through the 'quality assurance' and 'total quality management' movements, food safety plans, and the specification of international standards on quality assurance provided the means by which government's role in food safety could be reoriented, and probably reduced (NFA 1994). If the same level of food safety could be obtained at a lower cost through less government intervention, economic efficiency would be enhanced.

Analytic techniques

Food safety as 'value adding'

'Safety' is only one attribute of food. In figure 8.5, the curve AB is the 'production possibilities frontier' for the 'safety' and 'non-safety' attributes of the foodstuff. Food safety is the probability of non-contamination (a function of both risk and hazard). Food could be provided at level of safety A and no other attributes, or B level of other attributes and zero safety, or combinations of both. Consumers are indifferent between those combinations of food safety and non-safety attributes shown by the utility curve U. Given the production possibilities and consumer preferences for food safety indicated in figure 8.5, the consumer would choose to purchase this product with a0 units of food safety and b0 units of other attributes.[7] Consumers' attitudes to food safety are likely to be a function of other variables; for example, poor consumers may demand less food safety than wealthy consumers while the income elasticity of demand for food safety is probably high.

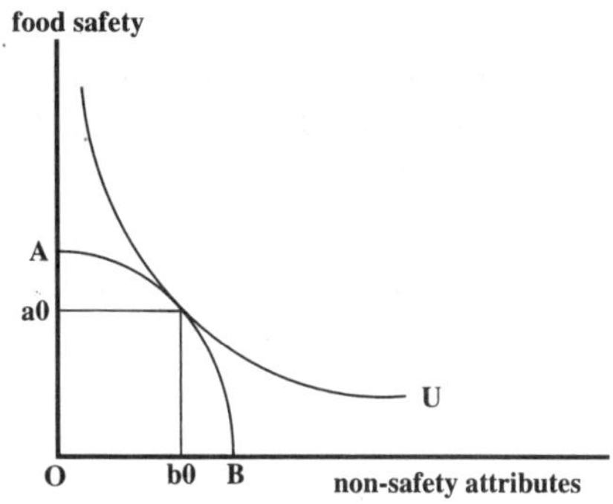

Figure 8.5 Safety—Other Attributes Trade-off

Food safety elements of production costs—incorporated in 'quality costs'—are summarised in figure 8.6. These costs have several dimensions: producing food safety ('preventive' costs); monitoring the level of safety produced ('appraisal' costs); and breakdowns in food safety production ('failure' costs).

- Total Production Costs
  - Process operating costs
  - Quality costs
    - Operating
      - Preventive
        - quality planning
        - quality auditing
        - subcontractor quality
        - process control
        - R&D quality improvement
        - R&D control equipment
        - quality training
        - quality insurance
      - Appraisal
        - design appraisal
        - receiving inspection
        - measuring equipment
        - inspection & test
        - materials
        - monitoring
        - approvals & endorsement
        - field testing, commissioning
        - stores evaluation
      - Failure
        - Internal
          - scrap
          - replacement, rework, repair
          - failure diagnosis
          - test/inspection
          - downtime
          - downgrading
          - administrative review
        - External
          - customer complaints
          - warranty claims
          - returned products
          - concessions
          - loss of sales
          - recall costs
          - product liability costs
    - Capital
      - depreciation
      - interest
      - opportunity

Source: Brandolini (1994) (after Taylor 1989)

Figure 8.6 Total Costs of Food Production

Ideas about the production and consumption of food safety can be combined in figure 8.7. For simplicity, the supply of the raw material (Sr in the right-hand quadrant) is assumed to be infinitely elastic at price pr. The left-hand quadrant shows the demand for food safety attributes per unit of final consumption. The supply curve Ss shows the social marginal cost of an additional unit of food safety. The curves Ds1, Ds2 and Ds3 show different consumers' demands for food safety attributes—group 1 has a relatively low valuation of food safety and group 3 has a relatively high valuation of food safety. A low quality final product could be supplied as Ss1, a medium quality final product as Ss2, and a high quality final product as Ss3. If the aggregate demand for food of a specified safety standard were related to group size, and group size were inversely related to income, the demands for each of the food groups might be D1, D2 and D3 respectively. Amounts q1, q2 and q3 of the low, medium and high quality products would be produced and consumed respectively.

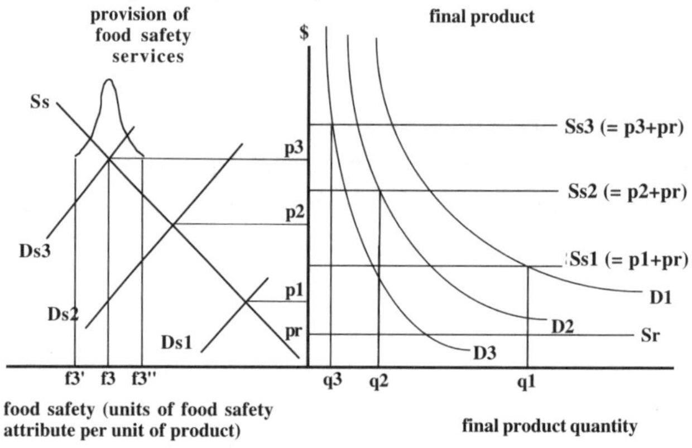

Figure 8.7 Markets for Food Safety

Stochastic food safety

The food safety problem is not simply deterministic. Consider a commodity reflecting a high level of demand for food safety (i.e. with demand Ds3 in the left-hand quadrant of figure 8.7). The level of food safety services embodied in each unit of this product is ostensibly f3. While processors, wholesalers, retailers and consumers intend the product to have a food safety level corresponding to f3, this level cannot be guaranteed—food safety is a random variable, with some distribution about the mean level f3 (cf. Swinbank 1993, p. 88). The actual level of food safety embodied in any particular sample of the product may be as high as f3′ or as low as f3″. Firms providing processing services which affect the level of food safety will therefore not only be concerned with the mean level of safety conferred by the processing, but also other characteristics such as the variance of safety, and perhaps even its skewness.

A simple stochastic model of food safety is illustrated in figure 8.8. Consumers prefer higher levels of food safety and lower levels of variability, giving rise to a family of utility curves. At low levels of expected food safety, the variability (measured by its standard deviation) is likely to be high. As increasing resources are devoted to food safety, its variability is likely to fall at the same time as the expected level of safety rises. Eventually, however, decreasing variability of food safety may only be achieved by concentrating resources on reducing variability rather than increasing the level of safety. These tradeoff characteristics give a 'risk frontier' (cf. figure 8.8). The optimal level of food safety is given by e0 expected food safety, and s0 variability of food safety. If there is a monotonic tradeoff between variability of food safety and its expected level—i.e. the risk frontier of figure 8.8 is everywhere downward sloping—and the consumer's utility function is as shown in this figure, the optimal level of food safety occurs at the maximum level attainable and at zero variability.

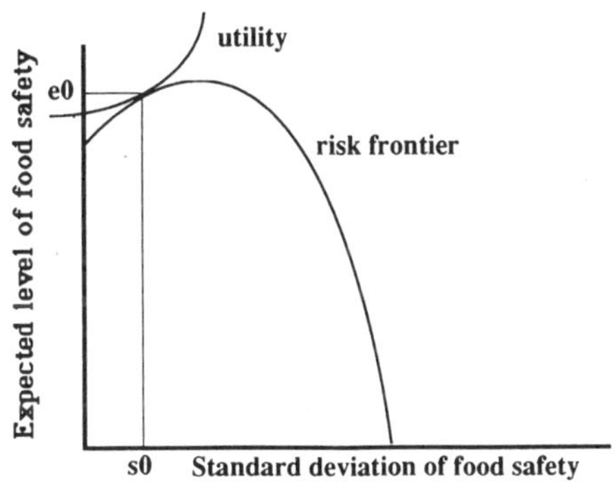

Figure 8.8 Simple Stochastic Model of Food Safety

Clean and green

The above analysis implies extensive inputs to ensure the production of safe, healthy, and presentable food. Extensive inputs are required to produce food at the farm level, including a wide variety of chemicals to promote plant growth and to protect against scourges (diseases, pests and weeds). Extensive inputs are subsequently required to detect produce that has been overdosed with chemicals and to either eliminate the contaminants or divert the contaminated produce into non-food uses. An alternative strategy is to reduce or eliminate the use of potential chemical contaminants in both farm and processing, and thus eliminate the need to detect and eradicate them. Reducing or eliminating chemicals in food production characterises the organic or 'green food' movement. If chemicals are not used to protect against pests or diseases then—to the extent that pests or diseases affect foodstuffs—other means of protection would be required, or higher levels of non-chemical contamination or damage would need to be accepted. Other means of protecting against contamination could include shorter storage and shelf life, and thus higher stock turnover and faster throughput from farm to consumer, or the use of non-contaminating chemicals (e.g. use of carbon dioxide in stored grain to prevent pest growth).

Rationalising food safety intervention

The effect of rationalising food safety regulation in Australia through the adoption of the National Food Code can be represented schematically in figure 8.9, with the simplifying assumption of food safety in a deterministic framework and a required level of food safety given by Ds. Before the mid-1980s, each state and territory had its own unique food safety legislation. National food producers wishing to sell their products nationally were faced with the problem of either manufacturing products consistent with all separate legislations, or marketing separate products tailored to the food safety requirements of a particular market (cf. similar problems with packaging and labelling, IAC 1989a). Whichever option they chose, manufacturers were faced with higher costs than if they only had to satisfy a single national food safety standard (food safety supply curves Ss1 and Ss2 respectively in figure 8.9). With higher food safety costs, the combination of raw material supply and a cost for food safety services which was invariant to the level of food production would result in a final product supply curve like Sf1. By harmonising food safety requirements among the states and territories, represented by a fall in the industry's costs of food safety compliance to Ss2, the marginal cost of food safety at equilibrium would fall from c1 to c2, and the supply curve of the final product would fall from Sf1 to Sf2.

A similar analysis could be presented concerning direct government regulation versus establishing standards implemented via industry self-regulation and government auditing. Should this approach lead to a similar lowering of the costs of obtaining food safety (e.g. from Ss1 to Ss2 in figure 8.9), a similar reduction in the final costs of retail food could be anticipated (Sf1 to Sf2). Since government's own costs of direct regulatory supervision of food production and marketing would need including, the food safety supply curves Ss1 and Ss2 would have to incorporate these. Again, the actual reduction in the costs of supplying food safety, resulting from modifying the form of food safety regulation, requires empirical analysis.

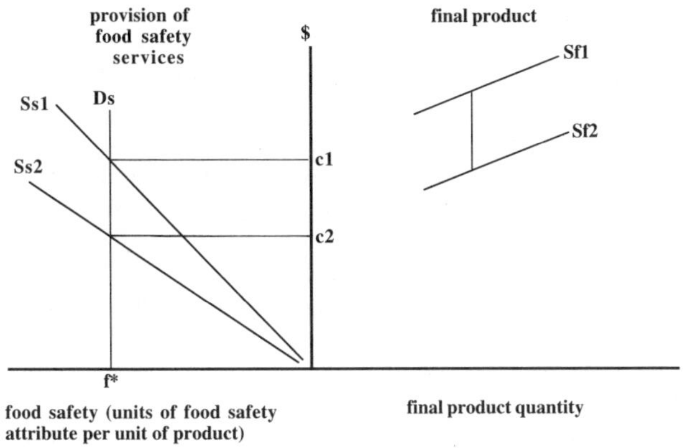

Figure 8.9 Effect of regulatory changes in markets for food safety

ISSUES FOR DISCUSSION

1 Comment on the following argument from an Australian politician:
'Forests are not logged for woodchips. They are logged for timber for houses, furniture and many other uses. Woodchips are a by-product processed from unmillable timber (hollow logs and off-cuts) that would otherwise rot on the forest floor or be burned, adding to greenhouse problems.

Woodchips are worth $80 per cubic metre, whereas unseasoned structural hardwood is worth $400 a cubic metre, kiln dried hardwood beams $800 a cubic metre and kiln dried appearance timber between $1000 and $2500 per cubic metre.

There is therefore no incentive or reason for millers to want to depend on woodchip production. Quite the opposite.'

2 Is agriculture a 'sunrise' or a 'sunset' industry? Why?

3 There have been major breakdowns of food safety (e.g. Australia: pesticide contamination of cattle feed in 1994, HUS in 1994–95; UK: mad cow disease in 1996) and quarantine (e.g. Australia: papaya fruit fly and rabbit calicivirus in 1995). What can or should be done to ensure these episodes are never repeated?

4 Comment on Swinbank's (1993, p. 88) argument that:
'. . . for the supplier, in most instances, a less safe food product is not intentionally supplied. The problem often lies in accidental contamination in the production process, not the conscious supply of a 'less safe' product; and thus the problem the consumer might face is judging not the intentions of the supplier but his [sic] quality assurance.'

Endnotes

1 Synonyms for value adding include 'downstream processing', 'forward linkages' and 'elaborately transformed manufactures'. Value-adding industries are often denoted 'sunrise', in contrast to supposed 'sunset' industries with which agriculture is often erroneously associated.

2 Thanks to Jim Johnston and Bill Porter for assistance with this section.

3 O'Keefe (1968, pp. 1–11) presented a history of food contamination and adulteration in England from the thirteenth century; Walker and Roberts (1988) analysed similar Australian issues since 1788.

4 cf. Swinbank (1993), Henson and Traill (1993), Kinsey (1993).

5 cf. analogy with falling vaccination rates in Australia for 'childhood' communicable diseases because previous vaccination campaigns have dramatically reduced the incidence of these diseases.

6 e.g. the rapid drop in Australian consumption of smallgoods following deaths from contaminated mettwurst in the summer of 1994–95; reaction to this event was compounded by the fact that children were most vulnerable. See also effects on beef consumption throughout Europe following the 'mad cow disease' (bovine spongiform encephalopathy) controversy over British beef production in March–April 1996.

7 This analysis assumes that, if the level of food safety a0 is chosen—e.g. the conjugate probability of food contamination risk and hazard is some value P—then this desired or *ex ante* value is actually delivered by the production system and is observed *ex post*. If, however, P is a random variable, then some model of decision making under uncertainty will be required (e.g. Choi & Jensen 1991).

Part III

Intrasectoral Issues: Structural Change

While having a direct impact on the profitability of agriculture, R&D is also one of the dynamic forces that are continually reshaping the agricultural sector. These forces of constant reconstruction, and their effects on the structure of agriculture, form the theme of Part III. There has previously been over-emphasis on the structure of *farming*—the size, capital intensity, enterprise selection, and ownership forms of farms. But, as outlined in the Introduction, the agricultural sector is not just its farms but also their relationships to the natural environment and firms producing farm inputs, as well as the sectors marketing agricultural produce (cf. figure 1). Thus, *structural change* in agriculture comprises changes in the structure of farming, together with changes in relationships between farming and the rest of the economy.

Structural adjustment in farming is evaluated in Chapter 9. Structural adjustment in the rural sector as it involves Australian Aborigines is evaluated in Chapter 10. In Chapter 11, there is a discussion of the relationships between structural change in farming and structural change in the surrounding rural economy.

9 Structural adjustment: European farming

Like other economic sectors, agriculture has been subject to profound changes. The number of farms has fallen, as has the total farm workforce and employment of labour; average farm size has risen, as has female employment, capital invested in agriculture, and the ratio of outputs to inputs (cf. table 9.1). Some of these changes have occurred as activities previously performed on-farm, such as breeding and caring for livestock as the primary power source, have been replaced by non-agricultural industries producing and maintaining tractors of increasing technical complexity—i.e. as agriculture has 'industrialised'.[1]

Table 9.1 Structure of Australian Farming, 1950–90

	1951–52	*1961–62*	*1971–72*	*1981–82*	*1991–92*
Farms ('000) (a)	203	202	188	174	124
Total area of farms (m ha)	441	475	500	491	466
Workforce ('000)—total	477	453	407	380	374
— employers/self employed			240	236	221
— unpaid family			23	12	23
— wage & salary			145	132	130
Female workforce	31	42	42(b)		
Capital stock (excl. unimproved land value)	41 473	57 888	92 597	99 122	89 872
Output volume (1989–90 = 100)	35	52	74	85	97
Gross output value (1987–88 =100) (c)	80	85	90	100	85
Net output value (1987–88 =100) (c)	222	168	149	109	39

Sources: ABARE (1994), Powell & Milham (1990)

Notes: a: several changes in the definition of 'farm' from the mid-1980s reduced farms counted
b: 1970–71
c: gross and net output value respectively, divided by Consumer Price Index

Policy issue

Policy issues concerning structural adjustment in farming involve both distribution and efficiency. Efficiency aspects concern whether or not there are market failures or imperfections in the adjustment process that could lead to an inefficient rate of adjustment, or an eventual inefficient disposition of resources. Distributional questions are also important: is the rate of farm adjustment 'too much' for individual producers or the agricultural sector to bear? Is the 'end-point' of structural adjustment socially desirable?

The issue of farm adjustment to long-run economic pressures for change is often confounded with short-run pressures from commodity price downturns and 'natural' disasters, especially drought. Ideally, Australian farmers should recognise the high variability of their physical and economic environments, and plan accordingly. If the valuation of the fixed land asset reflects these variabilities, and investment also reflects variability and long-term profitability, then the farming sector—and, indeed, dependent up- and down-stream sectors—would be financially structured to cope with farming variability. In some situations, however, such planning is exceptionally difficult. New entrant farmers with limited capital reserves are extremely vulnerable to adverse physical or commodity conditions until they establish sufficient reserves. New entrants who immediately face a run of good seasons and prices are more likely to survive than those who do not. Similarly, as the experience of the eastern Australian drought of 1991–95 has shown, farmers who are debilitated by one form of adverse conditions (e.g. falling wheat prices during the second half of the 1980s, and the wool price crash of 1990) are poorly placed to endure subsequent adversities.

Policy intention

Australian governments have responded to perceived problems of farm adjustment by evolving measures to facilitate on-farm adjustment to economic change. Kerin and Cook (1988, p. 61) noted that 'The Rural Adjustment Scheme (RAS) has been established with the objective of assisting rural industry structural adjustment, easing adjustment pressures (including prevention of a loss of potential long-term viable farmers from the rural sector) and assisting farmers to leave the industry if they are no longer viable.'

The Senate Rural and Regional Affairs and Transport References Committee (1994, pp. 1–8) briefly summarised the 1992 rural reconstruction scheme, and earlier schemes.[2] Reflecting developments in the deregulation of the Australian economy in the 1980s, the objectives of the 1992 scheme were 'to foster the development of a more profitable farm sector that is able to operate competitively in a deregulated financial and market environment, and to improve the competitiveness of the farm sector in a sustainable manner' (Senate Rural and Regional Affairs and Transport References Committee 1994, p. 5).

The 1992 Scheme was designed to provide support through grants to subsidise interest payable on loans and associated costs of loans, whether the loans are provided by a state or by another entity; grants for farm training, planning, appraisal, support services and rural adjustment research; or grants to assist farmers who do not have prospects of sustainable long-term profitability to leave the farm sector (Senate Rural and Regional Affairs and Transport References Committee 1994, p. 5).

Through the 1980s and early 1990s, government confronted two particular problems of adjustment assistance: the relationship between support for long-run adjustment in agriculture as opposed to support during intermittent crises; and the separation of adjustment assistance ostensibly for efficiency reasons from the more general provision of temporary social security support for farmers.

Assistance to farmers suffering from drought was administered by the federal and state governments from 1971 to 1989 under Natural Disaster Relief Arrangements. Arising from the Drought Policy Review Task Force (1990), the Federal Government established a National Drought Policy in 1992, whose objectives were to encourage primary producers and other sections of rural Australia to adopt self-reliant approaches to managing for climatic variability; to maintain and protect Australia's natural and environmental resource base during periods of extreme climatic stress; and to ensure early recovery of agricultural and rural industries, consistent with long-term sustainable levels (Lembit & Kingma 1995, p. 2).

The 1992 Rural Adjustment Scheme also included a provision for declaring 'exceptional circumstances' under which interest rate subsidies may be granted. Four exceptional circumstances were declared up to December 1994: the fall in wool prices, unseasonally heavy rain in eastern Australia, severe drought in Queensland, and severe drought in New South Wales (Senate Rural and Regional Affairs and Transport References Committee 1994, pp. 6–7). Additional measures were implemented in September 1994 (Senate Rural and Regional Affairs and Transport References Committee 1994, pp. 56–7), and again in December 1994 (Keating & Collins 1994). Previous policies to protect farmers during periods of severe stress have included one-off wool deficiency payments (1969–70) and underwriting arrangements in the wheat marketing scheme (see Chapter 7).

From a social security perspective, there are considerable difficulties in providing 'safety net' income support for farming. Traditional income support for employed workers uses simple criteria like 'does the worker have paid employment?' and 'what are the individual's assets?' in order to test whether or not the individual is eligible for income support. Additional tests may also be implemented, such as 'is the individual actively seeking work?' But farmers are self-employed and highly immobile, have high inter-annual income variability, and are often 'asset rich and income poor'. Traditional tests of eligibility for income support are difficult to apply: 'The provisions of the Social Security Act are not intended to provide income support to self-employed people, whether primary producers or other small business people, while they are still running their businesses. To do so would distort the economics of farm and business operations' (Commonwealth of Australia 1986, p. 78).

Constraints affecting farmers' receipt of temporary income relief via social security policy have been slowly relaxed, and this process was accelerated during 1994 in response to the continuation of drought in eastern Australia (Senate Rural and Regional Affairs and Transport References Committee 1994, pp. 56–7). The

Department of Social Security provides financial support for farm households, job search allowance and family payments (Senate Rural and Regional Affairs and Transport References Committee 1994, pp. 7–8).

Some efficiency and distributional aspects of this policy are difficult to disentangle. Social security payments are normally considered as distributional policy because they primarily involve transfer payments. However, since social welfare payments maintain human life and dignity for individuals who may re-enter the workforce, they clearly have important efficiency implications. Conversely, while payments to assist adjustment may improve efficient resource use over time, such payments may also have important distributional effects.

Area of agricultural sector

Adjustment assistance primarily targets farm businesses and farm households. However, changing the 'natural' rate of farm adjustment affects the rate of adjustment in the demand for inputs and the supply of outputs. Policies that succeed in affecting the rate of adjustment thus have potential impacts on the rate of adjustment in industries servicing agriculture. Similarly, maintaining cash flow to the farm household for living expenses during adversity may moderate the degree of stress suffered by rural firms supplying basic necessities to farm families.

A key long-run characteristic of farm adjustment has been the substitution of land and farm machinery for labour. If agricultural adjustment policies succeed in accelerating the rate of farm adjustment, the displacement of labour may be more rapid than it would otherwise have been. While some attention has been given to assisting the exit of farm families from the industry, relatively less attention has been paid to the exit of employed labour. Adjustment assistance may thus have assisted in mitigating the effects of farm poverty, but has done little for the poverty of employed farm workers (cf. Musgrave 1982, pp. 306–7).

Policy orientation

The principal orientation of adjustment policy is domestic. While the Commonwealth has no obvious constitutional role in this area, since the 1960s it has assumed an increasing and now major coordinating and funding role for farm adjustment. Although adjustment policy can be given an economic rationale, policy appears primarily oriented to distributional concerns. The Commonwealth appears to be especially concerned that farmers in all states have similar access to assistance. The states have been willing to accept Commonwealth involvement because it reduces claims that would be made on their finances if each state addressed adjustment policy individually.

Policy mechanism

The principal policy mechanism is the Federal Government's *Rural Adjustment Act 1992*. This Act, the outcome of an agreement between the Commonwealth and states, runs from January 1993 to December 2000. The Commonwealth determines the purposes and objectives of the scheme, and specifies outcomes, and the states and territories are advised of policy directives and guidelines by the Commonwealth. The Commonwealth funds 90 per cent of the scheme's costs, except for 'exceptional circumstances' (Senate Rural and Regional Affairs and Transport References Committee 1994, pp. 9–11).

The Rural Adjustment Scheme Advisory Council, established under the Act, comprises one representative from the Commonwealth, one on behalf of the states, and a representative from the National Farmers' Federation. Four other representatives are appointed for their special expertise in economics, financial administration, banking, sustainable agriculture, farm management or training. The council recommends an ongoing three-year plan to the Minister; proposes an annual program and budget; provides advice to the Minister; and has a particular role with respect to exceptional circumstances and possible changes to the Rural Adjustment Scheme in response to these circumstances (Senate Rural and Regional Affairs and Transport References Committee 1994, p. 8).

The Rural Adjustment Scheme (RAS) is administered by the states and territories—for example, in Victoria by the Rural Finance Corporation—where individual farmers' applications for assistance are considered and decisions are made. Considerable executive discretion may be exercised by state rural adjustment authorities within the guidelines. Because of this, the Senate Rural and Regional Affairs and Transport References Committee (1994, pp. 21–78) has evaluated several concerns, principally about different interpretations of the guidelines generally and particularly with respect to 'exceptional circumstances', and different approval rates among the states.[3]

Institutional structure

From the perspective of the supporting governments, the RAS ought to be one from which they can only gain benefits. It is a small program (less than $200 million per year in the early 1990s), it has no large opportunity costs for the government, and the beneficiaries might be expected to be grateful for financial assistance in a form that is not generally available to any other industry. The states are generally supportive of the 1992 scheme—as well they might be since they agreed to it initially and the Commonwealth provides nearly all the funding. Their principal criticisms related to the timing of the scheme's introduction, following low prices for some major commodities in the late 1980s and early 1990s and the beginning of the severe drought in eastern Australia during the early 1990s, and some administrative matters (Senate Rural and Regional Affairs and Transport References Committee 1994, pp. 13–14, 33–6).

Beneficiaries, however, appear to be generally extremely critical of the scheme—at least judging by evidence presented to the Senate Rural and Regional Affairs and Transport References Committee (1994, pp. 13–16, 21–78). The beneficiaries' principal concerns were that:

- the scheme was not applied uniformly across Australia;
- Commonwealth funding was too low and falling;
- farmers whom it was intended to assist were not obtaining assistance;
- much of the assistance was related to additional debt where the principal problem was high debt levels exacerbated by high interest rates in the late 1980s and early 1990s;
- there were inequities in obtaining assistance to leave farming;
- exceptional circumstances provisions were too difficult to access and not available equitably;
- farm household support was a total failure; and
- states were not doing a sufficiently good job of promoting and administering the scheme.

Policy performance evaluation

Apart from the recent report of the Senate Rural and Regional Affairs and Transport References Committee (1994), there have been relatively few official public evaluations of adjustment assistance. The Industries Assistance Commission (1984) evaluated rural adjustment, concluding that it doubted whether efficiency gains to the community outweighed the costs of the scheme. The principal justification for having a rural adjustment scheme was that it discouraged governments from implementing even less-appropriate policy responses when confronted by relentless demands for relief during intermittent crises or in response to longer run problems (cf. Lloyd 1986, p. 63). In evidence to the Senate Rural and Regional Affairs and Transport References Committee (1994, p. 17), Musgrave asserted 'that RAS is "perhaps the least of a range of policy evils"' (cf. Musgrave 1982, 1990). Lloyd (1986, chapter 6) provided a detailed analysis of the RAS with particular reference to its administration in Victoria and the role of the Rural Finance Commission.

Economic issues and analytic techniques

In the 1950s and 1960s, agricultural economists expended considerable efforts analysing the reasons for farm structural change. Their analysis remains relevant to issues of farm structural change and adjustment (cf. Longmire 1995), but agricultural economists have largely ignored structural adjustments in the relationship between agriculture and the rest of the economy. This analysis has recently been taken over by geographers (e.g. Sorensen & Epps 1993) and regional economists. There is, however, a major role for economists to play in continuing to explain and develop further understanding of the relationships between farm structural changes and changes in agriculture's structural relationships with the rest of the economy.

General equilibrium analysis

The 'farm problem' that has been used to justify the need for adjustment assistance may be illustrated by the following simple general equilibrium analysis. Consider a simple two-input, two-output closed economy which is transiting from an undeveloped to a developed state (figure 9.1). The economies being compared are, say, the 'undeveloped' Australian economy of 1860, and the 'developed' Australian economy of the 1990s. The curves labelled U and D are production possibilities frontiers for this economy in its undeveloped and developed states respectively, given input endowments and available production technologies for producing wheat and machines. Curves SWu and SWd show the highest levels of social welfare (for simplicity, defined directly over wheat and machines) that are attainable in the undeveloped and developed states respectively. In an undeveloped society, preferences are highest for the means of subsistence (i.e. wheat) and there is a relatively lower demand for machines; hence the SW curves are closer to the wheat axis. In the undeveloped state (U), Wu wheat is produced—and consumed, because a closed economy is assumed—and Mu machines are produced and consumed.

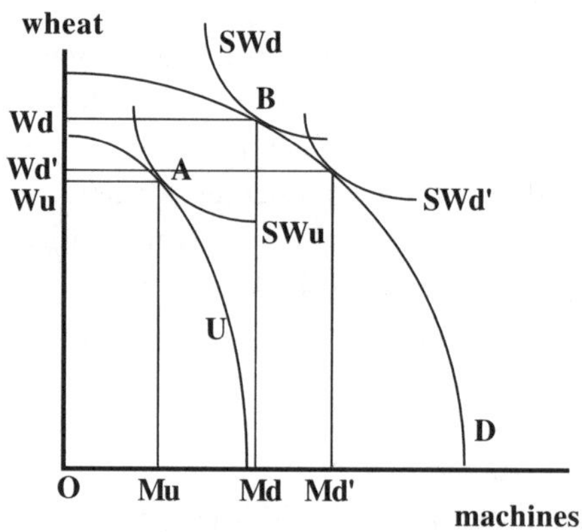

Figure 9.1 Simple Economy in Transition

The principal characteristic of economic development is that production possibilities expand: thus, with development, production opportunities expand from U to D through capital accumulation and technological change. Production opportunities for non-basic goods (e.g. machines) expand faster than for basic commodities (e.g. wheat). Thus, the proportional expansion of the production possibilities curve from U to D along the machines axis is greater than the corresponding expansion along the wheat axis. Even if relative preferences for wheat and machines remained unchanged—i.e. SWd is a proportional expansion of SWu, the growth in the production and consumption of machines (Md/Mu) is greater than the corresponding growth in the production and consumption of wheat (Wd/Wu).

As the economy develops, relative consumer tastes also change, and the growth in the demand for non-basic commodities like machines is likely to exceed the growth in the demand for food. Thus, in the developed economy, social preferences for wheat and machines in the developed economy may be SWd' rather than SWd. In this case, production and consumption of wheat has grown relatively little compared to the undeveloped economy (Wd′/Wu), while machine production and consumption has grown considerably (Md′/Mu).

Changes also occur on the input side (figure 9.2). The isoquant U shows the small aggregate output possible with given quantities of labour and capital in the undeveloped economy. In the undeveloped state, the relative price of labour to capital is represented by line u (labour is cheap and capital is expensive) determined by the quantities of capital and labour available (Lu and Ku respectively), and e is the optimal point. As the economy develops, the relative price of capital declines (now given by line d). Without any other changes, the new optimum would be at point b, utilising inputs Lu′ (and Lu′– Lu labour would be unemployed) and Ku′. However, economic growth is accompanied by growth in the capital stock (from Ku to Kd in figure 9.2) and so the optimum for the developed economy is at c on isoquant D, employing the Kd capital and Ld labour. This analysis could be extended to allow for technological change which would affect the location of isoquants (e.g. the comparable level of output D could be produced with less labour and/or capital than Ld, Kd), and the shape of the isoquant might be biased towards the use of capital.

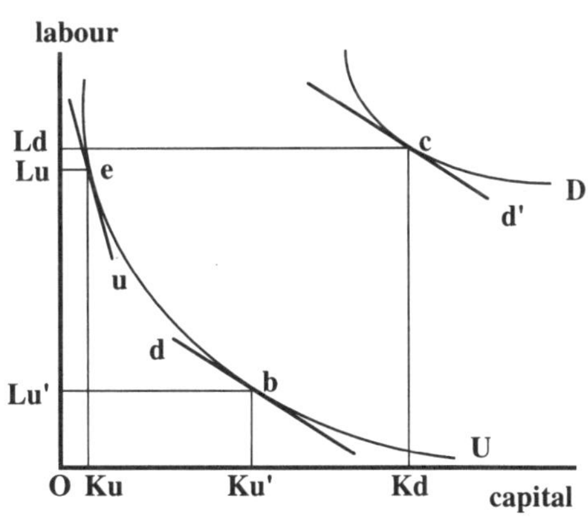

Figure 9.2 An Illustration of Economic Development in Production

The changes that occur in the aggregate economy will also occur in each of its constituent sectors unless there are economic barriers to adjustment, or unless policy barriers are erected to slow or prevent the effects of change being transmitted into particular sectors. Thus, for example, relatively slow growth in the consumption of basic commodities compared to faster growth for non-basic commodities will affect output and profitability in the basic commodities sector unless it is insulated by policies such as subsidies to produce these commodities (e.g. through the Common Agricultural Policy of the European Union). Similarly, changes in relative input prices will affect all sectors unless labour force policies insulate some sectors.

This analysis can be extended to an open economy to allow for international trade (figure 9.3). With a *closed* economy and production possibilities frontier of U in the undeveloped state, and with social welfare SWu1 defined over wheat and machines, the economy would optimally produce c wheat and p machines. With international trade possibilities represented by the international relative price line z, economic activity is optimised by producing (c′,p′), and selling (c′–b) wheat on the international market and buying (q–p′) machines to attain social welfare SWu2.

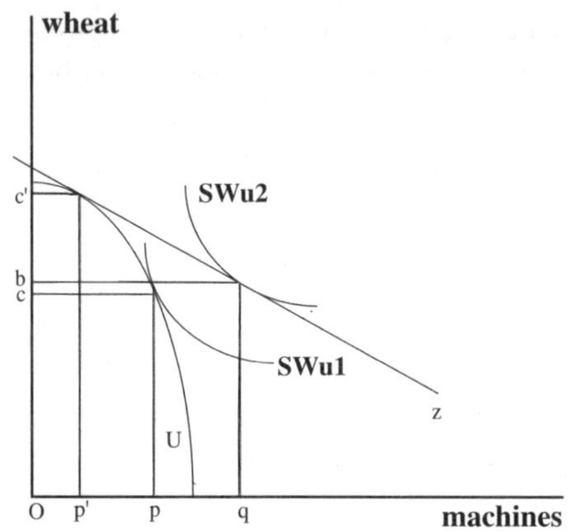

Figure 9.3 Undeveloped Open Economy

Suppose development in this economy is represented by an outwards shift in aggregate production possibilities from U to D in figure 9.4a. Suppose also that international prices remain unchanged (z′ is parallel to z) and social welfare SWd is a proportional shift from SWu2. In the developed economy, production of d′ wheat and e′ machines is optimal, and (d′–d) wheat is exported to purchase (e–e′) machines. How the economy actually changes as development occurs will depend on the nature of the shift of the production possibilities curve D, whether or not welfare preferences change with development (e.g. favouring machine consumption), and whether or not relative international prices for wheat and machines change as this development occurs. In figure 9.4a, development results in higher domestic production of both wheat and machines, and a higher consumption of both after the wheat-for-machines trade. Conversely, in figure 9.4b, growth favours the wheat industry—for example, in Australia, over many decades, the high commitment to agricultural R&D and the relatively poor technological performance of a largely protected and domestically oriented manufacturing industry biased growth in favour of agriculture. In this case, production of machines changes little (from p′ to e′), whereas there is a large growth in wheat output (from c′ to d′) (figure 9.4b). The final welfare solution results in extensive wheat exports to purchase machine imports.

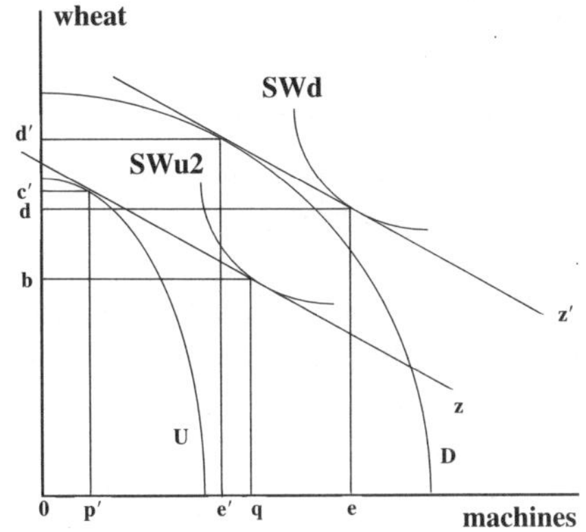

Figures 9.4a Machine-biased Development in an Open Economy

Other variations could also be examined, such as a combination of wheat-biased growth and a fall in the price of wheat relative to machines (figure 9.4c). While wheat output still expands, and wheat is still exported to purchase machines, less wheat is produced than if wheat prices had remained higher. The lower output of wheat, coupled with the lower international wheat price, means that fewer machines can be imported and hence social welfare is lower than if the wheat price had not fallen (cf. figures 9.4b and c).

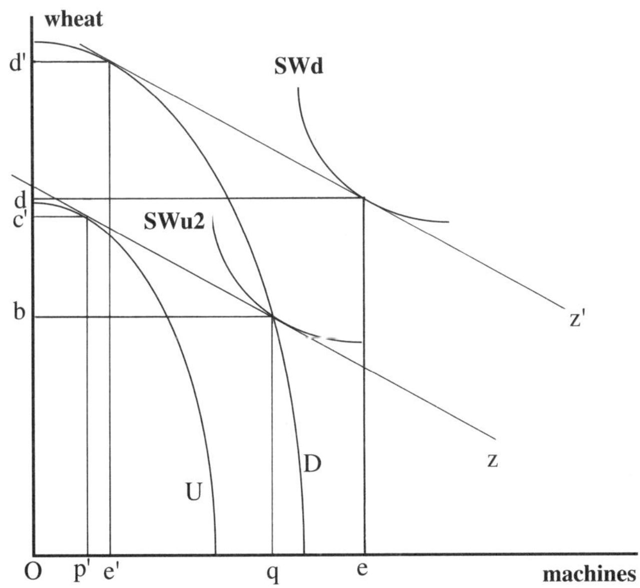

Figure 9.4b Wheat-biased Development in an Open Economy

Partial equilibrium analysis

Similar changes can be seen in a partial equilibrium context (figure 9.5). In the undeveloped economy, supply is Su and demand is Du, with equilibrium at price Pu and quantity Qu. With economic development, and assuming a slow population growth, the growth in domestic demand for wheat is small (from Du to Dd); however, the growth in supply (e.g. through technological change and lower prices for machinery) is large, from Su to Sd. While the equilibrium level of output grows substantially from Qu to Qd, the equilibrium price falls from Pu to Pd. Although industry output has increased, the fall in output price may be sufficient to require adjustment in the industry other than simply the adoption of the new technology. Compare the falling commodity price in figure 9.5 with its impact on individual firms as development occurs (figure 9.6). Suppose a firm in the undeveloped economy has a long-run average cost curve LRACu, and its long-run supply curve is LRMCu. The firm is in long-run equilibrium at price Pu and output q1. Now suppose economic development is associated with a reduction in the firm's average costs from LRACu to LRACd1; its long-run supply curve is now LRMCd1. If this kind of cost shift characterises *all* firms in the industry, a supply curve shift of the form illustrated in figure 9.5 will eventuate. If output price falls from Pu to Pd because of the new technology and the supply curve shift (figure 9.5), then all firms are at long-run equilibrium at the lower price Pd and the same level of output q1.

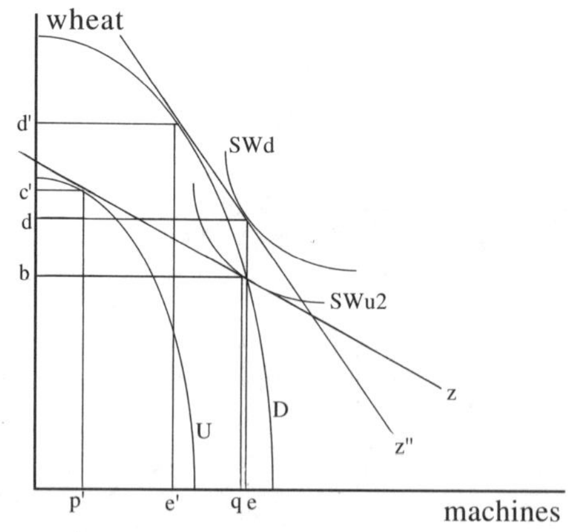

Figure 9.4c Wheat-biased Development with Falling International Wheat Price

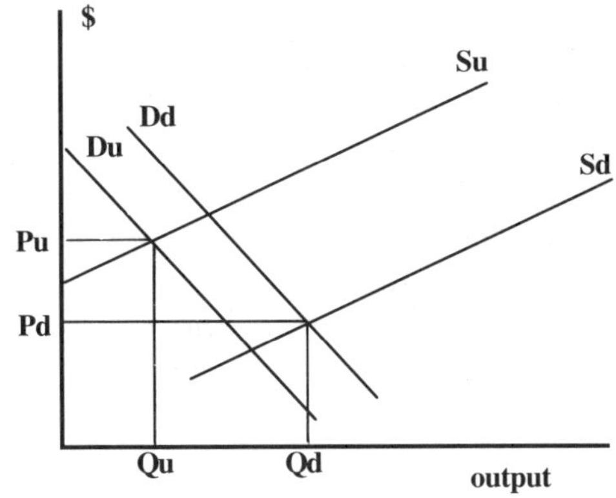

Figure 9.5 Economic Development in Partial Equilibrium—Wheat Industry, Closed Economy

But suppose not all firms are able to adopt the new technology to reduce their costs from LRACu to LRACd. Then the supply curve shift may not be as great as in figure 9.5, and price may only fall to Pd′ in figure 9.6. Firms able to reduce their costs to LRACd1 will expand their output to q3. However, firms that are not able to adopt the new technology and are forced to remain with the old technology represented by LRACu will make losses and will eventually become bankrupt. Structural adjustment will therefore ensue, as the bankrupt firms either sell their land to successful firms (whose long-run costs may therefore come to look like LRACd2) or the land will relocate to some other industry such as hobby farming, urban development or tourism.

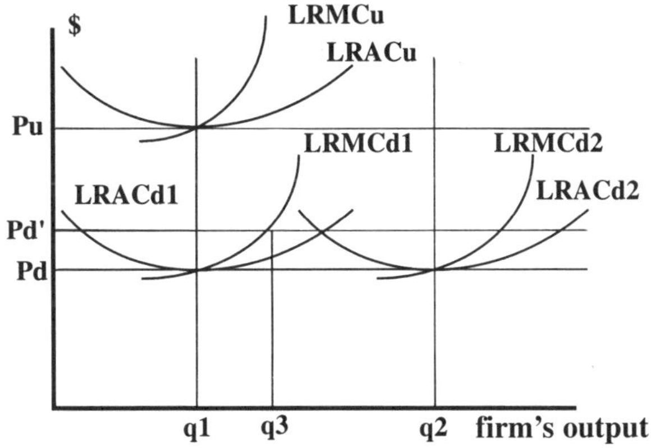

Figure 9.6 Individual Wheat Farm with Economic Development, Closed Economy

The technological change that created the shift from LRACu to LRACd1 was assumed to be scale preserving because long-run optimal output was unchanged at q1. However, technological change may not be size neutral—for example, more advanced machinery may be bigger and be most efficiently employed on larger-scale units. Thus the technological change may take the form of shifting costs from LRACu to LRACd2 directly; the efficient implementation of this technology would require a simultaneous increase in the size of innovating farms. If vacant land was not available, or land was not available from other enterprises, then the efficient adoption of the technology would require a simultaneous structural adjustment with the exit of some existing farms, which would be amalgamated with the innovating farms.

With an open economy, the results can be somewhat different, and these differences have been vitally important for the Australian agricultural sector. In figure 9.7 it is assumed (for simplicity) that the international price for wheat is Px when the economy is undeveloped, and also when it is developed. Instead of the undeveloped closed economy producing and consuming Qd, the undeveloped open economy produces Qut, consumes qu, and exports (Qut – qu). As the supply curve shifts to Sd, the wheat price remains constant at Px (by assumption), wheat output expands to Qdt, domestic wheat consumption expands to qd, and wheat exports expand to (Qdt – qd). Conversely, suppose the export wheat price fell as this country developed—as would be probable because of general economic development throughout the world, and the application in other countries of the technology that caused the shift of the wheat supply curve from Su to Sd. As long as the world price did not fall as far as Pd, this wheat industry would remain an exporter, and wheat production would be more profitable than if exports were not feasible. Many Australian agricultural producers in the past have not suffered the intensity of the cost–price squeeze that has afflicted other countries, because of the low-cost nature of much Australian agricultural production and the availability of export markets. Of course, the response of other countries' governments to the cost–price squeeze on their own farmers—by implementing massive subsidy schemes such as in the European Union, the USA and Japan—has diminished the potential benefits to Australian farmers from economic development.

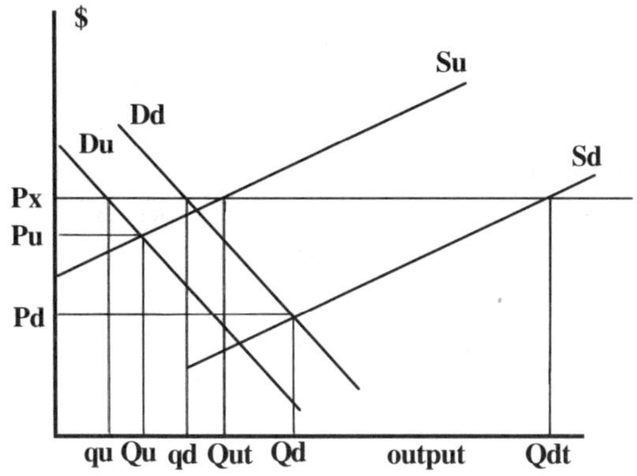

Figure 9.7 Economic Development in Partial Equilibrium—Wheat Industry, Open Economy

Barriers to adjustment

As relative prices change in the aggregate economy, or as agricultural technologies change and induce changes in the relative profitability of farm enterprises, change will be required in the farm sector. These changes may be short-run—e.g. with the collapse in wool prices following abandonment of the Reserve Price Scheme in 1990, producers moved into other enterprises such as cattle and cropping. Alternatively, changes may be of longer duration, such as the substitution away from wheat towards a variety of oilseed and legume crops in the past two decades. Governments may (attempt to) erect barriers to oppose or limit the intensity of the economic forces driving this adjustment. There are also, however, economic barriers to this adjustment.

Technological change in Australian (and world) agriculture in the period since the Second World War has been substantially driven by the mechanisation of farm operations. Working horses had been in decline since the 1920s, and had disappeared by the end of the 1950s. Tractors up to the early 1950s were capable of ploughing about 0.8 hectares per hour. Diesel tractors of the late 1950s were capable of up to 1.6 hectares per hour. Two-wheel drive diesel tractors reached their size limit of 120–150 drawbar horsepower by the late 1960s, capable of about 4 hectares per hour. These tractors also had major qualitative improvements of hydraulics and air-conditioned cabs. Four-wheel drive tractors of the 1970s reached sizes of 250–400 drawbar horsepower and were capable of 8–12 hectares per hour.[4] Because initially farms in Australian agriculture were small (table 9.1), the efficient use of new technology required growth in average farm size. Since good quality land had already been occupied, this could only occur by amalgamation. However, while optimal average farm size may grow relatively steadily over time as machinery size and capability increases reasonably steadily, correspondingly steady growth in actual farm size is not an option readily available to farmers. While leasing and sharefarming may provide short-term opportunities for growth, such opportunities are only available with the agreement of a second party. A permanent increase in business size requires the purchase of additional land.

Efficient growth would normally occur by an increase in size contiguous with the existing property to minimise travelling time and thus maximise technical efficiency.[5] Contiguous expansion may, however, present difficulties. Adjoining farmers may not be prepared to sell their landholdings—indeed, they may themselves be seeking to expand. Even if they are willing to sell, the value they place on their holdings may reflect a desire to realise on assets that would be redundant to a proposed purchaser (e.g. homestead, machinery sheds, shearing shed), and it may thus be difficult to conclude a satisfactory bargain. Landholdings may only be listed for sale intermittently—for example as a deceased estate or at bankruptcy. Conversely, the purchasing farmer may not wish to acquire all the adjoining property, as it may be beyond a prudent financial limit or might require hired labour to work effectively. The purchaser might be reluctant to use hired labour, or the additional labour required to work the expanded holding might require less than a single labour unit whereas labour may only be available in discrete units. Thus, there may be serious imperfections in local land markets that seriously constrain opportunities for business expansion.

Closely related to the problems inherent in land markets and asset fixity are problems of operator fixity. Figure 9.8 illustrates the key economic events that determine the fate of farm firms. If the firm has a long-run average total cost curve like LRACt, and the average output price remains above P1, the firm is economically viable in the long run. Corresponding to LRACt is a long-run average cost curve that covers only part of total costs, say all those except capital replacement (LRAClm); if average output price is between P1 and P2, the firm is viable in the 'medium' run—i.e. until capital replacement is required. Below LRAClm is the average variable cost curve (AVC), where only variable costs are covered: between average output prices P2 and P3, the firm might continue on operation in the short run, although the farm household may suffer severe poverty (cf. Standen 1972). At prices below P3, the firm cannot even cover variable costs and will become bankrupt.

Farmer adjustment

Once a farm household realises that it has no long-term future in farming, it faces a discrete choice as to whether or not to abandon farming. If it takes the decision to leave farming, the household then faces the decision when to exit. However, just because the farm firm is, or will become, bankrupt, it does not necessarily follow that farming will be immediately abandoned. The farm household's problem is that it derives a set of benefits from farming which contribute to household utility (HU, measured in annual terms), and these benefits must be compared to the potential benefits of leaving farming (opportunity cost of farming (OC), again annualised):

HU = pecuniary returns from farming + off-farm earnings + non-pecuniary farming returns

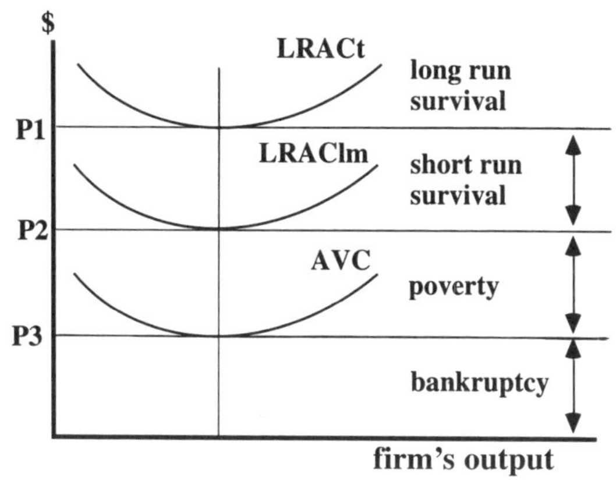

Figure 9.8 Farm Firm Financial Viability

Pecuniary returns from farming comprise the household's net income from continuing farming. Off-farm earnings might include activities such as share cropping (which might not be available if the farm business were dissolved) and paid employment (that might still be available if the business were liquidated). Non-pecuniary farming returns are the 'psychic' value obtained from rural living and, in particular, from the farm household living on its own farm and being an independent business. The 'household utility' from farming is sometimes expressed as the farm household's 'reservation price'.

The opportunity costs of farming (OC) are defined as:

OC = non-farm income + investment return from farm asset sales + non-pecuniary returns from non-farm living – adjustment costs (+ government assistance)

Non-farm income might include urban employment (or even rural employment if this were obtainable). Depending when the farm business is liquidated, alternative non-farm income may include social security payments. If the farm household has positive assets remaining after liquidating the farm business, some investment return may be thereby obtained. There are also non-pecuniary benefits from non-farm living, which may well be low or negative if the farm household is strongly attached to farming. Adjustment costs are those that would be incurred if the business were liquidated and the farm firm relocated; these costs would include the annualised cost of liquidating the business, finding alternative housing, and obtaining an alternative source of income (including possible retraining). Where government assistance is available for the firm or farm household if it abandons farming, this may be seen as either reducing adjustment costs or increasing the opportunity cost of farming.

The farm household will voluntarily abandon farming if OC > HU. In 'reservation price' terminology, the farm household will quit farming if the opportunity cost of farming exceeds its reservation price. However, if the difference between the non-pecuniary benefits of farm and non-farm living are very great, the household may remain in farming despite being in material poverty. Similarly, if non-farm pecuniary income is low and especially if the household workforce is nearing retirement age, or if its net assets are low, the farm household may not adjust, at least until permanent social security benefits are available from retirement pensions.[6]

When a farm business is liquidated, the next best use of farm land is generally agricultural. The farm holding may continue as an independent unit if another farmer has access to lower cost technology or has smaller household income requirements, or if it becomes a part-time or hobby farming unit. Alternatively the farm holding may be absorbed into one or more surrounding farms. Outside farming, the farm unit may be retained intact for recreation, or be used for forestry, recreation, mining or urban purposes.

The human dimension

Some analysts may be quite dispassionate about the implications of the preceding argument, and a good understanding of economic problems often requires detached analysis. The agricultural policy analyst must understand the circumstances which yield differential economic and financial results for farm firms. But the analyst must also be sensitive to the human dimensions of the problems being studied. In the context of figure 9.8, any firm receiving average prices below P1 must eventually become bankrupt, although the process may be spread over decades. For the individuals involved this is a human tragedy, and cannot be glibly shrugged off as the consequences of 'market forces'. The consequences of economic failure may not only be catastrophic for the adult generation but the financial implications of poverty may also extend to subsequent generations through impacts on health, education and personal relationships. Thus, even from a purely economic perspective, there are dynamic efficiency costs of farm poverty (and poverty generally) which—*independently of social welfare arguments*—might justify government intervention.

Land degradation

The consequences of farm firm financial trauma may also extend to the land base. Drynan (1995, pers. comm.) has argued that a profit-maximising firm which faces additional constraints—e.g. a minimum income constraint—may generate a supply curve that is downward sloping at low prices (figure 9.9). The rationale is that, above price P1, the firm's minimum income constraint is satisfied and the firm simply maximises profit as reflected by its supply curve S. Below P1, however, the firm does not earn sufficient income to satisfy its minimum requirements—e.g. the minimum income required for the firm's variable costs and normal household expenses. Thus, the firm attempts to increase its income by increasing its output. The firm can only achieve this outcome by changing its technology, for example by neglecting good husbandry practices that it would otherwise employ in prosperous times. The firm's new supply curve is S′. At P2, for example, the firm produces q2 rather than q1 which would have been produced if the firm were profit maximising. The firm may neglect good soil husbandry practices, making land susceptible to erosion or acidification; it may increase grazing pressure, thus reducing long-run pasture profitability and also contributing to the potential for erosion. Land may be cleared cheaply, leaving it vulnerable to on-site degradation, and increasing the possibility for off-site effects such as dryland salinity. The firm may neglect the control of pests, diseases and weeds, increasing its own long-run costs and possibly imposing external costs on others. Decision makers may be less cautious in using agricultural chemicals, and allow physical spillovers that impose external costs on other landholders or the community at large. Such changes in technology may be rationalised as disinvestment in natural resource capital stocks.

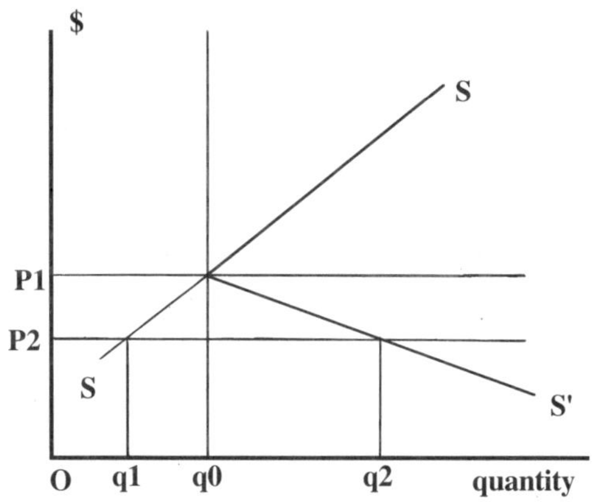

Figure 9.9 Downward-sloping Supply Curve

Policy implications

The preceding argument suggests that a purely financial analysis of farm adjustment is likely to be of limited relevance in guiding appropriate policy making. Financial analysis may reveal whether or when the farm is likely to bankrupt in the long or short term, but this analysis will not necessarily be a good predictor of the process of adjustment because the latter is intimately connected with non-financial issues. A good example is provided by the dairy industry on the North Coast of New South Wales: by 1960 it was realised that this industry was unlikely to be viable in the long run. Despite desperate searches for improved dairy technologies, and providing these producers with access to the more lucrative metropolitan wholemilk market, it was realised that the industry would probably not be viable. There was, however, no stampede out of the industry. Standen (1972) showed that the best option for many producers was to remain in dairying—and that this option coincided with the efficient use of society's resources, although the welfare consequences were unpleasant for many farm families. Adjustment occurred steadily through the 1960s and 1970s as farm operators reached retirement age and became eligible for social security, because some profitable albeit limited alternative enterprises were discovered (e.g. sub-tropical orchard fruits), and because urban Australia discovered the delights of rural living and rural recreation.

Conclusion

As an economy grows and international markets change, profound changes may occur in the optimal allocation of a nation's resources. Investment in an industry like agriculture is relatively long-term and may not respond to short-term price changes in either inputs or outputs. The long-term pattern of price changes both between agricultural industries, and between the agricultural sector and other sectors such as mining, manufacturing and services, may take many years if not decades to discern. If expectations about future price changes and the relative rates of technological change are different from those that eventually emerge, the need for structural adjustment may take a long time for factor owners to discern. If the changes that actually emerge are themselves modified by government intervention—e.g. by tariff protection—the adjustment process will be complex. Where the intervention is modified, there will again be a complex interaction between modification of the intervention and the underlying economic forces. Structural adjustment in the economy is a slow, often episodic, process often driven by crises such as major falls in commodity prices, droughts, the collapse of government intervention, or international crises.

Problems of farm structural adjustment, and associated farm poverty, are only one aspect of rural adjustment and rural poverty. Agricultural economists have often ignored those in the rural community who have borne most of the brunt of structural adjustment during the expansion of commercial farming in Australia, and who still form the largest impoverished rural group. Australian Aborigines thus are the focus of Chapter 10.

ISSUES FOR DISCUSSION

1 Do external changes (such as economic growth in the rest of the economy) or agricultural crises (such as drought or the present wool marketing problems), justify continued rural adjustment assistance by Australian governments?
2 Agricultural production and marketing are unstable activities. Discuss whether or not this instability justifies government intervention to mitigate its effects.
3 At the opening of the 1996 Sydney Royal Easter Show, the President of the Royal Agricultural Society was also reported as saying that, in agriculture, 'no longer do the good years make up for the bad years, and therein lies a problem which the nation must address'. Comment.

Endnotes

1 Interpretation of these statistics concerning structural change in farming must sometimes be undertaken with extreme care, as apparent changes may actually be artefacts of the statistical collection process, particularly associated with the definition of a 'farm'.

2 For description and analysis of the earlier schemes, see Musgrave (1982, 1990) and Lloyd (1986). Differences between the 1988 and 1992 Rural Adjustment Schemes were reviewed in Senate Rural and Regional Affairs and Transport References Committee (1994, pp. 12–13).

3 Lloyd (1986, pp. 62–3) gave a nice example concerning how early repayment of loans by farmers to state RAS authorities could be accumulated by states and used for other purposes (e.g. in Victoria, these funds were used for general rural lending).

4 Bob Batterham kindly provided information about machinery in this paragraph.

5 In some cases, other objectives such as spreading risk across different environments, or obtaining land of different qualities, may indicate that non-contiguous expansion is desirable.

6 A 'reservation price' for farming is not unique to farmers; other independent business people or wage and salary earners may also have 'reservation prices' which need to be exceeded in order to induce them to change their employment, their household location, or both.

10 Structural adjustment: Aborigines in the rural sector

Aborigines are not peripheral to agricultural and resource policy. In just over two centuries of European occupation, the most significant 'rural adjustment' has affected Aborigines. White Australia's treatment of Aborigines has been, where not actually totally antipathetic to their interests, generally characterised by a spectacular failure to accord them a dignified role in Australian society. Even where white society has attempted to improve Aborigines' standards of living, these attempts have been more characterised by failure than success (e.g. section 10.1). The High Court's *Mabo* native title judgement, the fear campaign that was whipped up in its wake, and its implications for the distribution of natural resources in the Australian rural sector are examined in section 10.2.

10.1 GOVERNMENT FAILURE—THE VILLAGE OF TOOMELAH

Toomelah is a small New South Wales Aboriginal village on the banks of the Macintyre River close to the Queensland–New South Wales border. In 1987, some of its residents allegedly rioted in the neighbouring Queensland town of Goondiwindi. New South Wales refused requests from the Queensland Government for Toomelah residents allegedly involved to be extradited to Queensland. The Commonwealth's Human Rights and Equal Opportunity Commission conducted an investigation into conditions at Toomelah, later supplemented by the New South Wales Ombudsman's investigation (1992). Aboriginal well-being at Toomelah was so far below standards generally applying in Australian society that justice demanded its redress.

Policy issue

Contemporary Aboriginal society is generally—although not totally—characterised by unsatisfactory social indicators: extremely poor health conditions (ABS 1994b, p. 415); poor education availability (ABS 1995); high unemployment rates and income distribution skewed towards low levels (ABS 1993, pp. 16,18); and poor housing (Jones 1994a).

Toomelah is a microcosm of contemporary Aboriginal experience throughout Australia and is a graphic example of forgotten people in Australian rural society. Wootten found in Toomelah in 1970 a '"community of 250 people . . . had been left for two weeks without a water supply and the situation is continuing. The houses . . . average 10 people to each two-roomed house"' (quoted in Human Rights and Equal Opportunity Commission 1988, para. 1.28). Nearly two decades later, the Human Rights and Equal Opportunity Commission (1988, para. 1.22) concluded that 'the people of Toomelah still suffered living standards far below those experienced by the vast majority of non-Aboriginal residents of New South Wales'.

Australia is an unjust society if gross inequalities are permitted to flourish and citizens are denied access to adequate health, education, housing and transport.[1] Society's resources cannot be regarded as efficiently allocated if some people are so inadequately supplied with basic services that their chances of equal participation in economic activity are severely impaired.

Policy intention

Early attitudes to Aborigines varied between active and passive extermination, followed by a period when it was confidently expected (or at least hoped) that Aborigines would simply disappear and 'soothing the dying pillow' was the appropriate policy. By the 1950s policy was 'assimilationist', where the intention was 'that all persons of Aboriginal descent will choose to attain a similar manner and standard of living to that of other Australians and live as members of a single Australian community' (quoted in Human Rights and Equal Opportunity Commission 1988, para. 2.3). The assimilationist idea was replaced in the early 1970s—except in Queensland where it remained policy until the late 1980s—by 'self-determination'. From the mid-1970s, 'self-determination' was replaced by 'self-management' where 'the emphasis was on Aboriginal responsibility rather than power or control . . . [but where] Effective control remained with the bureaucracy . . .' By the late 1980s, contemporary federal policy was to affirm both self-determination and self-management (Human Rights and Equal Opportunity Commission 1988, paras. 2.5, 2.8, 2.11).

Specific policy intentions for Toomelah, if they can be said to have existed, epitomised those endured generally by Aborigines.

Area of agricultural sector

Australian Aborigines were traditionally rural dwellers and approximately 70 per cent of Aborigines and Torres Strait Islanders still live outside capital cities (ABS 1993, p. 3). Since the early 1970s, Aborigines have become increasingly urbanised. At the time of the Human Rights and Equal Opportunity Commission's 1988 Report, Toomelah was an Aboriginal community of about 500 inhabitants about 18 kilometres from Boggabilla, a similarly sized, predominantly non-Aboriginal community. Just as

Toomelah was the poor relation of its neighbour, Boggabilla was a poor relation of Goondiwindi just across the Queensland border.

Policy orientation

Until 1967, Aboriginal 'affairs' were a state government responsibility. By the 1967 referendum amending the Commonwealth Constitution, the material in square brackets below was deleted:

- s.51: The Parliament shall, subject to this Constitution, have power to make laws for the peace, order, and good government of the Commonwealth with respect to:

  . . .

  (xxvi) The people of any race [other than the aboriginal race in any State] for whom it is deemed necessary to make special laws:

  . . .

- [s.127. In reckoning the numbers of the people of the Commonwealth, or of a State or other part of the Commonwealth, aboriginal natives shall not be counted.]

Since 1967, there have been six federal governments and a series of policy swings. During the Hawke–Keating Labor governments, the Aboriginal Affairs Ministry appears to have been a political Siberia generally reserved for ministers from the ALP's minority left wing.

Policy mechanism

Australian legislative structures supposedly exist to ensure that policy disasters like those at Toomelah should not happen. Australia has both the resources and the infrastructure to enable all Australians access to adequate public infrastructure (e.g. public health, education, transport) and basic necessities (food, clothing and shelter). The principal failure at Toomelah was the non-delivery of services which most non-Aboriginal Australians take for granted. Toomelah residents' low living standards compared to the rest of the rural community resulted from significant policy implementation failures by federal, state and local governments. The three levels of government especially failed to coordinate their responses to the problems evident at Toomelah.

Institutional structure

In a public choice context, governments supply services to social groups that are both visible and important. Aborigines have not traditionally formed an interest group sufficiently strong that governments—whether federal, state or local—have been forced to consider their interests. The Human Rights and Equal Opportunity Commission (1988, para. 5.6) commented with regard to Toomelah that 'Groups such as Aborigines who have little political influence are, therefore, disadvantaged'. This situation appears to be changing, particularly as a consequence of the development of Aboriginal political expertise in managing the land rights debate following the High Court's *Mabo* decision (cf. section 10.2).

Policy performance evaluation

Aborigines are a much studied group in Australian society (cf. Human Rights and Equal Opportunity Commission 1988). The Royal Commission into Aboriginal Deaths in Custody (1991) was a mammoth report which located the problem of Aboriginal deaths in custody in a series of policy failures having a more general social context. The concerns and problems of Australian Aborigines also formed part of wider social inquiries such as the Australian Government's Commission of Inquiry into Poverty (1975).

There is also the opportunity for regular scrutiny of policy as it affects Aboriginal Australians through annual budget appropriations by state and federal governments for specifically Aboriginal services, and to review Aborigines' access to services that are provided to the community generally. The annual budgets of local councils provide the same opportunities for scrutiny of local services provision.

Economic issues

At Toomelah, the provision of infrastructure was totally inadequate:

> . . . [in January 1987] [Race Discrimination] Commissioner Moss found the housing to be quite unsatisfactory both in quality and quantity. . . . the water supply . . . was rationed and dispensed twice a day for fifteen minutes at a time; the sewerage system was completely inadequate; the roads were unsealed dirt tracks. Access to and within the settlement was impossible after heavy rain. There was no drainage or street lighting, no regular garbage collection, and no store (Human Rights and Equal Opportunity Commission 1988, para. 1.6).

The following discussion focuses only on water supply and local government services.

Water

Water supply is crucial to the survival of any society. Without clean drinking water, disease will be endemic, and the population will be in a chronic state of poor health and susceptible to non-water-borne illnesses. Water is central to the disposal of human wastes in areas of population concentration; if these wastes are not satisfactorily disposed of, their putrefaction will contribute to the spread of disease.

Toomelah was identified as having a contaminated water supply in the late 1960s. Its water supply—or, more accurately, its water *non*-supply—in 1987 is documented in Box 10.1.

Box 10.1 Water Supply at Toomelah, 1987–88

When the Inquiry visited Toomelah in July 1987, water was being rationed. It was available twice a day for fifteen minutes at a time, as it had been for years. (para. 4.1)

Shortage of water is not just an inconvenience. It is a major crisis which . . . the community health worker sees as the basic cause of many health problems at Toomelah . . . The clinic often must function without water and there is no hot water . . . residents most frequently complain of ailments related directly to the lack of water and poor sewerage disposal: infected sores, gastroenteritis, growth failure, respiratory, ear and skin infections, and stress. (para. 4.3)

The Toomelah Primary School has air-conditioning and wet rooms for artwork, but the lack of water means they cannot be used. The shortage of water has also hindered the development of urgently needed facilities such as sewerage and roads. The ADC [Aboriginal Development Commission] advised the Inquiry that . . . the water and sewerage improvements must be in place before roadworks can begin. (para. 4.4)

Lack of water at one time also forced the abandonment of plans for a new preschool. After the Toomelah community successfully lobbied the New South Wales State Government for money to establish a pre-school, the Moree Plains Shire Council put the application at the top of its priority list. However, the Council then refused to approve the development plans because of the lack of water and sewerage. The Council would not supply them itself. (para. 4.5)

In the period 1981 to 1987, the Department of Aboriginal Affairs provided approximately $500 000 in an attempt to secure an adequate water supply for Toomelah. In 1983, the Department initiated its first major program to increase the water supply by sinking bores. Within twelve months the community was experiencing shortages. Within two and a half years, two of the three bores had dried up. The second program was commenced in 1986. The Inquiry was told that a bore sunk late in 1987 has now been capped, and pumping stations and storage units have been installed. (para. 4.7)

. . . if the Toomelah community obtained a pump and a water licence (costing $100 for a ten-year licence), the required supply of water from the Macintyre would be made available on demand. Such water would probably not be suitable for drinking without further treatment, but it certainly would be suitable for sewerage and other domestic purposes.

The people of Toomelah were left ignorant of the reason for the sudden drying up, some years ago, of the Macintyre River in their area. Until that time, it had been flowing freely and supplying water to Toomelah for decades. The reason, of course, was the damming of the river and the controlled release of only sufficient water for the towns and properties *licensed* to make use of it. None of the relevant authorities, neither those responsible for servicing Toomelah nor those responsible for regulating water supply, had ever informed the Toomelah community either of the reason for the decision to rely in future on bores or of their right to apply to have water released to them by the Water Resources Commission. (paras. 4.11–12, emphasis added)

There is considerable disagreement as to whether local government authorities have responsibility for the provision of such services to Aboriginal communities. The Department of Aboriginal Affairs insists that water and sewerage provision is the responsibility of the local government authority and the Department of Local Government agrees. However, the Local Government and Shires Associations' representative told the Inquiry:

> I find it extremely unusual that there is even contention about this. It [Toomelah] is a private property as far as the Local Government Act is concerned [see Box 10.2] and the council cannot expend funds, its own funds, in the area. (para. 4.15)

Source: Human Rights and Equal Opportunity Commission (1988)

For want of simple notification and follow-up to ensure the advice was understood and acted upon, a village of 500 people only had water for two 15-minute periods per day when Australia generally was enjoying an economic boom.

Local government services

The serious government failure revealed by the Toomelah water supply debacle was only one aspect of a more general failure to supply local government services (cf. Box 10.2). The refusal of local government authorities to supply services to Toomelah was bizarre since their funding was partly based on the existence and needs of local Aboriginal communities. Further, since Commonwealth and state governments subsidised the provision of some council services, failure to provide these services to Toomelah excluded its residents from obtaining access to funding to which they, as Australian citizens, were entitled.

Box 10.2 Government Failure in General Provision of Services

. . . there is considerable dispute and confusion as to the responsibility and, indeed, the power of local government authorities to expend funds on most Aboriginal land in New South Wales. ('Aboriginal land' is land owned in freehold title by Local Aboriginal Land Councils constituted and incorporated under the *Aboriginal Land Rights Act 1983*.). The Local Government and Shires Associations of New South Wales and the Department of Aboriginal Affairs (in its written submission) asserted that such land is private land and that, therefore, the Local Government Act prohibits a local government authority expending funds within its boundaries. Neither was able to explain how the [Moree Plains] Shire Council was able to rate Toomelah per house while treating the entire community as a private settlement.

. . . the New South Wales Department of Local Government . . . stated that a local government authority is responsible for the provision of services throughout its area . . . we cannot understand the attitude of the Federal Aboriginal Affairs agencies. Although established for the sole purpose of servicing the needs of Aborigines, these agencies took a view of this situation which defeated the most basic interests of the people at Toomelah. (paras. 6.13–14)

It is at best ironic that local government bodies in New South Wales refuse to service Aboriginal land while receiving general purpose grant funds which, although not specifically allotted for that purpose, are assessed in part on the existence and needs of an Aboriginal community. The 1985 Report of the National Inquiry into Local Government Finance states:

> . . . local councils throughout Australia do not generally cater adequately for Aboriginals although their general purpose grants are in part calculated on the basis of Aboriginal populations residing within their boundaries. (para. 6.15)

. . . the involvement of a local government authority as agent for the Commonwealth or State is at present essential to secure certain specific purpose grants, including Commonwealth grants for roads and state grants for water supply and sewerage works. A local government authority which refuses to provide services to an Aboriginal community which it provides to non-Aboriginal communities within its boundaries effectively prevents funds which may be available under these programs from being expended in that Aboriginal community. (para. 6.16)

Not only does the Moree Plains Shire Council refuse to supply any services to Toomelah and argue that it is prohibited from expending its funds within the community; it has even failed to supply water and sewerage services to the community's boundary. The Shire President . . . told the Inquiry:

> Council sees Toomelah as a private settlement which, therefore, differs in comparison to Boggabilla and other shire villages, where Council has a responsibility to provide and maintain various services . . . Council is not responsible for the provision of services at Toomelah . . .

Not content to rely on its interpretation of the Local Government Act as prohibiting servicing of private land and its labelling of Toomelah as a 'private settlement', the Shire Council further told the Inquiry:

> We have no financial capability of providing these services [to Toomelah].

The term 'private settlement' was, as the Shire Council readily conceded, invented by the Council itself. It has no legal meaning under the Local Government Act, or for that matter under any other legislation. The Council has thus described no other place within its area and does not know of any other such creature in Australia. We consider that the Council has so described Toomelah in order to avoid its obligation to service the residents in the same way as all other residents of the Shire are serviced. (para. 6.17)

With five hundred people, Toomelah could qualify for declaration as a 'town' . . . Assistant Secretary of the Department of Local Government, told the Inquiry he believed that Moree Plains Shire Council has not nominated Toomelah for town or village status because:

> . . . it was clear to me that they [Moree Plains Shire Council] felt that giving Toomelah the status of a village would carry various connotations in things like what you would normally expect to see in a village. Would you expect to see a school there? Would you expect to see various other provisions of services which you might not expect to see in an isolated settlement? That seemed to be the attitude they were taking. (para. 6.19)

When the Inquiry visited Toomelah in July 1987, part of the area was in Moree Plains Shire and part in Yallaroi Shire. Apparently unknown to either shire council, and certainly unknown to the residents, the border between the two actually bisected one of the Toomelah houses. The Aboriginal community seems to have been ignored altogether in the creation of the shire boundaries, almost as if no-one in the responsible department knew they were there. Since our July 1987 hearings, the boundary has been moved so that Toomelah is now entirely within the Moree Plains Shire area. At our December 1987 hearing the Shire President . . . expressed some disquiet about this development:

> What I was suggesting to you in the hearing, and to the powers that be, was that before that was undertaken, let us clean up the whole mess that is the Toomelah out there at the present time; that Moree Plains, before it accepted unreservedly that Toomelah would be in Moree Plains shire—we wanted the water systems, the sewerage systems, all upgraded, and the general living conditions of the people much better than they were.

In the light of the fact that the Moree Plains Shire Council not only has not exercised its discretion to expend its own funds in its part of Toomelah but in fact claims that it is prohibited by law from doing so, we can only speculate as to why the council was concerned to have conditions there upgraded before the remainder of the area came within its jurisdiction. (para. 6.32)

Source: Human Rights and Equal Opportunity Commission (1988)

It is tragic that, while Toomelah embodied an objective which many Aborigines had sought for decades—the recognition of their own land—it should have been achieved while denying its inhabitants the rights to services that most other Australians (except many fellow Aborigines) take for granted. For policy analysts, the tragedy of Toomelah warns that creation of policy structures must take account of potential interactions with other policies and institutions. Without such consideration, desirable policy may not achieve its intentions, however honourable, and may even result in noble intentions being perverted into their precise opposite.

Analytic techniques

Edwards and Watson (1978, p. 192) argued that 'economic policy making often hinges on simple figuring of the key economic magnitudes involved'. Empirical observation and appropriate analysis, not complex theory, are often all that is required to analyse many policy problems. No sophisticated economic techniques are required to conclude that neither economic efficiency nor a just distribution of welfare was achieved for Toomelah. But, more than all of these, policy analysis which is intended to resolve human tragedy demands thoughtful and compassionate implementation, absent from government responses to conditions at Toomelah.

Conclusion

Toomelah was not unique amongst Aboriginal settlements. Not far from Toomelah, in the Clarence River valley, was Baryulgil, where asbestos tailings 'were used extensively for a variety of purposes . . . including use in road construction and surfacing, in land fill and topdressing around private homes and in "sand" pits for the local school' (Human Rights and Equal Opportunity Commission 1990, pp. 4–5). Despite identification of these problems by 1983, residents were still complaining of inadequate policy response in 1990. Mornington Island in Queensland also exhibits policy responses which mirror the ineffective responses at Toomelah and Baryulgil (cf. Human Rights and Equal Opportunity Commission 1995). As shown in the Royal Commission on Aboriginal Deaths in Custody, the social and living conditions for many Aborigines throughout Australia are far from satisfactory.

Blainey (1976, pp. 253–4) concluded his *Triumph of the Nomads* by noting that

> on the east coast the white sails of the English ships were a symbol of a gale which in the following hundred years would slowly cross the continent, blowing out the flames of countless campfires, covering with drift-sand the grinding stones and fishing nets, silencing the sounds of hundreds of languages, and stripping the ancient aboriginal names from nearly every valley and headland.

It is unconscionable that, more than two centuries later, this gale continues to separate indigenous Australians from their rights as citizens.

10.2 EXOGENOUS STRUCTURAL CHANGE: THE HIGH COURT'S 1992 DECISION ON NATIVE TITLE[2]

The High Court's decision in the *Mabo* case of 3 June 1992 re-asserted Aboriginal common law rights to title in land (*Mabo and Others v State of Queensland*; Mabo 1992). The central issue in the judgement was that 'the High Court of Australia held in *Mabo v Queensland* that the common law of Australia recognises a form of native title' (anon. 1992, para. 1). The case was *common law*—i.e. judge-made law rather than parliamentary statute—and not based on an interpretation of statute or the Commonwealth Constitution. The decision was controversial because many Australians believed Aborigines had never owned land in a way recognised by English law and, further, even if Aborigines ever had owned their land they had long since lost possession by right of 'settlement' or 'conquest'.

The High Court's decision prompted both federal and state governments to legislate for Aboriginal land rights. The approach of the Western Australian Government was inconsistent with that of the Federal Government and the former, together with South Australia, challenged the federal legislation in the High Court. This challenge was dismissed in March 1995.

This section is chiefly concerned with the High Court's judgement, which illustrates the emergence of very significant institutional change affecting agriculture and natural resources via the judiciary, rather than the legislature or the executive. The judgement also illustrates the importance of property rights in a policy context. Finally, the judgement is important because it prompted non-indigenous Australians to confront the consequences of European settlement for the continent's original inhabitants.

Policy issue—the *Mabo* judgement

The majority of judges decided that 'upon the acquisition of sovereignty by the Crown it acquired a "radical" title to the land' but that this '"Radical" title was subject to existing native title, the elements of which are to be determined according to the traditional laws and customs of those people with whom the title resides' (anon. 1992, Appendix, paras. 2–3, 4). The Court thus 'rejected the notion that Australia was *terra nullius* (land belonging to no one) at the time of European settlement and the resulting implication that the absolute ownership of land vested in the Crown' (anon. 1992, Appendix, para. 2). However, 'The High Court's rejection of the proposition that Australia was *terra nullius* at the time of settlement does not mean that Aboriginal people can claim sovereignty under international law' (anon. 1992, Appendix, para. 41). The Crown's sovereignty did confer title for 'those areas of the continent that were "truly uninhabited"' (anon. 1992, Appendix, para. 4), however native title claim would probably never be denied on these grounds since, at the time of the settlement in Sydney in 1788, it was unknown whether or not areas other than Sydney's immediate vicinity were actually uninhabited. While the Crown obtained sovereignty, 'the acquisition of sovereignty did not itself effect the automatic extinguishment of native title, where it existed, although certain actions by governments or the loss [by Aborigines] of a traditional connection with the land . . . will have extinguished the [native] title in respect of particular areas' (anon. 1992, Appendix, para. 4).

The majority of judges accepted that extinguishment of native title resulted from grant of freehold title (anon. 1992, Appendix, para. 10) although the judges differed over what other actions might have extinguished this title (anon. 1992, Appendix, para. 11). Where native title possibly exists, part of the test of its continued existence depended on 'whether an interest in the land granted by the Crown, or the Crown's actions as an occupier, were inconsistent with the continuing right to enjoy native title, and therefore extinguished that title' (anon. 1992, Appendix, para. 11). Judges' views also varied as to whether or not native title was extinguished by grants of leasehold (anon. 1992, Appendix, para. 12). The Crown's appropriation of land for an existing use probably extinguished native title—'roads, railways and buildings, such as post offices, were mentioned' but appropriation for some future use would not extinguish native title (anon. 1992, Appendix, para. 13). Land set aside for a national park or wasteland might not extinguish native title, whereas declaration of a wilderness area might be inconsistent with continuing native title (anon. 1992, Appendix, para. 13). 'Native title might not have been extinguished by the granting of authorities to prospect for minerals', but a mining lease probably would extinguish it (anon. 1992, Appendix, paras. 14, 19).

While the direct intent of the *Mabo* judgement affected *land* title, the generic resource 'land' has many economic dimensions. From the Court's judgement, it was unclear as to 'whether the scope of any title might extend to the natural resources of the land, including minerals' or whether native title rights might have only been restricted to resources traditionally used by Aborigines (anon. 1992, Appendix, para. 20). For example, Justice Brennan (with justices Mason and McHugh concurring) held 'The nature and incidents of native title must be ascertained as a matter of fact by reference to the traditional laws acknowledged by and the traditional customs observed by the indigenous inhabitants (Mabo 1992, p. 2 (iv) and p. 42). Justices Deane and Gaudron argued that 'The traditional law and custom is not, however, frozen as at the moment of the establishment of a Colony' (Mabo 1992, p. 83; cf. Brennan et al. in Mabo 1992, p. 44). It was thus unclear whether or not native title only granted rights to exploit resources with traditional technologies (anon. 1992, Appendix, para. 28). Other issues not directly canvassed in the *Mabo* judgement included whether or not native title extended to water resources (surface, sub-surface, marine), littoral (foreshores, beaches, sea-bed), or fisheries (e.g. anon. 1992, Appendix, paras. 6, 19–25; Council for Aboriginal Reconciliation 1993, p. 9).

Policy intention

The High Court, by recognising native title, might be considered extraordinarily generous to Australian Aborigines. However, the Court was actually extremely generous to *non*-Aboriginal Australians. The Court refused to consider the issue of sovereignty. The Court found that native title was only recognised by common law (which could be readily overcome by statute) rather than as an implied guarantee in the Constitution (which could not) (cf. *Australian Capital Television v Commonwealth*, handed down less than four months after *Mabo*). The Court refused to specify the content of native title, which provided the opportunity for it to do so subsequently. As the existence and content of native title could only be assessed case-by-case, legislatures obtained time to extinguish native title and the Court itself time to reconsider the practical effect of its judgement. Further, since *Mabo* did not directly raise mining issues, no conclusion was reached as to whether or not native title encompassed mineral rights. Through its judgement, the Court provided a 'statute of limitations' on native title claims by indicating that validly alienated land was only subject to compensation after enactment of the Racial Discrimination Act in 1975. Wootten (1993a) rightly commented that 'the Mabo decision . . . was the minimum that could have been given with any decency.'

Many Aborigines saw *Mabo* as offering little. By specifying that valid extinguishment had occurred by the granting of freehold title, by the dedication of Crown lands to a specified use, and probably by the grant of leasehold title—except where, principally in Western Australia, leasehold tenure permitted continuing Aboriginal interest—the judgement effectively limited potential native title claims to unutilised low value Crown land. Since some of these areas have been claimed, or are amenable to claim, under land rights legislation, the judgement offered little *additional* opportunity for Aboriginal land claims except in Western Australia. By linking compensation to the *Racial Discrimination Act 1975*, the High Court effectively foreclosed any opportunities for compensation claims for extinguishment of native title *prior* to 1975, which would have affected the country's more valuable land.

Contrarily, many clearly felt that at best Aborigines deserved no more secure access to land than they enjoyed prior to the *Mabo* decision. Thus, if *Mabo* offered opportunities for obtaining better title to *any* land, it offered too much. Particularly as there were unresolved issues in the High Court's judgement which could have provided future benefits to Aborigines, the decision was seen by some as lying on a spectrum between highly undesirable and disastrous. These views were partly exacerbated by some Aboriginal land claims, ostensibly under *Mabo* principles, which even supporters of the *Mabo* decision described as having had 'a disastrous and pointless effect on race relations in many parts of rural Australia, and given such comfort to those seeking to mobilise opinion against Aboriginal land rights' (Wootten 1993b; cf. Godden 1994 on the economics of 'ambit' claims).

Area of agricultural sector

A perspective on the potential for lands to be affected by *Mabo* is provided in table 10.1. Only land in the last column—i.e. where there is a continuing and traditional association of Aborigines with the land—is susceptible to *Mabo*-style claims. Further, land is only susceptible to claim if this land falls into *some* title/land use categories. This table provides a perspective on some of the sillier and/or more provocative claims about potential consequences of the *Mabo* judgement—such as peoples' backyards being at risk.

Table 10.1 Coverage of *Mabo*

	Aboriginal Association	
Non-Crown land	Lapsed	Continuing and traditional
• alienated	no claim	n.a.
• leasehold—pastoral	no claim	governments intend to legislate away—compensation payable
• mining lease		
— exploration	no claim	potential claim
— extraction	no claim	potential claim
Crown land		
• dedicated use	no claim	no claim
• non-dedicated use	no claim	claim

Table 10.2 presents some idea of the area of land potentially claimable under common law native title; however, lumping 'Pastoral and vacant Crown land' together is unhelpful. Auslig (1992) presented existing Aboriginal tenures, freehold title, pastoral leases and various types of Crown lands in map form. The Australian Chamber of Commerce and Industry (1993) overlayed similar data with major common law native title land claims up to 22 June 1993. Except in Western Australia, only a small area of Australia's land was likely to be susceptible to native title claims under common law.

Table 10.2 Australian Land Ownership by Type

Tenure classification	*Australian area affected (%)*
Aboriginal land or land under claim	16.4
National parks and nature reserves	7.5
Private and other land	(est.) 8.0
Pastoral and vacant Crown land	(est.) 68.1

Source: Ewing (1992, p. 9)

Another aspect of the judgement's coverage concerned how many Aborigines stood to benefit. Since urban Aborigines are unlikely to have a continuing traditional association with land required under the *Mabo* principles then, as revealed in table 10.3, at least 57 per cent and perhaps as many as 75 per cent of Aborigines were highly unlikely to satisfy the High Court's 'traditional association with land' criterion for grant of native title. Perhaps as few as 5 per cent of the Aboriginal population—Aborigines on 'outstations'—may qualify for native title under the High Court's criteria. Concern was also expressed *because* only a few Aborigines stood to benefit from *Mabo* (cf. Godden 1994)

Table 10.3 Distribution of Aboriginal Population, 1981

	no.	*%*		*no.*	*%*
Major urban	28.4	19.6	Cities	63.9	41.3
Other urban	54.6	37.8	Small non-Aboriginal towns	53.1	34.3
Rural	61.6	42.6	Aboriginal towns	30.3	19.6
			Outstations	7.6	4.9
Total	143.4			154.9	

Source: Fisk (1985, tables 1.1, 2.1)

Policy orientation

The High Court judged the *Mabo* case because some Aborigines believed the political system had largely failed to deliver land rights justice. Except in the Northern Territory, where land rights were granted by Commonwealth legislation in 1976, and South Australia where progress had been made in recognising land rights in that state's north, no effective land rights had been recognised, even where land rights legislation existed (e.g. NSW). Except in the territories, control over land was the province of the states unless the Commonwealth acquired land under 'just terms'. The policy challenge posed by the *Mabo* judgement was not merely related to land tenure, but also to race, where the Commonwealth had clear jurisdiction (by s.51(xxvi) of the Constitution). Some High Court judges also acknowledged that Australia's international reputation was involved with its limited and belated recognition of the rights of the descendants of the continent's original inhabitants.

Policy mechanism

Because the policy mechanism in the *Mabo* case was a High Court judgment, much of the controversy concerned whether the Court had the right to, or should, deliver such a judgement.

Lawyers' reactions to the High Court's *Mabo* decision were as varied as in the wider community (cf. Godden 1994). A major concern of some legal commentators was that the Court—quite explicitly—overturned the supposedly widely accepted law of property on colonisation, and the judgement was seen as inventing new law on land title. But the High Court explicitly argued that in determining the status of land title it was applying English common law at the time of European settlement of Australia. The Court's view, reflecting historical evidence (e.g. Reynolds 1992), was that the common law applying at the time of European settlement had been so widely misinterpreted that justice required a radical reassessment of native title. *Mabo* is thus seen as recognising that native title always existed and has only been validly extinguished in particular circumstances. Some of the High Court judges were also criticised for lack of 'judicial restraint' for carrying the *Mabo* judgement beyond the Murray Islands (cf. Godden 1994).

Because appeals to the UK's Privy Council had effectively ended from all Australian legal jurisdictions by the 1980s, the Australian High Court is the final arbiter of all Australian law. Since the High Court's *Mabo* decision could not be appealed, it was unassailable except to the extent that native title was *validly* extinguished by state and/or Commonwealth government statutes, with compensation for extinguishment after 1975; the Commonwealth's *Racial Discrimination Act 1975* was repealed or amended; the Commonwealth Constitution were amended to permit extinguishment of native title without compensation; and/or a subsequent case enabled the High Court to 'distinguish' the *Mabo* case from other cases. State and territory legislation would also be subject to the *Racial Discrimination Act 1975* (anon. 1992, Appendix, paras. 51–8).

Two broad policy responses to the *Mabo* judgement were possible. The policy implications of the judgement could be accepted, and legislation developed to establish a comprehensive policy framework for native title, rather than relying on the accumulation of future case law. The Commonwealth and most of the states followed this path. Alternatively, policy implications of the judgement could be rejected, and a policy framework developed which sought to wholly or partially negate the principles enunciated in *Mabo*; the Western Australian Government's legislative attempt to follow this path was defeated in the High Court.

Institutional structure—public choice aspects of High Court's judgement

Induced institutional innovation

The exogenous institutional change of the High Court's *Mabo* decision led to major perceived changes in the value of land. These changes help explain the lobbying of groups such as miners to preserve their existing titles, by Aborigines to claim land under *Mabo* principles, and attempts by both groups to extend or modify these principles. There was also deliberate distortion of the likely changes arising from the judgement: those perceiving they could be detrimentally affected deliberately engendered fear to facilitate construction of a grand coalition to persuade governments to nullify this judgement.

The lobby group most directly affected by the *Mabo* decision, the Aborigines, initially welcomed it as a belated recognition of Aborigines as Australia's first settlers. The decision had a symbolic significance to Aborigines greater than its capacity to deliver practical land rights. This symbolic significance was reinforced in late 1992 by the Prime Minister's Redfern Park speech (Keating 1992). In the ensuing political debate, however, Aborigines appeared by mid-1993 to have become very disillusioned about the prospects for *Mabo* either initiating a new enlightenment in non-Aboriginal Australia's relations with Aboriginal Australians, or delivering real advantages through the practical implementation of native title. The physical dispersal—even fragmentation—of Aboriginal Australia contributed to a fragmented response until the Eva Valley and subsequent Canberra meetings. The fragmented Aboriginal voice—and thus high costs of united decision-making in contrast to the unity of the other principal interest group, the miners—contributed to Aborigines' failure to ensure that the policy response process proceeded at a pace suitable to their interests. The lack of control over pace ensured that, once the claim that non-Aboriginal groups would suffer serious losses under native title became widely accepted, it became very difficult to prevent a policy response that did not result in native title being effectively emasculated. After a torrid and very public disagreement between the Prime Minister and Aboriginal negotiators in October 1993, a compromise was reached. Aborigines subsequently became very conscious of the need for public unity, even while thrashing out internal differences.

Although very clear differences remained within the Aboriginal community over the appropriate response to *Mabo*, Aborigines are probably a much more effective political force than they were prior to the *Mabo* debate. Their emergence as an effective political force may be the most enduring legacy of the *Mabo* debate. If this 'cohesion despite differences' is maintained, Aborigines are less likely to be steamrolled by the political process and thus will have a better chance of ensuring an equitable future for themselves.

Once the farmers' groups had become convinced that native title was extinguished by freehold or leasehold under the original *Mabo* decision, or accepted governments' assurances that they could secure agricultural and pastoral leasehold under appropriate legislation, the farming lobby appeared to become relatively indifferent to the native title issue. The principal exception was in pastoral areas of Western Australia where, because of visible Aboriginal populations and large areas potentially claimable, pastoralists became extremely agitated over the prospect of native title claims. These fierce reactions were probably deliberately engineered for political purposes.

The National Farmers' Federation played an honourable role during 1993 in negotiating an acceptable legislative response to *Mabo*. The Federation's Executive Director, R. Farley, also a member of the Federal Government's Council on Aboriginal Reconciliation, was particularly influential. The NFF was acutely aware of the fragility of any possible compromise on native title; its President warning at one stage 'that "if the Coalition pushed the Government too far and can't get any sort of compromise, then we'll end up in the hands of the Democrats and the Greens"' (Chamberlin 1993, 5 November). During the torturous Senate debate on the legislation, Farley was reported as saying that he felt 'betrayed, bitter and foolish' after a government amendment to clarify some rights relating to some renewable mining and pastoral leases was defeated, despite three National Party senators crossing the floor (Davies 1993, 20 December).

Of all the interest groups, the mining lobby was most active. Their activities covered the entire spectrum available to lobby groups—they were involved in discussions with governments, and they publicised their case widely (e.g. in their own publications, and extensive advertising in the daily press).

Non-Aboriginal responses to the *Mabo* decision were often concentrated at the apoplectic end of the hysteria spectrum. The Prime Minister's plea to 'ignore the isolated outbreaks of hysteria and hostility of the past few months' (Keating 1992, p. 4) was, with few honourable exceptions, largely ignored by non-Aborigines. One of the alleged problems of *Mabo* was that it created uncertainty about land title in Australia: many responses appeared deliberately designed to engender further uncertainty, possibly to exacerbate further tensions and divisions in the community. One example was questioning whether state forests were susceptible to native title claim, when clearly they were a *dedicated* use rather than vacant Crown land, and it was highly improbable they would have been affected by the native title judgement.

Many historical 'analyses' were trotted out to support arguments that Aborigines could not and should not have, and never could or should have had, access to land title. Such 'analyses' had already been comprehensively demolished by Reynolds (1992, pp. 31, 53–4).

Policy performance evaluation

Insufficient time was available for calm and measured responses to the *Mabo* decision. To his credit, Prime Minister Keating apparently decided that the *Mabo* judgement provided one of those rare opportunities when politicians can demonstrate statesmanship—even though the Federal Labor Government had previously fudged an opportunity to enact land rights legislation. Despite the politics of fear and hostility that characterised the native title debate throughout 1993, Keating was relatively unwavering in pursuing a successful legislative response.[3]

Federal Government

The Federal Government circulated a draft analysis of the *Mabo* judgement by September 1992 (cf. anon. 1992) and began consultations in October 1992. In December 1992, Keating addressed the *Mabo* issue in his 'Redfern Park' speech (Keating 1992). A preliminary identification of lands susceptible to native title claims was released in late 1992. A White Paper, released in June 1993, evaluated the legal position resulting from the High Court's decision, proposed 33 principles that the Commonwealth believed should underlie any government response to *Mabo*, and proposed a system for identifying native title and resolving possible conflicts between native title and existing titles, and dealing with others' uses of resources falling under native title (Commonwealth of Australia 1993a). The White Paper also dealt with wider issues of the consequences of Aboriginal dispossession, and with reconciliation. A special meeting of the Council of Australian Governments in early June 1993 failed to resolve any substantive issues. Federal Cabinet decided on the form of its proposed *Mabo* legislation in late July, proposing tribunals at Commonwealth and state levels to attempt to overcome the states' opposition to federal dominance. The Commonwealth then ran into considerable opposition from Aborigines, arising from the Eva Valley and subsequent Canberra meetings, and a bitter dispute with the Queensland Government over the Wik people's claim to land around the Weipa bauxite deposits.

The Commonwealth released an outline of its proposed legislation on native title in September 1993 (Commonwealth of Australia 1993b). After tortuous negotiations in late October, the Federal Government achieved a general compromise with 'official' Aboriginal negotiators, the National Farmers' Federation and miners, based on the September 1993 proposals. This agreement formed the basis of the draft Commonwealth Native Title Bill 1993. Since the government did not have a Senate majority, the success of its Native Title Bill depended on either support from one of the parties in the Coalition or from both the minor parties (Australian Democrats and Western Australian Greens). As the Coalition remained implacably opposed to the principles of the government's native title legislation, the government was hostage to the Democrats and the Greens. Following the October agreement, parties dissatisfied with the compromise sought to improve their positions by further bargaining. Before the Federal Government's legislation was finalised, the Western Australian Government's *Land (Titles and Traditional Usage) Act 1993* came into effect on 2 December, extinguishing all native title in that state.

Upon its introduction to the Senate, the Native Title Bill was diverted for two weeks to its Standing Committee on Legal and Constitutional Affairs, with threats of diversion to a Senate Select Committee, which could have substantially delayed it. The Standing Committee divided along party lines in its report. During the subsequent long debate, unusually including a Saturday sitting, the Prime Minister was reported as declaring that Parliament would sit—including Christmas Eve and Boxing Day—until the Bill was passed. The Bill finally passed the Senate during the night of 21 December, and re-passed the House of Representatives on 22 December to ratify the Senate amendments. The Act was proclaimed on 24 December 1993, and calls for registrations of interest in land affected by native title were advertised early in January 1994. By early 1996, there had been no successful land claims under the legislation.

Federal Opposition

The Federal Opposition had difficulties in developing a response to *Mabo*. McLachlan (1993, pp. 6–8) noted 'a most extraordinary decision', 'an adventure indulged in by the High Court of Australia' which failed 'to take account of the immense damage it would do to the rights other Australians thought they had'. McGauran (1993, pp. 1, 2) argued that 'the High Court of Australia rocked the foundations of the social and economic future of the nation' and had 'generated unprecedented uncertainty across the community'. Fischer (1993, p. 4) attracted publicity for observing that 'At no stage did Aboriginal civilisation develop substantial buildings, roadways or even a wheeled cart as part of their different priorities and approach';[4] he also observed that 'Those in the guilt industry have to consider that developing cultures and peoples will always overtake relatively stationary cultures' and 'All I am saying is that these things are relative and the horrors of the past were not caused by this generation of Australians'.[5]

The Federal Opposition established its *Mabo* position on 28 June. The first substantive paragraph of the Media Release accompanying its '*Mabo* Issues Paper' provided the theme of its reaction: 'The implications of the *Mabo* case present one of the most important challenges *to Australia's investment climate* for decades' (Hewson 1993, para.2, emphasis added). Aborigines' preferences did not loom very large in these priorities. Subsequently, both the Opposition Leader, and the Coalition's spokesman on native title, reiterated that 'The overriding objective at the present time should be to secure existing titles and remove any doubt as to the validity of those titles' (Hewson 1993, para.7). The 'existing' titles to be secured did not, however, include *native* title.

Once the Native Title Bill entered the Senate, the Opposition commenced a filibuster with the apparent objective of causing maximum delay. The Leader of the Opposition commented, after agreement was finally reached in the Senate to pass the Native Title Bill, that it was 'a day of shame for the Australian people' (Taylor 1993).

Little-effect states

Tasmania, New South Wales and Victoria were likely to be relatively lightly affected by native title for two reasons. First, except for the Western Division of New South Wales, most private land in these states is freehold and most Crown land has a dedicated use (e.g. forestry, wilderness) and therefore is not susceptible to native title claims. Second, the pressure of settlement on Aboriginal populations in these states had been intensified by the extensive resettlement of Aborigines on lands outside their traditional clan or nation areas. Thus, even where Aborigines in these states had a current association with land, it would have been extremely difficult for them to prove a *traditional* association. Perhaps because history had largely 'solved' the native title problem for non-Aborigines, New South Wales and Victoria were quicker than other states to develop legislation in response to the *Mabo* decision.

Unclear states and territories

While South Australia and the Northern Territory had large areas potentially claimable under native title, they had a relatively small exposure to native title claims. Much of the land that could have been claimed under native title had already been granted to Aborigines under various forms of title. To minimise the potential impact of native title legislation, the South Australian Premier opposed the states' sharing the burden on *Mabo* compensation, and also opposed granting Aborigines rights to object to land development.

Although Queensland has a large Aboriginal population, its Aborigines have been largely confined to relatively small reserves and thus the area of land claimable under native title appears to be quite small. As revealed by the extraordinary fracas over the Wik people's claim in the Weipa area, Queensland appears to be more susceptible to land rights claims under breach of fiduciary duty where reserves were created with the state as a 'trustee' for the Aboriginal population (cf. Mabo 1992, p. 48 per Brennan, J.). The Native Title (Queensland) Bill 1993 was drafted on the assumption that the Commonwealth's Native Title Bill 1993 would be enacted by the Federal Parliament.

Large-effect state

Western Australia has the largest area of land open to claim under the *Mabo* decision. This state was settled later than others, much of its land is unsuitable for agricultural and pastoral uses and therefore freehold or leasehold grants were not made (about 40 per cent of the state was *vacant* Crown land), and its pastoral leasehold often retained access rights for Aborigines.

The Western Australian political response to *Mabo* was three-fold. The first response was to deny that the High Court had the right to make the decision it did. The second response was to argue that, since the Court erred in its *Mabo* decision, it was up to the people to rectify this error in a referendum. The Western Australian Government's third response was to pass the *Land (Titles and Traditional Usage) Act 1993* which extinguished all native title in the generally understood senses of 'ownership' and possession, and replaced it with 'statutory rights of traditional usage'. Court (1993, p. 1) asserted that sovereignty implied 'the Crown acquired the absolute ownership of all land within the state' and thus 'no title to land in Western Australia has existed other than title granted by or under the Crown'.[6]

The principles on which the Western Australian Government based its response to the High Court's *Mabo* have Aboriginal interests tacked on as an afterthought. These principles were to ensure the dominance of the state government in the 'policy and administration' of Western Australia's natural resources; to ensure 'certainty of land title for existing and future title holders' (but not for native title holders since their interests were forthwith extinguished); and to integrate with strategies to improve Aboriginal living standards. Nothing in the Bill dealt with the last of these principles.

Economic issues and analytic techniques

Property rights in land

Some commentators were concerned that the type of title granted under *Mabo* principles was 'inferior' title because it was not unrestricted freehold (e.g. Brunton 1992, p. 10). These and similar arguments can be seen in several contexts:

1 Efficient economic outcomes depend on the preferences of *all* members of the community. Aborigines themselves are best placed to choose their preferred form(s) of land tenure.

2 Most of Australia's land is held as communal title (e.g. Crown land and leasehold), which some commentators argued is inappropriate for Aborigines. Since both the *Mabo* judgement and land rights legislation principally envisage the conversion of Crown land to some form of communal Aboriginal tenure, the result in property rights terms is no worse than currently. If it is true that 'forms of inalienable tenure ... destroy the value of the land' (Brunton 1992, p. 10), the land to which Aborigines might expect to have access via native title has *already* had its value 'destroyed' because of the form of its existing tenure. Ironically, much of Australia's wealth is also held in communal forms of title—e.g. all social infrastructure (see above); all private property held by joint-stock companies; and all property held by not-for-profit firms (e.g. churches, trades union and professional associations, philanthropic organisations, cultural associations).
3 The *Mabo* decision did not preclude Aboriginal groups who gained native title from subsequently seeking conversion to some other form of title by ceding such land to the Crown. Aborigines seeking such a conversion would want to ensure that their deal with the relevant government was watertight! Exactly this possibility was canvassed in the *Mabo* judgement (Mabo 1992, per Brennan, p. 3; see also Commonwealth of Australia 1993b, para.29 and the Commonwealth's Native Title Bill 1993 s.20).
4 NSW Farmers (1993, p. 3) argued that 'communal ownership ... denies Aboriginals the ability to upgrade and efficiently use their land'. This argument rests on an odd interpretation of neoclassical economics where private, individual property rights are a *sufficient* condition for Pareto efficiency. However, such property rights are not a *necessary* condition for efficiency.

Urgings by non-Aboriginal Australians against the type of land right granted via *Mabo* or land rights legislation might best be seen as the attitudes of participants in the political process who objected to the principle of Aboriginal land rights. Native title under *Mabo* principles may not be, in the long run, the optimal form of title for successful grantees. In the short run, however, native title was likely to be far superior to the alternative of no title at all. *Mabo* principles offered a route to future title which may be superior in the long run—and this may have been the real foundation of the objection to the *Mabo* judgement.

Comparative statics—ignoring time and uncertainty

The allocation of Australia's resources between Aboriginal and non-Aboriginal Australians may be illustrated using conventional Pareto welfare economics analysis. In a two-person world, the original possible utility combinations are represented by the OB0 segment of the vertical axis (figure 10.1). Aborigines had a refined production technology, well-adapted to exploiting Australia's natural resources and, prior to European colonisation, only Aboriginal preferences mattered. The relevant social welfare function was that of Australian Aborigines (e.g. line b0b0), and the Pareto optimal allocation of resources is the point B0.

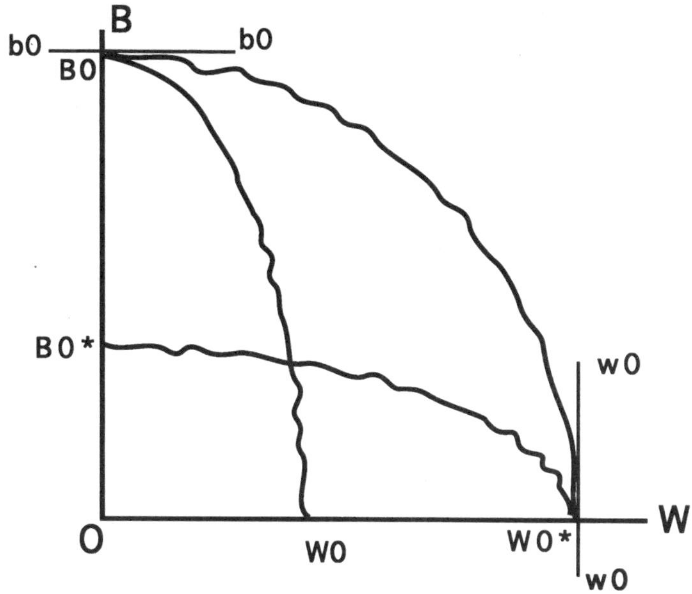

Figure 10.1 Welfare Implications of White Settlement

European Australians initially understood production conditions poorly, and produced little. Combined Aboriginal and European production might, for example, have offered utility possibilities in early years of European settlement represented by the utility frontier B0W0 (figure 10.1). As Europeans learned to better exploit the land, their production possibilities improved. Conversely, as Aborigines were dispossessed of their land, resources and culture, their command over resources declined and even their production technologies depreciated; over time, the utility possibilities frontier contracted to B0*W0*. Once Europeans had effectively colonised Australia and had dispossessed Aborigines from the better-watered areas, the only relevant preferences in the implied social welfare function were those of non-Aborigines (e.g. line w0w0). The optimal allocation of resources—ignoring Aborigines—was therefore point W0* (figure 10.1).

The economic effect of making native title available to Aborigines is depicted in figure 10.2, where utility is assumed to be a function of services provided by land and money income. Because a grant of native title for land under *Mabo* principles presupposes a continuing and traditional association of an Aboriginal group with this land, native title is simply unavailable to non-traditional Aborigines. These Aborigines' land asset is generally zero, and their non-land non-transfer payment income is generally low (e.g. point Z in figure 10.2a). Their welfare level is commensurately low. But the unavailability of native title to non-traditional Aborigines means that there are no economic welfare implications for them—nor for non-Aborigines with whom they have economic relations—of the *Mabo* decision.

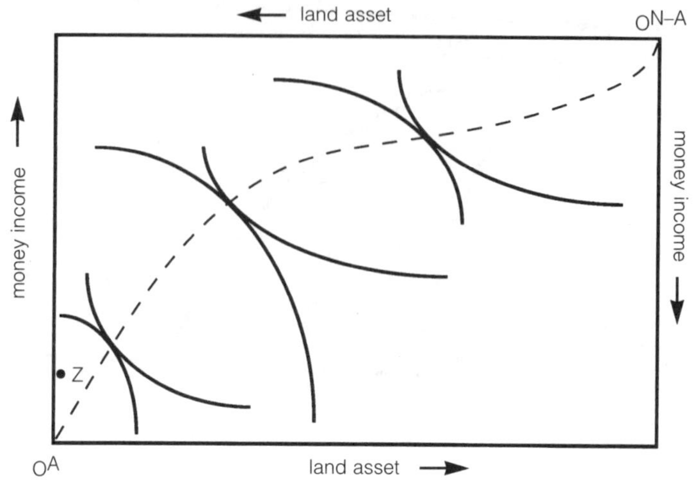

Figure 10.2a *Mabo* and Non-traditional Aborigines

The economic situation of 'traditional' Aborigines, for whom native title became available under the *Mabo* decision, is depicted in figure 10.2b. Suppose non-Aborigines and traditional Aborigines currently have access to initial allocations of land and money income corresponding to point Z. Voluntary exchanges of land and money income could lead to efficient outcomes on the segment *ab* of the contract curve. Traditional Aborigines' land access is, however, not secure: Aboriginal experience following European settlement has been that, as non-Aborigines want additional land, they take it. Except in the Northern Territory and South Australia, Aborigines' continued access to lands available to them prior to the *Mabo* decision was very insecure. Aborigines had every reason to expect that, if non-Aboriginal Australia found a valuable use for their land, Aborigines would lose their current rights to this land (cf. Godden 1993, 1994). Without secure property rights, Aborigines had every reason to anticipate future losses of land access, such as represented by point Z′ in figure 10.2b, with corresponding access through voluntary bargaining only to segment *mn* of the contract curve, yielding lower utility to Aborigines than obtainable via bargaining from their initial allocation at Z, and higher utility to others.

Voluntary bargaining which permits attainment of the segment *ab* of the contract curve CC in figure 10.2b provides Pareto-safe economic change. Pareto-safe changes comprise that sub-set of welfare improvements where, without specification of a social welfare function or welfare improvement criteria, it can confidently be concluded that welfare has increased. Other changes may be welfare-increasing but not Pareto-safe; but such changes require explicit welfare criteria (or a social welfare function) to be specified to decisively conclude that welfare does, indeed, increase. Pareto-safe changes are most assuredly identified by voluntary transactions. By contrast, where private force or

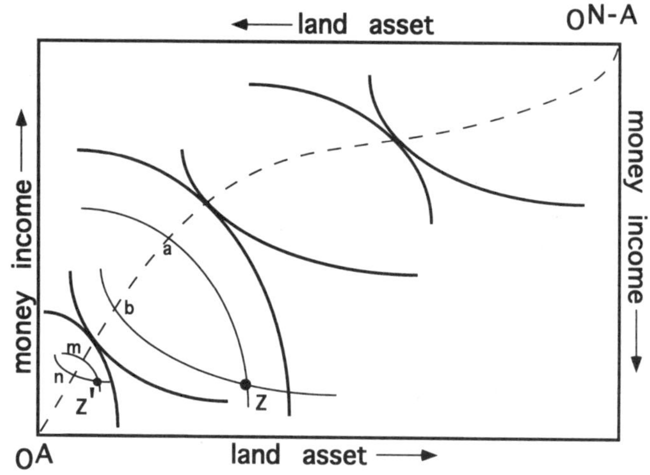

Figure 10.2b *Mabo* and Traditional Aborigines

the coercive powers of the State are used to effect economic change, there is considerable doubt as to whether or not the change is Pareto-safe. Particularly where land expropriation affected groups identified solely on the basis of race—as with Aborigines' historical land losses—and without compensation being paid, the most reasonable conclusions to draw are that:

i because the change was involuntary, it was not Pareto-safe;

ii because compensation was not paid and the disadvantaged group was poor, the change was not welfare-increasing; and

iii because the change was coercive it probably was not even a potential Pareto improvement or potentially welfare-increasing.

Where the *Mabo* decision provided *secure* property rights, native title would prevent *future* expropriation of Aboriginal land, and a non-voluntary reallocation of resources comparable in the shift Z to Z' (figure 10.2b). If coercive land transfers are neither Pareto-safe, welfare-increasing nor potential welfare improvements, then the High Court's decision to secure native title via the common law is likely to have prevented subsequent reductions in social welfare.

The likely social welfare implications of the *Mabo* decision for traditional Aborigines are illustrated in figure 10.3. With the utility possibility frontier B0W0* corresponding to the contract curve in figure 10.2b, and assuming the existence of some weakly egalitarian social welfare function,[7] the highest level of social welfare attainable is SWF_4 (given by Z* in figure 10.3). The segment *ab* of the utility possibility frontier B0W0* (figure 10.3) corresponds to the segment *ab* of the contract curve in figure 10.2b, and represents the level of social welfare (SWF_3) attainable by traditional Aborigines with secure access to land prior to *Mabo*. With expropriated property rights implied by Z' in figure 10.2b, only segment *mn* of the utility possibility frontier B0W0*—corresponding to segment *mn* of the contract curve in figure 10.2b—would then be attainable by voluntary trade (figure 10.3). As long as the segment *ab* of B0W0* is to the 'right' of Z*, the level of social welfare implied by *mn* is lower than that implied by ab (e.g. $SWF_2 < SWF_3$). From the preceding argument, a *prima facie* case can be made that expropriation of title from a relatively poor group like traditional Aborigines would be welfare-reducing.

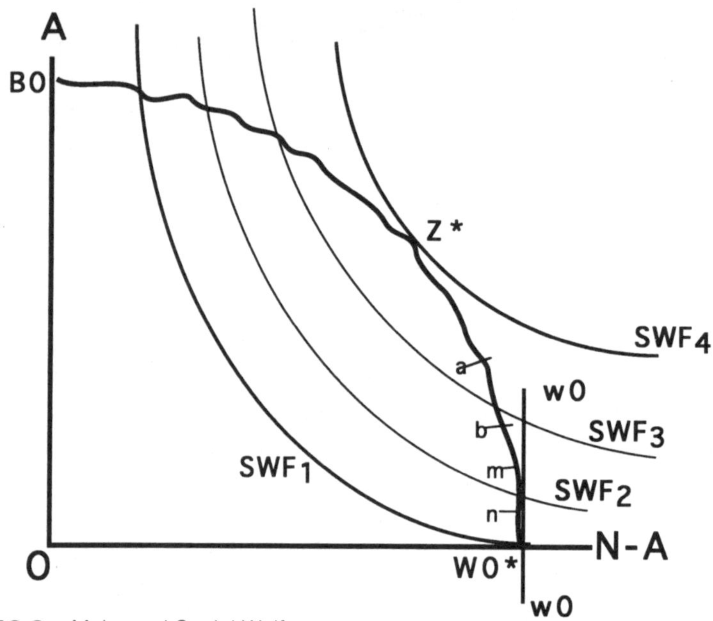

Figure 10.3 *Mabo* and Social Welfare

The social welfare indifference curve w0w0 is undemocratic since it was constructed on the assumption that the welfare of Aborigines is irrelevant. Any egalitarian social welfare function has indifference curves that asymptote to the axes. Thus the level of welfare implied by a putative 'social' welfare function whose indifference curves are like w0w0 and imply a social welfare maximum at W0*, actually corresponds to a level of social welfare consistent with the social welfare indifference curve SWF_1. The High Court's *Mabo* decision is consistent with the establishment of a system of secure property rights for at least some Australian Aborigines that would prevent achievable welfare from changing from segment *ab* to segment *mn*, and ultimately W0* (figure 10.3); and with the assertion that the welfare function cannot be of the form given by w0w0.

The High Court's judgement implied a decision to prevent social welfare falling from some level like SWF_3 to some level like SWF_1 (figure 10.3). Of course, *absolute* certainty about the relative welfare effects of voluntary land trades versus expropriation would require an empirical analysis involving, *inter alia,* specification of either an acceptable social welfare function or acceptable criteria for evaluating relative welfare changes. While it cannot be asserted with absolute certainty that the High Court's

Mabo decision was welfare-increasing (in the sense of preventing future reductions of social welfare), increased welfare was highly probable.

The preceding analysis is useful for examining the coercive transfer of a private good (i.e. one which is both rival and price-excludable) because the benefits can be analysed solely in terms of the utility effects on two individuals. For non-Aborigines, who are primarily interested in land for its economic potential, land is a resource that yields private goods whose benefits to the community are substantially appropriated by the land owner. However, with traditional Aboriginal land (and possibly also already alienated land for both traditional and non-traditional Aborigines), the land resource is closer to a public good because, for cultural purposes, it is utilised in a non-rival way by the group which owns, previously owned or had an entitlement to the resource. Thus, the welfare impact of expropriating land from Aborigines, or the effect of the High Court's native title in preventing such expropriation, must be assessed in terms of the welfare effects for the affected Aboriginal *group* compared to the effects on the individual(s) to whom this title might be transferred.

Resource effects vs wealth transfers—effects on output

Implicit in the foregoing argument about native title is the assumption that the level of output is independent of the initial assignment of property rights. Suppose, as in figure 10.4, the initial assignment of land property rights is to Aborigines and the corresponding level of money income is given by the vertical height of the smaller Edgeworth–Bowley box ($O^AA = O^{NA}B$). The contract curve O^AO^{NA} indicates the efficient bargaining possibilities. With the initial assignment of rights at Q, Aborigines and non-Aborigines have available the segment *ab* of the contract curve O^AO^{NA} for voluntary trade.

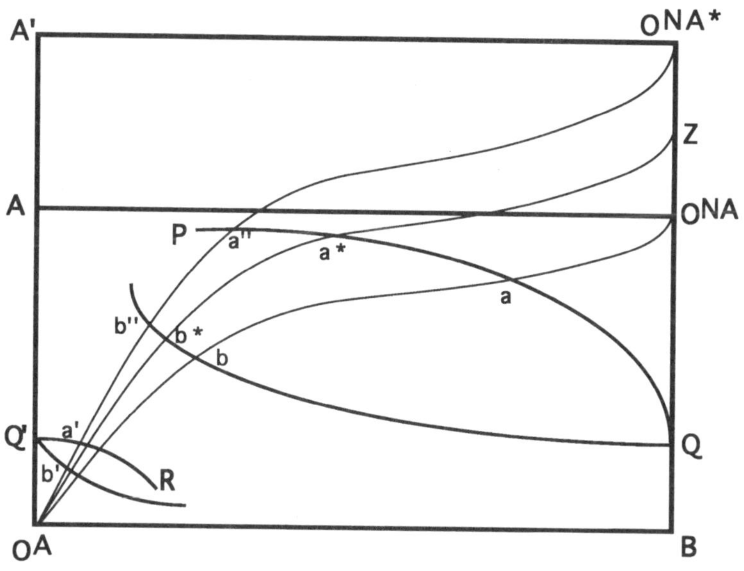

Figure 10.4 Growth in the non-land income with change in land ownership

Now suppose that non-Aborigines could, if they controlled the land resource, increase money income (e.g. by mining). Then, with an initial assignment of rights Q′ (i.e. Aborigines have no rights to land) the relevant Edgeworth–Bowley box has vertical height $O^{A}A' = O^{NA*}B$ (figure 10.4). The relevant contract curve is then $O^{A}O^{NA*}$ and, for the initial assignment of rights Q′, segment *a′b′* of contract curve $O^{A}O^{NA*}$ is feasible for voluntary trade. Non-Aborigines' indifference curves PQ and Q′R represent higher levels of utility with the expanded box having origin O^{NA*} than the same indifference curves in the box with origin O^{NA}. The Aborigines' indifference curves represent the same levels of utility for both boxes as they have the same origin O^{A}.

Possible welfare implications are illustrated in figure 10.5. Corresponding to the contract curve $O^{A}O^{NA}$ is the utility possibility curve B0W0*, and corresponding to the contract curve $O^{A}O^{NA*}$ is the utility possibility curve B0*W0**. The initially available voluntary trades from Q are contained in the segment *ab* on the utility possibility curve B0W0*. With the expanded box and the land expropriation represented by Q′, Aboriginal welfare has unequivocally fallen, and non-Aboriginal welfare has unequivocally increased (segment *a′b′* on curve B0*W0**). The expropriation is thus clearly not Pareto-safe, although it is possible that social welfare might have increased, decreased, or remained unchanged. In figure 10.5, for example, the change from Q to Q′ is depicted as non-Pareto-safe, but with social welfare unchanged. However, the expropriation implied by Q′ is not the only way that land could become available to non-Aborigines for non-traditional uses. For example, voluntary trades between Aborigines and non-Aborigines that commenced at Q would allow efficient solutions to be realised along the segment *a″b″* on the contract curve $O^{A}O^{NA*}$. Pareto-safe exchanges resulting from Aboriginal/non-Aboriginal trades that corresponded to segment *a″b″* on the contract curve $O^{A}O^{NA*}$ in figure 10.4 would occur on the segment *a″b″* of B0*W0** in figure 10.5,[8] and these changes would unequivocally increase social welfare. By a similar argument to that in the preceding section, the expropriation implied by the initial starting point Q′ is most likely to result in a lower level of social welfare than would have been available via voluntary trades—i.e. the segment *a′b′* will be at a lower level of welfare than *a′b′* in figure 10.5. Moreover, the segment *a′b′* is most likely to occur at a lower level of social welfare than the initial level of welfare corresponding to segment *ab* on the frontier B0W0* because, if the change to Q′ did *not* diminish Aboriginal welfare so much that social welfare was diminished, coercion to achieve the trade would have been unnecessary.

As previously argued, absolute certainty about the relative welfare effects of voluntary land trades versus expropriation would require an empirical analysis which, *inter alia*, involved specification of an acceptable social welfare function or explicit criteria to identify an improvement in welfare.

The voluntary trade in land beginning from Q (figure 10.4) might not result in all economically valuable land becoming available to non-Aborigines. For example, traditional Aborigines might not be prepared to trade land of religious or other cultural significance. In this case, the potential increase in money income in figure 10.4 is less than

O^AA', and the resulting contract curve for efficient trades lies between O^AO^{NA} and O^AO^{NA*}. Corresponding to this contract curve there will be a utility possibilities curve UV lying between B0W0* to B0*W0** (figure 10.5), and the available voluntary trade segment with the initial assignment of property rights at Q will be some segment *a*b** of the utility possibilities curve UV 'north-east' of the initial segment *ab* (figure 10.5). With voluntary trade, therefore, Aborigines would be unequivocally better off if the increase in non-land income occurred via a trade in property rights. Since the effects on Aborigines' welfare is severe if the increase non-land income occurs via expropriation of Aborigines' property rights, increasing non-land income by allowing Aborigines to trade voluntarily with non-Aborigines to effect this income change will almost certainly increase *social* welfare more than if the increase in non-land income occurs via expropriation.

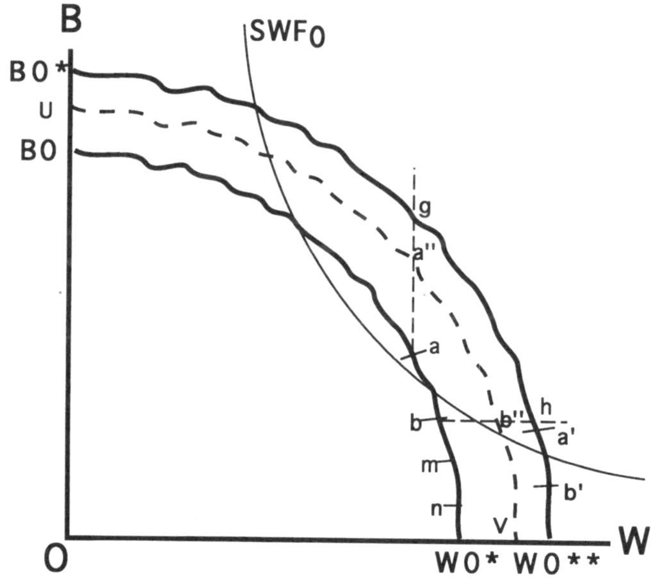

Figure 10.5 Income-increasing Land Reallocations

Risk and uncertainty—*whose* uncertainty?

Not only did non-Aboriginal Australians work 'themselves into positive paroxysms of guiltlessness' in response to the *Mabo* judgement (Wootten 1993a), they also hyperventilated on uncertainty (cf. Godden 1993, 1994). Uncertainty following *Mabo* should be seen in its proper context. Prior to *Mabo*, and except where land rights grants had been made with certainty of tenure (e.g. Northern Territory), Aborigines occupying Crown land such as reserves had no certainty of tenure. Aborigines' settlements could be, and were, relocated at the whim of government, especially where alternative land uses became attractive, such as the exploration or exploitation of mineral resources (cf. Godden 1994). Thus, prior to *Mabo*, resource developers were able to obtain a high degree of resource security from government, and thus ultimately non-Aboriginal society had certainty in the use of land. The cost of this certainty was, however, total uncertainty for Aboriginal occupiers.

Traditional Aborigines'—and non-Aborigines'—utility possibilities illustrated in figure 10.4 should, therefore, be seen in a risky context. Traditional Aborigines' land title was not, in fact, a single point such as Q but rather a probability distribution around their current level of land access. The feasible voluntary trade segment of the contract was, similarly, not an identifiable segment such as *ab*, but a probability distribution of such segments along the contract curve $O^{A}O^{NA}$. Because traditional Aborigines' access to land under the High Court's native title decision was confined to tracts of land where native title could be deemed not to have been extinguished, there was no possibility of expanding traditional Aborigines' existing access to land under common law native title. Given continuing demands for land for non-agricultural purposes (e.g. mining), traditional Aborigines' expectations about their likely future land holdings prior to the High Court's *Mabo* decision must have been that they would continue to lose land. They would rationally have expected that eventually they would lose *all* their land. Thus, the pre-*Mabo* burden for traditional Aborigines was that their long-run expected land holding would be zero (i.e. Q′ in figure 10.4). Except in the Northern Territory and northern South Australia, traditional Aborigines must have expected that, without native title that vested them with land rights, eventually they would have been left with none of the land resource which is central to their culture, and a low living standard essentially based on charity or social security.

Conclusion

Since *Mabo* principles enhanced some Aborigines' ability to access future welfare possibilities superior to their existing expectations—and did so at the cost of at most minor reductions in the welfare of non-Aboriginal Australians—the *Mabo* decision was more likely to increase Australia's welfare than to reduce it. But results of the 1996 federal election suggest that the native title issue is far from settled.

ISSUES FOR DISCUSSION

1 Should the High Court have made a determination on native title in the *Mabo* case, or should this decision have been left to the parliaments of the individual states?

2 'Structural adjustment in the Australian rural sector should be endless for European agriculture but structural adjustment should eventually come to an end for rural Aborigines.' Discuss.

3 The Prime Minister recently failed to persuade the Council of Australian Governments to adopt a common national approach on the High Court of Australia's *Mabo* decision on 'native title'. Discuss economic issues resulting from the *Mabo* decision which relate to agricultural and resources policy.

Endnotes

1 Of course, not everyone has the same attitude to distributional issues; in a reported conversation involving the Australian Head of State: 'Somewhat superfluously, Bob Hawke told the Queen that Mr Howe [Minister for Social Security] was particularly concerned with the problems of poverty. "Yes, it's such a drain on the taxpayers", Her Majesty replied' (*Sydney Morning Herald,* 20 June 1988, p. 1). Thanks to an agreement between the British Government and one of her predecessors, the Queen (of the United Kingdom) until recently has paid no income tax, despite a reported wealth in billions of pounds and an annual salary from the Civil List in the tens of millions.

2 This section is a condensed version of Godden (1993, 1994) in which further detail and references are available.

3 The Prime Minister's commitment to Mabo—reported as 'If the Government did nothing in this term of Parliament but get the response to the Mabo case right "it would be worth it"'—was apparently not shared by others: 'the questioning of Keating's preoccupation with the Mabo issue has become more and more hostile' and 'The Labor Caucus . . . has become impatient about the perplexing inability of the Prime Minister to get the Mabo debate under control and alarmed that he was neglecting the political management of other issues . . . And to what end? some asked. Mabo was never going to be an issue which would win Labor votes. It was more likely to be a vote-loser' (Kitney 1993, 3 September).

4 In an elegant demolition of Fischer's argument, Wootten (1993b) observed, *inter alia*, that it probably wouldn't have made much sense for Aborigines to develop a wheeled cart since they had no draught animals. He also argued that 'Justice is not a prize for being the most powerful, most technologically advanced or even the most self-righteous: rather it is the tribute the powerful will, if they are genuinely righteous, pay to the less advantaged, the more so because they have often built their wealth and power at the expense of the latter.'

5 Some of these 'past' horrors are however within the memory of living, mainly Aboriginal, Australians; there are also horrors of the very recent past and the present (cf. section 10.1 and Godden 1994). Wootten (1993a) commented that a feeling of 'guilt' over such injustices is not necessary to justify action to eliminate them: 'In what other sphere do we regard guilt as the only reason for action. Should Granville Sharp and Wilberforce have ignored slavery because they had not caused it?'. Hulme (1993, pp. 56–7) argued that 'guilt' about Aborigines was being over-emphasised and should be put in context: 'I read again and again that we must not blame present-day Germans for what Hitler did in the war . . . One wonders whether Deane and Gaudron JJ. would say that Germany bears a national legacy of unutterable shame, for the Holocaust.' But contemporary Germans do continue to grapple with just these issues of guilt and national responsibility. In November 1993, on the combined anniversaries of 'Crystal Night' (1938) and the demolition of the Berlin Wall, the Speaker of the German Bundestag said: 'Whoever believes [he can] escape the history of his nation by saying [he did] not . . . participate will be wrong. He wants to break with things he cannot break with. History does not discharge anybody. We are all involved and have to take it over as a heritage and at the same time as a charge. The present generation has not made itself guilty of former crimes. But today we would fail if we would not take the responsibility for those days. What matters is to learn from the past in order not again to become guilty some day . . . ' (Süssmuth 1993). The German President had previously said: 'All of us, whether guilty or not, whether old or young, must accept the past. We are all affected by its consequences and held responsible for it . . . Whoever does not wish to remember inhumanity becomes susceptible to the dangers of new infection . . . As human beings, we seek reconciliation. Precisely for this reason we must understand that there can be no reconciliation without memory' (von Weizsäcker 1986, p. 48).

6 The Land (Titles and Traditional Usage) Bill, however, recognised that the Crown was not 'the sole source of title' to land since its section 7 extinguished 'any native title to land that existed immediately before' commencement of the Act.

7 'Weak' egalitarianism implies that indifference curves of such a social welfare function asymptote to the utility axis for each individual in the economy; i.e. as one individual's utility tends to zero, another individual's welfare must tend to infinity to hold social welfare constant. Australia's social welfare institutions are consistent with 'weak' egalitarianism since sufficient income transfers are made available to prevent absolute destitution.

8 In figure 10.5, Pareto-safe improvements made available by land development from segment *ab* in the absence of land development are represented by available welfare points denoted by area *gabh*.

11 Rural services

Farm structural change occurs when capital is substituted for land and labour, resulting in substantially larger farm units and an outflow of farm labour, both employed and self-employed. The labour exodus from farming reduces the farm population and its demands for services from small country towns, as well as reducing the farm labour supply living in those towns. The substituted farm capital tends to be highly specialised, and requires specialised services for its distribution and maintenance. Because suppliers of these services require economies of scale in their own activities, they tend to consolidate in larger country towns and cities. Capital substitution thus places a double squeeze on small rural towns from both the supply and demand sides. Larger country towns tend to grow at the expense of the smaller, and a small number of larger regional centres dominate this growth. Thus there are changes in the size and distribution of the rural population.

Changes in the aggregate demand for agricultural output have accentuated changes in the structure of the export sector of the Australian economy. Agricultural output has increased in industrialised countries with highly protected agricultures (especially Western Europe and the United States) with only slow import liberalisation in some potential importers of Australian agricultural products (e.g. Japan). These exogenous demand changes have reinforced the pressure for quite substantial structural change in the farm sector, which has flow-on effects into related economic activities in rural and regional areas.

These changes produce additional flow-on effects. As population contracts in the smaller rural towns, it becomes increasingly expensive to maintain the social infrastructure. Where this depends on local taxation, such as government rates, it has been increasingly difficult for the local community to raise sufficient revenue to fund the existing level of services, or the additional services demanded as individual affluence grows. As state and federal governments have become increasingly conscious of the need for efficiencies in their delivery of services, the flow of funds to small communities, where there are no or at best limited economies of scale in the provision of social services, has been reduced. Reduction in services creates a downward spiral as population moves to the larger towns and cities where these services are provided, creating even greater problems for the local delivery of services in smaller towns.

Concurrent exogenous changes have both offset and exacerbated these changes. Improving technology in motor vehicles and aircraft has reduced the costs of travel to firms and services located in larger centres, and has enabled larger centres to compete even more effectively with services remaining in small towns. These same reductions in costs have, however, also expanded the potential markets for some firms operating in small rural centres.

In this chapter, some policy aspects of changes in the relationships between the farming sector and the rest of the economy are examined. These include changes in the demand for private (or sometimes social) infrastructure to 'add value' to farm products, and changes in demand for social infrastructure. Of particular interest is the interaction between changes in the farm sector, associated changes in production infrastructure, and changes in rural social infrastructure.

Policy issue

The key policy issue is the appropriate provision of services—both private and public infrastructure—to the rural community. This issue has both efficiency and distributional aspects. The efficiency aspect primarily concerns the efficient public provision of 'public' goods—given the demands for public goods in rural areas, what is the efficient level of provision? The distributional aspect concerns providing services in rural areas comparable as far as possible with those enjoyed by urban Australia. There are, however, two rural Australias. In the large country cities and their surrounding areas, the provision of infrastructure—medical services, public health (water, sewerage, electricity), education (school and, in many cases, tertiary), public security and legal services, sporting and cultural facilities (including access to the media)—is similar to that enjoyed by metropolitan Australia. In smaller urban areas, however, local access to these services is, and is likely to remain, seriously restricted.

Several competing policy threads can be discerned relating to the efficient provision of services which are public goods or publicly provided; the just distribution of social services in a community whose generally egalitarian values are prepared to tolerate some sub-optimality; public choice dimensions where social provision of infrastructure is an important political currency; and the importance of social infrastructure for achieving the 'decentralisation' to which most Australian governments at least pay lip-service.

Policy intention

Rural services are provided by federal, state and local government, their provision reflecting the objectives of decision makers at each level of government. These objectives have rarely been crystallised into a single statement. The Federal Labor

Government identified four principles underlying its policies for rural and regional Australia:

1 Social justice, which means that all Australians, regardless of their background and where they live, have fair access to necessary services.
2 Maximising opportunities for all, so that all people have the chance to achieve a personally fulfilling life.
3 Economic management, which ensures that Australians can not only get jobs but have opportunities to improve their job skills and rewards.
4 Consultation, which is important to ensure that the views of all people are considered in developing policies which affect them (Hawke 1989, p. 5).

Each of the three levels of government continually juggles its limited resources with some semblance of an intention to provide better services in rural Australia. The degree to which this intention is actually implemented varies, partly as a consequence of the political complexion of government at any particular time. Different governments also emphasise the provision of different types of services to rural Australia, while services themselves may be quite defective for some groups in rural Australia, even where the overall level of services is satisfactory (cf. Chapter 10).

Area of sector

Services are important to all parts of the rural sector. Public infrastructure is important in:

- servicing agricultural input markets—providing water, transport and electricity;
- regulating input markets, such as agricultural chemicals, and professional services;
- servicing the farm sector—providing specialist management services via state agriculture departments, soil conservation agencies and federal/state Landcare arrangements;
- servicing natural resource industries—regulating stream flow and fisheries, forestry production, managing national parks and wilderness areas, managing indigenous and feral wildlife; and
- servicing the marketing chain—providing transport infrastructure and marketing facilities.

As well as production infrastructure, local, state and federal governments provide somewhat similar social infrastructure to rural and urban communities. Except in rural communities which are close to major urban areas, however, the extent to which social infrastructure is provided is limited by economies of scale and the difficulty of providing some services over considerable distances.

Policy orientation

The increasing power of the Federal Government has been reflected in an increasing willingness by federal governments of both political persuasions to provide rural services in addition to general social services. Especially since the Labor Government of the early 1970s' push into regional development and attempts to link directly with local government, federal governments have continued to exert their influence over rural service provision.

The Federal Government has engaged in establishing national priorities for rural and community infrastructure, and has partially achieved these objectives via s.96 grants and/or through cooperative federal–state agreements on establishing and funding infrastructure projects. The inclusion of a representative of local government on the Council of Australian Governments marked a major change in the role of local government in the interaction between the Commonwealth and states (cf. Chapter 3). However, providing and operating public infrastructure—whether specifically for the agricultural sector or more general community facilities—is largely the responsibility of state and local government.

Policy mechanism

Major spending activities are likely to require legislation, and thus parliamentary scrutiny, but executives of Commonwealth and state governments have considerable flexibility in selecting the form of specific government activities. At the Commonwealth level, major decisions affecting rural infrastructure may emanate from departments involved in education (e.g. TAFEs and universities), health (funding for hospitals), transport (e.g. funding of the recent completion of the Perth–Brisbane standard gauge railway, rural airports), and primary industries (e.g. irrigation schemes). At the state level, departments of agriculture, water resources, national parks and wildlife, forestry, conservation, and soil conservation provide services. Local government provides services such as roads and water. There are also many quangos, which deliver services such as electricity and pest control.

A key element of rural service provision is the commitment of capital works funding by federal and state governments. Such funds are formally controlled by Budget legislation, but particular decisions are developed administratively and via behind-the-scenes bargaining. In New South Wales, for example, the Capital Works Sub-Committee of Cabinet plays the major role in ordering and recommending spending priorities. Sometimes major decisions can be implemented administratively, particularly with regard to measures like the geographical location of government offices.

Institutional structure

Lobby groups influencing the provision of rural services include political parties, especially the National Party in rural and regional Australia, farmer groups, and local government. The search for lobbying efficiency has led local government to organise into regional groupings; in New South Wales there are also associations of country mayors (representing larger country towns and cities) and shires (encompassing smaller towns and areas of agriculture, forestry and national parks) (Sorensen 1994, p. 34).

Policy performance evaluation

Sorensen (1994) argued that a resurgent interest in regional development had not been consistently served by recent reviews of regional development problems and proposals for policy development. The Industry Commission (1993) recommended increasing the flexibility of regional labour markets, mitigating the efficiency consequences of social security measures, reducing regulatory impediments to regional development (including environmental and transport), shifting the emphasis from targeting specific regions to making regional location in general more attractive to business, and requiring more rigorous and transparent efficiency analysis of regional infrastructure investment proposals (Sorensen 1994, pp. 37–38). Sorensen (1994, pp. 38–40) scathingly reviewed the Kelty report on regional economic development, arguing that it was naive in its dealings with federally oriented policies on finance, transport, labour markets and industry policy. Principally, it lacked analysis of the costs and benefits of the proposed policies. McKinsey and Co. released a discussion paper based on a survey of business views about impediments to regional investment and job creation. Sorensen (1994, pp. 40–2) noted the report's finding focused on two issues: rejuvenation of regional leadership; and the growth of existing regional businesses, especially those focused on export opportunities, which are internationally competitive. Sher and Sher (1994) provided an overseas perspective on the needs for regional development policies in Australia (cf. Godden 1995).

Economic issues

Local processing

How does agriculture interact economically with its local community? At its simplest level, this interaction may be conceived of as the addition of post-farmgate processing services to an agricultural product. An idealised representation of such activities is shown in figure 11.1. The right-hand panel shows the supply of raw material (Sr) from the farm sector. The left-hand panel shows the market for post-farmgate activities—this market comprises the assembly, storage, handling, transport, processing, wholesaling and retailing of product. These activities could occur at the local level, elsewhere in the domestic economy (national), or overseas (international). Some of these activities are location-specific: assembly, the collection of raw material from the farm, is a price-inelastic activity that occurs at the local level; conversely, wholesaling and retailing occur near the final consumer; transport obviously occurs between farmgate and consumer, but the quantity and type of transport services provided partly depend on where processing occurs. Storage, handling, and processing could occur at the local level or elsewhere in the national or international economies. For convenience, these services are assumed to be homogenous.

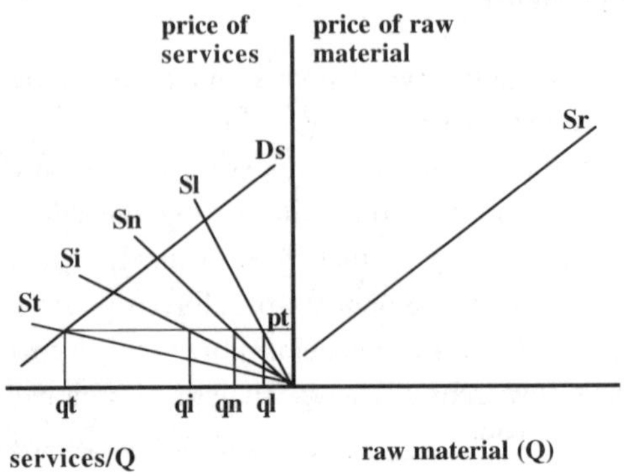

Figure 11.1 Spatial Provision of Post-farmgate Services

The horizontal axis of the left-hand panel of figure 11.1 shows the quantity of services added to each unit of the raw material whose supply is shown in the right-hand panel. The vertical axis of the left hand panel is the cost per unit of supplying each service. The local, national and international supplies of post-farmgate services are represented by curves Sl, Sn and Si respectively. The 'total' supply of these services is the horizontal aggregation of these three curves (St in figure 11.1). These curves show the marginal cost of different levels of services added to each unit of the raw material prior to final consumption.

For a given final demand for services to be added to the raw material (Ds), the equilibrium aggregate quantity of services added to each unit of raw material is qt, at price pt. At price pt, qi services are added in the international market, qn services are added in the non-local national market, and ql services are added to the raw material in the local market.

With this model, it is possible to explore the consequences of changes throughout the economy for the local provision of services to this commodity.

1 Transport costs

Improved transport throughout the national economy reduces the cost of transport services. The cost of transporting bulk raw materials to other locations in the national economy, or to the international economy, falls. The supply curves Sl, Sn and Si—and the total supply curve St—rotate towards the quantity (horizontal) axis (figure 11.1). Because Sn and Si involve the transport of *bulk* rural commodities, whereas Sl incorporates transport costs of lower bulk finished commodities, falling transport costs reduce cost curves Sn and Si more than Sl. As long as the demand curve Ds is not perfectly inelastic, the total quantity of services added per unit of the raw material will rise. The output of services added in each of the three markets will rise, but the level of services added in the national and international markets will rise relative to the provision of local services because of the falling per unit cost of transport. If Ds is perfectly inelastic, production of services in national and international markets will rise and local production of services will fall.

The degree to which changes in transport costs affect the location of processing depends on the importance of transport costs in final product price. Transport costs are less important for high valued or non-perishable commodities. Wool was an important export commodity during Australia's initial economic development because it had a high value relative to its bulk, and was non-perishable. It was profitable to export wool over long internal distances with slow and expensive transport, and wool could even bear long delays in transport such as river boats being marooned for years during drought. By contrast, bulky or perishable commodities such as sugar cane or milk are sensitive respectively to transport costs and transport speed. Where commodities are highly sensitive to transport costs, a fall in local supply which causes a processing facility to close may force other viable producers of the commodity to cease production as alternative processing facilities may be beyond the reach of profitable transport. Where enterprises are highly sensitive to transport costs, increases in transport cost do not necessarily cause major problems if there are alternative profitable enterprises. In the case of the dairy industry on the north coast of New South Wales, which declined rapidly in the 1960s and 1970s, the principal alternative local enterprises were subtropical horticulture and part-time beef farming or hobby farming, which were markedly less sensitive to transport costs. The availability and profitability of these non-traditional farm enterprises contributed substantially to a largely successful adjustment process.

The development of good rail and road services led to the demise of local processing of bulk commodities, because it was more efficient to transport them to large-scale central facilities rather than reduce the bulk by local processing prior to transport. Good examples are the demise of local wool washing and scouring, and local dairy factories producing cheese and butter. It is not inevitable, however, that local processing will continue to decline because of changing transport costs. Consider, for example, current policy controversies over emissions of 'greenhouse' gases. Transport, dependent on fossil fuels, is a major emitter of the greenhouse gas carbon dioxide. If policy responses to the enhanced greenhouse effect, such as carbon taxes, result in rising transport fuel costs, then the marketing services supply curves in the left-hand panel of figure 11.1 will rise, and the increase in Sn and Si will probably be greater than that for Sl, and hence the level of local processing may rise relative to processing elsewhere.

2 *Technology changes*

Prior to the Second World War, most grain was moved from rural Australia in bags, which was labour-intensive. Since many grain-based products (e.g. flour) were (and are) also moved in bags, there was extensive local grain processing. As the cost of labour rose, bulk handling increasingly replaced handling in bags until, by the 1960s, bulk handling dominated. Control of vermin, pests and diseases was also easier with bulk handling. As bulk handling of grain replaced bagged handling, it became cheaper to transport grain to distant markets for processing. Thus a thriving local rural wheat-milling industry has largely disappeared except where tied to further specialised processing (e.g. gluten, and starch and its derivatives). Similarly, the market milk sector of the dairy industry moved increasingly away from metropolitan areas as refrigerated and bulk transport became available, and as the political process adjusted to these changes. A reversal of the local processing decline might result if transport costs rose, or if local processing costs fell relative to corresponding national or international costs.

3 Dependence on local output

Processing costs are affected by throughput via economies of size. Particularly at the local processing level, changes in local raw material production—through either a shift in demand for local production or a shift in the supply curve Sr in figure 11.1—will affect the efficiency of local processing. Thus, for example, if local output increases, the cost of local service supply is likely to fall (Sl rotates towards the horizontal axis in figure 11.1) and locally provided processing services are likely to rise relative to service provision elsewhere. Conversely, if local production falls, the cost of providing local processing services would rise, and local service provision would fall relatively to service provision elsewhere.

Again the dairy industry provides good historical examples. As dairying contracted throughout Australia after the Second World War, local small-scale processing became increasingly expensive relative to larger scale, non-local processing. Eventually, the costs of local processing became so high that local factories closed, and milk was transported over greater distances to larger scale processing facilities.

4 Local factor markets

Local processing depends on the availability of local factors, particularly labour. Factor demands for agricultural processing can often be seasonal. The rapid growth in the services sector of the economy, particularly occurring in urban areas, has put increasing pressure on local labour markets. Non-local processing, which may be less seasonal in nature or which provides a greater range of industries with complementary seasonal labour demands, can offer greater employment stability. To compete against non-local labour markets, local processors may have to offer higher wages. In this case, the local cost curve Sl would rotate towards the price (i.e. vertical) axis, implying higher cost local processing and lower levels of local processing at equilibrium (figure 11.1). Such problems are illustrated by the increasing difficulties of many agricultural industries dependent on hand-picking (e.g. fruit growing, grapes).

These changes would be exacerbated by other contemporaneous changes in the economy. Concentration of government and private services in larger regional centres has made living in these centres relatively more attractive, both because employment is more readily available and because of the high income elasticity of demand for services more readily available in the larger centres. These population shifts have also made it generally more difficult for local, small-scale processors to obtain labour without offering commensurately higher wages to offset the (perceived) disadvantages of living in more isolated areas.

Other changes

A similar analysis can be conducted for the provision of farm inputs. Increasingly, farm input supply has contracted, for efficiency reasons, to the larger rural centres where larger scale operations reap economies of scale by servicing larger hinterlands made possible by better transport and communications. These changes have been accentuated by rapid increases in the technological sophistication of farm inputs whose maintenance requires more highly skilled labour. The efficient servicing of these inputs may require maintenance operators to cover large service territories, sometimes as large as a single state.

Recent technological changes may partially offset some of the disadvantages of local processing suggested above. Advances in communications, related to computers, satellites and fibre optics—and, undoubtedly, future similar developments—have lessened dependence in some industries on physical propinquity for efficient production. Information-related services may be produced in any location and workers may sometimes cooperate in information-related tasks regardless of location. Workers and businesses in information-based industries, which previously could not operate in relatively isolated areas, may become less dependent on urban areas for efficient location. For those workers who value non-urban lifestyles, rapid developments in communications technology may herald a partial reversal of the 'drift to the city'. The extent to which urban drift can be reversed, however, will depend on the extent to which new country residents as consumers can obtain access to services that can only be efficiently provided in larger urban centres, such as regional or metropolitan cities.

Social infrastructure

The differential effects of economic change on public infrastructure are illustrated in figure 11.2. In panel (a), aggregate demand for a public good in a rural town (ΣMBit, summed over I individuals) is low, because of the small population. Because of the small scale of public good production, the marginal cost of providing the public good in the town (MCt) is high. The optimal provision of the public good in the town (Gt) is therefore low. The cost of providing the public good is 0dcGt.

In a rural city (panel (b) of figure 11.2), demand for the public good (ΣMCjc, summing over J individuals) is high because of the larger population, and the cost of providing the service (MCc) is lower because of larger scale operations. The optimal public good provision in the city (Gc) is much higher than in the town. Optimal production Gc in the rural city means that Gc is available to every resident, whereas non-mobile rural town residents only have ready access to the lower optimal level of services Gt. However, because public goods are non-rival, townspeople can enjoy the city's provision of those public good services whose consumption does not require physical location in the town (i.e. some part of Gc).

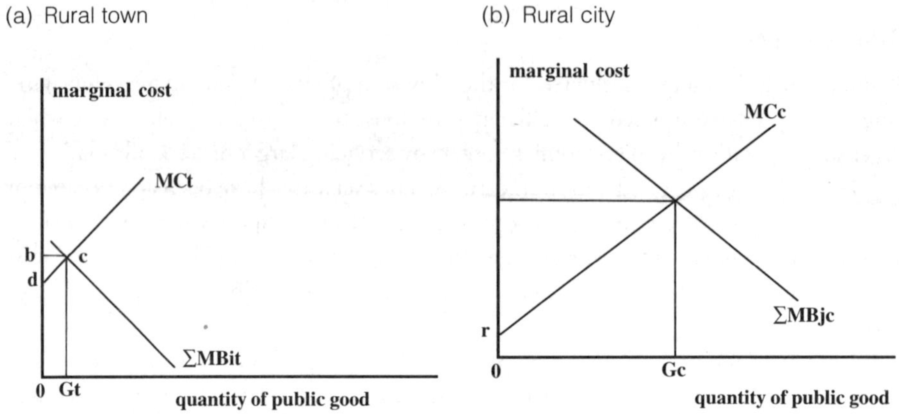

Figure 11.2 Public Goods in Two Rural Australias

For true public goods, which are publicly funded, or for any publicly provided goods that are not public goods (e.g. sporting facilities, education), there is a considerable difference between town and city provision if the cost of provision is borne by the whole community rather than solely by the local community. For the inhabitants of rural towns, goods that are publicly provided in the towns have low private access cost (e.g. low costs of travel and small travel times). The same goods when publicly provided in the city may have similar costs of provision, but access costs for town dwellers are high because they must travel to the rural city to enjoy the publicly provided goods, thus incurring both physical and time costs of travel. Town dwellers would prefer local provision of public goods (if the provision cost 0dcGt is [largely] borne by the community as a whole, not the local community, and access costs are low) rather than provision of the *same* services in the country city. Even if the provision cost is similar, the access cost is much higher *and is borne privately*. Rural objections to the closure of small country schools can be understood as significantly higher private access costs—although these higher costs may need to be traded off against any differences between the quality of education provision in rural versus rural city schools.

The dilemma for rural town dwellers comes from the higher *level* of provision of public goods, and publicly provided goods, in rural cities because of the higher level of optimal production (Gc in panel (b) of figure 11.2 compared to Gt in panel (a)). Continuing the school education example: although private access costs to rural city schools are higher for rural town dwellers, rural cities may offer a much higher level of publicly provided goods measured in either physical or quality terms (e.g. a greater range of subjects taught). Thus rural town dwellers must trade off higher access costs against access to better services. Even where private access costs are subsidised—e.g. school travel—these subsidies only offset the physical costs of travel and cannot reduce the opportunity costs.

Some changes in social infrastructure, even when they replace local services, have dramatically increased the availability of services across rural Australia. Aero-medical services, initially the 'Flying Doctor' for outback Australia and later medical retrieval services for more settled areas, have dramatically improved rural access to medical services. More recent developments, such as the upgrading of hospital facilities in larger regional centres, have dramatically reduced the need to transport rural patients to metropolitan hospitals for specialised services. Providing specialised medical services in major country cities has improved the social infrastructure of medical treatment by minimising patients' dislocation from their community. For rural towns, replacing local medical services by improved services in rural cities, and improved communications and travel opportunities to these centres, has dramatically increased the quality of service available to most rural town dwellers. Similarly, the recent rapid growth of the TAFE education sector and the growth of regional universities has substantially altered rural access to higher education services.

Effects on agriculture

The decline of smaller rural villages and towns which accompanies economic, technological and social changes also affects agricultural production. Smaller urban centres provide some services that contribute to the quality of life of farming families. Such services include marketed services and products such as local retailing, general practitioner medical services, banking, legal and accounting services; public services such as primary schools and perhaps junior high schools, police services, local courts and perhaps basic hospital services; and community services such as sporting and cultural clubs and community service organisations. Local population decline makes it increasingly difficult for private enterprise and government to justify providing marketed and public services at previous levels, or even to continue any local provision. Population decline also pressures locally provided community services because falling populations reduce the income and voluntary labour bases on which local community service provision depends. Better quality (and perhaps lower cost) transport provides access to more distant substitutes for market, public and community services.

Withdrawal of local services also makes farm living less attractive while the growing array of services offered by the larger, and especially growing, regional centres makes rural city living more attractive—for farm children, if not for existing farmers. Thus it may be increasingly difficult to entice future potential farmers to take up farming. If the farm population responds negatively to the withdrawal of local services, and thus farming becomes increasingly capital-intensive, the farm population will continue to fall, increasingly pressuring these local services.

The effects of isolation on farm families may partly depend on family wealth. Wealthy rural families have frequently been prepared to send their children to elite schools in metropolitan centres or large rural cities. Long travel times to school are thus replaced by boarding costs and separation from family. The children of lower income parents have endured long daily bus trips or any limitations of small local schools. Wealthy rural families have also always had ready access to the best transport available—which currently may mean their own aircraft—and so isolation because of high travel costs has always been less important. Good medical and retail services, for example, have always been readily available to rural elites. The combination of access to the best transport options and high demands for good quality services has meant that wealthy rural families have depended much less on local country services. If less-wealthy farm families become ever less prepared to accept lower levels of public and private services available in country areas and become increasingly reluctant to take up or continue farming, these social reasons will help to accelerate the agglomeration of farm holdings.

Political economy of rural infrastructure provision

The provision of social infrastructure in general, and rural infrastructure in particular, provides a neat illustration of the contrast between economic efficiency and public choice. Provision of social infrastructure is an important element of pork-barrelling, one means by which politicians persuade their constituents to continue their voting support for the local member. However, public infrastructure finance is always limited, and infrastructure projects are 'lumpy'. Demands for public infrastructure are many and there is no widow's cruse of finance. Capital spending is constrained by being allocated across electorates. Since all politicians need to demonstrate faithfulness to their electorates, some funds are generally allocated to all electorates—even to Opposition members' constituencies. Project construction for even relatively small projects may extend well beyond the minimum period in which the works could be constructed (cf. table 11.1).

Suppose there are ten identical projects, each costing \$A and each of which yields an annual return of \$B. Suppose the total finance available for construction in each year is also \$A and the minimum construction time for each project is one year. Two possible construction strategies are (i) to complete one project each year, and (ii) to allocate each project \$0.1*A for each of ten years. The net present value of construction costs for each strategy is identical, since \$A is spent each year. The difference between the two strategies is that, under (ii), benefits from any project are only available from year eleven. However, under strategy (i), benefits are available from each project sequentially, beginning in the second year. Under strategy (i), significantly more benefits are available—as long as the discount rate exceeds zero—because projects yield benefits sooner than they would under strategy (ii). At a 5 per cent p.a. discount rate, a project completed in year one that confers a constant annual value on completion, provides about 40 per cent of the discounted present value of its services in its first ten years of operation. Thus, if project construction was spread over ten years rather than being completed in its first year, 40 per cent of the discounted value of the project would be lost.

Table 11.1 Expenditure Horizons for Some Capital Works Projects in NSW

Project	*Start and finish dates*	*Total project cost ($m)*	*Expenditure 1988 ($m)*	*Expenditure*
Agricultural research complex, Camden Park	1984 1991	28.7	1.2	5.8
Balmain Public School, facilities for physically handicapped[b]	1986 1988	2.0	a	1.0
TAFE Rural Studies and residential facilities, Yallah[c]	1986 1989	1.9	a	0.2
New Ambulance Station, Macquarie Fields	1985 1989	0.6	0.1	0.5
Rankin Park Hospital (new)	1986 1991	160.0	25.7	41.1
Bayswater–Mt Piper 500 kv transmission line	1983 1987	100.1	97.4	2.7
Mt Piper-Wallerawang 132 kv transmission line	1985 1988	0.7	0.5	0.1
Shipley Reservoir, Blackheath, roofing	1986 1990	0.1	a	a
Split Rock Dam, Manilla, construction	1983 1989	52.6	43.7	8.1
Binalong water supply	1986 1989	0.4	a	a
Gosford/Wyong water supply	1976 1992	28.4	20.2	1.7
Redevelopment of Darling Harbour	1984 1989	521.5	337.2	184.3
Wahroonga–Berowra freeway	1983 1989	108.8	32.7	55.4
Hume Highway dual carriageway (9 km of road)	1976 1989	19.0	11.0	5.0
New Motor Registry, Gunnedah	1986 1987	0.4	a	0.4
New Motor Registry, Orange	1985 1988	0.6	0.1	0.5
Balmain Bus turning area	1986 1988	0.3	a	0.1

Source: NSW (1987–88)

Notes: a: less than $0.1m

b: of the 54 education department school projects listed in the source as work in progress, 24 per cent were scheduled for completion in 2 years, 43 per cent in 3 years, and 33 per cent were scheduled for completion in 4 years, even though the cost of each project was low (the largest project was only $12.7m)

c: of the 25 TAFE projects listed as work in progress, 16 per cent were scheduled for completion in 2 years, 32 per cent in 3 years, 48 per cent in 4 years and 4 per cent (1 project) scheduled for completion in 5 years; the largest project was $11.0m.

To politicians, however, it is not the economic return from the project that confers the principal benefits of government investment in social infrastructure. Capital works provide local politicians with other significant returns, including political support deriving from the announcement of the future commencement of a project, availability of funds for the project, its commencement, each annual allocation for the project's continued construction, completion of the project, and its official opening—especially if by a high profile public figure. Especially where factors are relatively immobile between rural towns and cities, annual expenditure on a project demonstrates the government's or the individual politician's concern for maintaining local employment. However, the rationing of government capital works across electorates means that completion times are much longer than is technically necessary. Long completion times—and deferring the realisation of benefits from project completion—appear to be a very high price to pay for treating electorates equally each year in the distribution of capital works funds. Economic efficiency might be better served by rotating the allocation of scarce capital works funds around electorates.

The use of government capital works as a political tool indicates an economic irrationality in government infrastructure investment. The *spending of funds* generates a project's return to the politician whereas, from an economist's perspective, *completion of the project* generates the return.

Decentralisation

Population outflows from agriculture and many villages and small towns in rural areas have been a symptom of the relative decline of the agricultural sector. A small number of regional urban centres has grown rapidly. One response to the relative decline of agriculture and much of its urban service areas has been to seek to reverse the population losses, especially by capitalising on the growth prospects of the 'naturally' growing regional centres.

Various strategies have been attempted, including subsidies to encourage private enterprise to relocate to regional centres. Government itself has shifted some of its operations to regional centres. Initially such shifts involved replacing metropolitan administration of rural staff with local administration. In the New South Wales Department of Agriculture, for example, this process began in the early 1950s and was largely completed by about 1980. Departments such as health and education also regionalised their activities. Once regionalisation was complete, continued expansion of government activities in rural areas required more dramatic shifts of government activities—particularly given the substantial rationalisation of government activities by the New South Wales Coalition Government in the late 1980s that had inadvertently targeted rural areas. Thus, for example, the Head Office of the agriculture department—with some 20 per cent of all departmental staff—was relocated to the regional centre of Orange in the late 1980s, and other smaller scale shifts also occurred. These shifts were copied elsewhere: the Victorian agriculture department was shifted to the regional centre of Bendigo, although this shift was eventually reversed following a change of government. Even the Federal Government followed suit: after decades of concentrating government departments in Canberra, the Industry Commission was to be relocated to Melbourne (only partially completed) and the Federal Opposition promised before the 1993 election to shift the Treasury to Sydney.

Attempts to accelerate the 'natural' rate of decentralisation have largely foundered because the private sector has realised that active decentralisation—i.e. relocating existing facilities to rural areas—is an investment decision. The discount rate is central to investment decisions: if substantial investment costs are required to enable the relocation, these costs may outweigh substantial future benefits if the discount rate is positive. Even if there are net *static* benefits of such relocation—e.g. land and building rental costs may be cheaper in rural areas for the employer, and housing costs are likely to be substantially cheaper for employees—relocation costs are often high. Some employees are reluctant to move and are more likely to switch employment in the original location rather than move. Changes to married female participation in the labour force have probably reduced the willingness of many couples to move because of reduced employment prospects for the partner not being relocated. Thus employers may face significant adjustment costs as the existing workforce is replaced by new employees willing to move, or employed at the new location. The higher the degree of specialised skills required of employees, the more expensive employers' adjustment costs are likely to be. Ongoing adjustment costs for some employees are also likely: where a career path involves employment by several employers, a large population centre may facilitate promotion because there may be no need for household relocation to take up new employment. Adjustment costs are likely to be lower for single employees, married employees without children, or households which rent rather than own their own homes, and are likely to be less important the higher the individual's personal discount rate.

Where firms act as if the discount rate is zero and/or where others bear the substantial portion of the adjustment cost, then the investment costs of relocation are unlikely to preclude a move. Governments are typical of such 'firms' but, in the case of New South Wales Agriculture's move to Orange, even the state Treasury—which might have been expected to have supported money-saving relocations—was apparently strenuously opposed (Godden 1991).

Analytic techniques

Two broad approaches are possible to regional economic analysis. The first is optimality analysis, seeking to identify the optimal allocation of resources to regions, or particular uses within regions. The second broad approach is benefit–cost analysis (BCA), in which possible ways of improving resource use in regions might be evaluated.

Regional BCA

In principle, BCA is relatively straightforward. A comparison is required between the benefits and costs of various policy or investment options. Analysis is likely to be complicated because costs are generally incurred before benefits, and some form of accounting for the time differences between benefits and costs is required (e.g. discounting).

Thus it is necessary to choose a discount rate. Many of the benefits (and often costs) of policy options are likely to occur as non-marketed goods and services and, in the absence of such markets, it may be difficult to assess their value. Finally, some criterion is required to discriminate between preferred and inferior policy options.

In regional policy analysis, all these problems will be manifest. There will also be the additional problem of accounting for 'indirect', 'flow-on' or 'multiplier' effects of policy —that is, the effects of policy on industries related to the policy target. At the level of an individual region, such multiplier effects may be of major significance—for example, a policy that affects timber availability for logging may have substantial effects in related industries such as transport and sawmilling, as well as in the logging industry itself. At the national level, multiplier effects probably have minor importance because they occur in all regions (although not necessarily at the same degree of intensity). However, multiplier effects might be distributionally important if it were desirable to favour one or some regions over others. One way of taking account of the multiplier effects of investment or policy is via input–output analysis.

Input–output analysis

In a closed economy, the output of an industry can be divided into that used as inputs to other industries, and that used in final consumption. This insight provides the basis for input–output analysis, in which the output of each industry is modelled as a function of final demands and the coefficients that link one industry's output with another industry's use of it as input (e.g. James & Throsby 1973, chapter 18). The 'impact multiplier' is the effect on an industry of a change in another industry's level of final consumption. The distribution of an industry's output to input purchases and payments to factors may be calculated. The effects of changes in industry output resulting from changes in final consumption levels may be tracked through to their impacts on employment levels and wage income, and profits.

The key assumptions of input–output analysis are that the input–output coefficients are exogenously given, as are the levels of final consumption. In practice, input–output coefficients are the consequence of relative input prices and relative input–output prices, the current technology, and the existing structure of firms. The levels of final consumption will be a consequence of relative output prices. The model also assumes that inputs are available in infinitely elastic supply.

An input–output model could be solved for a national economy, or for some section of a national economy such as a region. In the latter case, the parameters of the model could be further disaggregated:

- final consumption could be disaggregated into local final consumption and non-regional local consumption or 'regional' exports;
- investment could be disaggregated into local investment, and investment outside the region;
- exports would be exports from the regional economy; and

- imports would comprise imports into the region from elsewhere in the national economy and imports into the region from other countries.

Regional input–output models may be used to examine the effects of changes in non-regional consumption of regional output, and of changes in regional productivity (e.g. the levels of the input–output coefficients), including changes in government policy which affect regional productivity or the demand for the region's output (cf. Mules 1983, Powell & Jensen 1989).

IAC (1989b) drew attention to the limitations of input–output analysis as a consequence of assumptions about fixed coefficients technology; the implications of using average input coefficients when marginal coefficients are the key variables; and the implications of infinitely elastic supply and infinitely inelastic demand. Further, the quantitative impacts of input–output analysis are gross rather than net effects. The IAC indicated some ways around these problems, ultimately drawing attention to the potential importance of computable general equilibrium modelling (see below). The IAC especially noted that, at a policy level, input–output modelling evaluates the quantitative equilibrium consequences of simulated changes. However, input–output analysis by itself cannot indicate the economic desirability of changes in output and employment. The IAC also drew attention to the limitations of regional extensions of input–output modelling.

Spatial equilibrium models

A closer modelling approximation to regional reality might be achieved if the final consumptions in input–output analysis were demand functions rather than exogenously determined quantities. Spatial equilibrium analysis provides one mechanism for endogenising demand quantities and prices using demand schedules for internal and external markets for several regions simultaneously. For example, Batterham and MacAulay (1994) showed how farm level supply in several 'regions' represented by 'local silo' areas could be solved in the context of external demands represented by ports with different 'demands' resulting from different handling costs. This problem was solved in the context of examining possible implications from changing the storage, handling and transport arrangements for Australian grain—especially wheat. While solution with multiple products is possible, spatial equilibrium models are usually solved for a small number of products rather than for the entire range of a region's industries.

General equilibrium models

Input–output models of an economy can be extended in a variety of ways. One approach is 'computable general equilibrium' (CGE) modelling. In this family of models, the entire economy is modelled simultaneously at as great a degree of disaggregation as possible. Both the production and consumption sides of the economy are modelled endogenously: on the production side, both the selection of input combinations and the level of output are determined endogenously for a given production structure, and input availability may also be endogenously determined; on the consumption side, consumers choose the level of consumption given endogenous prices and given preference functions. One such model is the ORANI CGE model of the Australian economy (cf. Chapter 12 for details). This model may be further extended by considering the regional implications of aggregate economic change.

ISSUES FOR DISCUSSION

1 In opening the 1996 Sydney Royal Easter Show, the Governor-General was reported as saying 'things are changing in our rural areas and the whole face of rural New South Wales is scarred by unemployment, particularly youth unemployment, which is undermining family life and stability both in town and on the farms. [If these problems are not addressed] our nation may be irretrievably changed for the worse'. Comment.

Part IV

Intersectoral Issues

On the eve of completing his term as President of the National Farmers' Federation in 1994, Mr G. Blight was interviewed about the Commonwealth Government's Budget. Blight made little comment on rural measures in the Budget, concentrating on macroeconomic issues, especially the underlying growth assumptions. Invited to review his period as NFF President, Blight again focused primarily on macroeconomic issues, and was particularly critical of the Government's contemporary industrial relations reforms.[1]

Blight's interview partly reflected the NFF's internal structures—its peak level concentrates on sectoral and intersectoral relations, while its constituent councils are responsible for commodity policy. However, both this interview and the NFF structure also reflect a recognition of the importance to agriculture—and the resource industries generally—of relationships between the primary industries and the rest of the economy. As the share of agriculture in the aggregate economy has declined, so has its impact on the rest of the economy. Conversely, this relative economic decline has made the rest of the economy more important to agriculture. Economic events elsewhere in the economy shape agricultural activity and, *a fortiori*, so do government policies primarily directed at other sectors. For example, policies directed at managing macroeconomic activity by targeting inflation, interest rates, employment and wage rates, and the exchange rate directly influence agriculture. Because some 60–90 per cent of production is exported in the great extensive industries of wool, wheat, meat, sugar and dairy,[2] export conditions have important effects on agricultural profitability. Government policy which targets or selectively influences export activity or especially affects Australian agriculture's international competitiveness has a major impact on the economic health of agricultural industries.

In Part IV, attention is focused on how non-sectorally specific government policy affects the agricultural and resources sector. Key mechanisms linking agriculture with the rest of the economy are examined to determine the impact on resource industries of macroeconomic policy for other industries. The principal underlying theme is general equilibrium economic analysis, although there is also some focus on simplified versions of a general equilibrium model.

Because the discussion in Part IV primarily concerns the *effects* of policy, little attention is paid here to explaining why or how such policy is developed. The structure outlined in the Introduction (pp. 11–14) may also be used to explain the rationale for or genesis of general economic policy, but that explanation would be the focus of courses in macroeconomic policy, or policy for other sectors of the economy. The current focus is, however, on how the identified policy issue, the intended policy and the mechanism by which it is implemented actually feed through into the agricultural and resources sector. Attention is particularly focused on how these industries may be *differentially* affected compared to the rest of the economy: there may be opposite effects to those intended for the economy as a whole and even opposite or unintended effects for the agricultural and resource sector.

12 Agriculture in an economy-wide context

This chapter provides a framework for examining the effects of government economy-wide policy on agriculture and resource sectors. Beginning with simple partial equilibrium models of agricultural industries with aggregate variables added, the discussion develops analysis of the agricultural sector within a general equilibrium framework. The models constructed are then used to examine effects on agriculture and resources of structural change elsewhere in the economy.

12.1 RURAL INDUSTRIES WITH AN EXOGENOUS EXCHANGE RATE

Because exports comprise an important proportion of the income of Australia's major agricultural industries, these industries must be considered in an international context. For example, models of research evaluation in the context of exporting industries were based on the implicit assumption of a constant exchange rate (Chapter 5). Excess supply and demand schedules were evaluated in a common international currency by transforming the prices in each market to an international currency via constant exchange rates. However, for a single country like Australia, variable exchange rates have potentially major effects on exporting industries. Exchange rates may vary because of general government economic policy, or because of exogenous factors outside the government's policy control. It is therefore important to model the processes by which exchange rates affect these industries.

An exporting Australian agricultural industry with variable exchange rates is depicted in figure 12.1. In the bottom-left quadrant is depicted the excess demand for the product of this industry denominated in an international currency. Assuming this industry is internationally 'small'—i.e. it is a price-taker on international markets—the excess demand curve is the vertical line ED (in a conventional, upper-right quadrant price–quantity diagram, this excess demand curve would be horizontal). This industry could export any quantity (X) at the international currency price P^* (denominated in US dollars).

The domestic market is represented in the upper-right quadrant of figure 12.1. In this quadrant, the excess supply schedule of Australian producers is ES (cf. Chapter 5). In the domestic market, the excess demand schedule (EDd), denominated in domestic currency, is obtained by converting ED to domestic currency via the exchange rate. The exchange rate is shown in the upper-left quadrant. The line rx represents the relative value of US and Australian currency. The slope of rx shows the number of Australian dollars that are equivalent to one US dollar; conversely, the reciprocal of the slope shows the conventionally denominated exchange rate for Australia—i.e. the US dollar equivalent of one Australian dollar. The Australian dollar equivalent of an international currency price (e.g. US dollars) is derived by reading the vertical axis value of the curve rx corresponding to the US dollar price. At the time of writing, the exchange rate was approximately $US0.75 per $A1.00. Thus, if the price of the commodity depicted in figure 12.1 was $US0.75 per unit of the commodity, the corresponding price in Australian currency would be $A1.00. Conversely, if the commodity price were $US1.00, the Australian price would be $A1.33.

In figure 12.1, international currency price P* translates into Australian dollar price Pd*. The excess demand curve ED (in US dollars) translates into the demand curve EDd (in Australian dollars). Given the Australian excess supply schedule ES, the equilibrium level of exports is X* with a corresponding equilibrium export price in Australian dollars of Pd*.[3]

Suppose the Australian dollar depreciates, represented by clockwise rotation of the exchange rate curve from rx to rx′ (figure 12.1). One US dollar buys more Australian dollars (Australian commodities become cheaper to overseas buyers) or, alternatively, an Australian dollar buys fewer US dollars (imported commodities become more expensive for Australian purchasers). While the excess demand curve for Australian exports denominated in *international* currency remains unchanged, there is a shift in the excess demand curve denominated in Australian dollars. The excess demand curve ED, when transformed into domestic currency at the depreciated exchange rate rx′, results in a domestically perceived export demand curve EDd′ (figure 12.1). As the exchange rate depreciates from rx to rx′, the equilibrium export price in Australian dollars rises from Pd* to Pd**, and the equilibrium level of Australian exports rises from X* to X**. The price in international currency, however, remains constant at P*. Conversely, if the initial exchange rate was rx′, an *appreciation* of the Australian dollar means the excess demand curve denominated in Australian dollars falls from EDd′ to EDd, and the equilibrium level of exports falls from X** to X*.

The assumption of Australia being a small producer and thus a price-taker on export markets may not characterise some Australian agricultural products, notably wool. The excess demand schedule denominated in international currency is probably downward sloping (curve ED in the bottom-left quadrant of figure 12.2). For convenience, a 45-degree line has been added to the bottom-right quadrant; this line is used to 'map' export quantities in the international market into identical export quantities in the domestic market. For example, quantities X1 and X3 in the bottom-left quadrant are identical to corresponding quantities X1 and X3 in the upper-right quadrant.

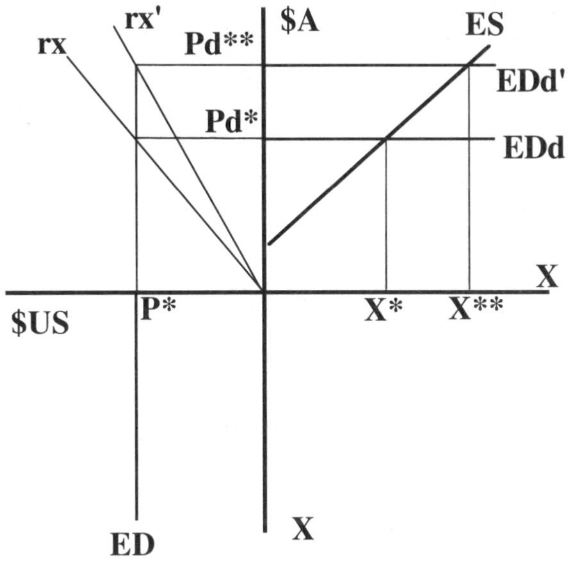

Figure 12.1 Exchange Rates with Small Country Assumption

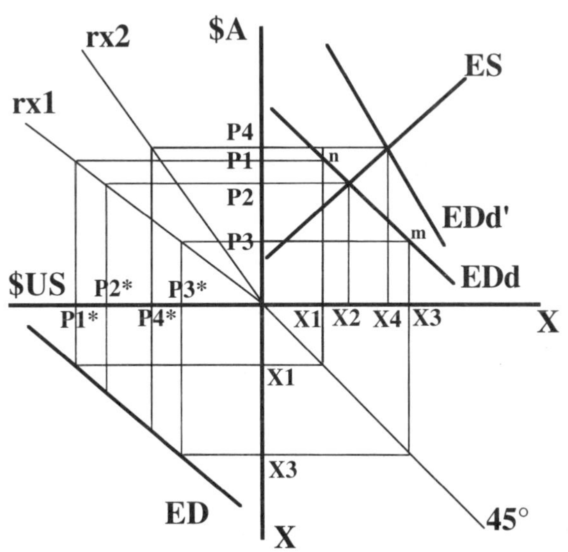

Figure 12.2 Downward-sloping Excess Demand

Excess demand denominated in international currency may again be transformed into an excess demand schedule denominated in Australian dollars. Consider ED in the bottom-left quadrant: at price P3* (denominated in $US), a quantity X3 would be demanded on international markets. Export level X3 is transformed via the 45-degree line in the bottom-right quadrant into exports X3 in the domestic market. The corresponding international price P3* for this level of exports is transformed via the exchange rate line rx1 into price P3 denominated in Australian dollars. The price–quantity pair (P3, X3) gives the point m on an excess demand schedule denominated in Australian dollars in the upper-right quadrant. Similarly, the international price–quantity pair (P1*, X1) maps the point n in the upper-right quadrant. With a linear ED (assumed for convenience), the excess demand schedule denominated in Australian dollars (EDd) is also linear and is derived by connecting points m and n. Given ED (and its domestic Australian equivalent EDd) and ES, the equilibrium level of exports is X2, selling at the international price P2* (figure 12.2).[4]

Depreciation of the Australian dollar from rx1 to rx2 causes an outwards shift and rotation of the excess demand curve denominated in Australian dollars from EDd to EDd′ (figure 12.2). With this depreciation, the equilibrium level of exports rises from X2 to X4, and the commodity price in domestic currency rises from P2 to P4. The international price of the commodity falls from P2* to P4* because of the increased optimal level of exports. Conversely, if the initial exchange rate was rx2, exchange rate appreciation to rx1 would contract excess demand denominated in Australian dollars from EDd′ to EDd, with a commensurate fall in the level of exports (X4 to X2) and the domestic price of exports (P4 to P2), but with an increase in the international price from P4* to P2*.

This model has been used to examine the effect of an exogenous change in the exchange rate on exporting industries whose level of exports has no effect on the exchange rate. The model may also be used to examine the effect on this industry of government policy which successfully induced a change in the exchange rate—or, indeed, government policy which actually affected the exchange rate whether or not this was intended.

This model may additionally be used to examine the effect of inflation on an exporting industry. In figure 12.3, the preceding model has been reproduced, with EDd corresponding to exchange rate rx. Inflation increases the Australian dollar price of production, so that excess supply contracts from ES to ES1. The domestic price corresponding to the initial equilibrium level of exports X2 (P2) has increased to P2′, but at P2′ and the initial exchange rate rx overseas buyers are only prepared to purchase X2′. The level of exports corresponding to P2′ can be read from the domestic currency excess demand schedule EDd, or the international currency excess demand schedule ED reading through the exchange rate line rx. However, exports X2′ at domestic price P2′ is not an equilibrium: at P2′ the export quantity (X2) exceeds international purchases (X2′); equilibrium occurs where EDd and ES1 intersect. The effect of inflation on exporting industries—or the effect of government policies that give rise to inflation or permit it to occur—is to decrease the level of exports in this industry *ceteris paribus*.

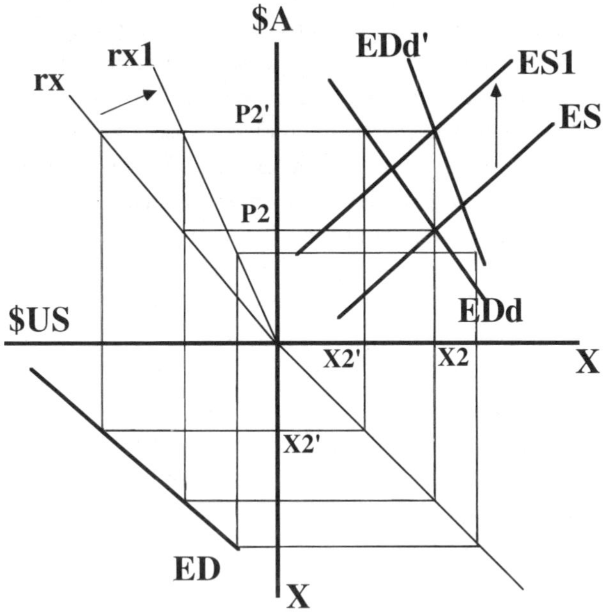

Figure 12.3 Inflation with Offsetting Currency Depreciation

Suppose, however, that domestic inflation was accompanied by—or government policies induced—a simultaneous depreciation in the exchange rate from rx to rx1 (figure 12.3). This depreciation would cause a shift in the excess demand curve denominated in *domestic* currency from EDd to EDd′. The shift in the domestically perceived excess demand schedule would, in figure 12.3, exactly offset the effect of the initial price rise. The initial level of exports remains the equilibrium level, the international price of the commodity remains unchanged, but the Australian dollar price of exports and domestic consumption has risen to P2′.

12.2 ENDOGENOUS EXCHANGE RATE

The model depicted in section 12.1 is useful for examining the implications of exogenous or government-induced changes in the exchange rate on a single exporting industry. But the exchange rate may not be exogenous for 'large' exporting industries, or aggregated exporting industries like 'agriculture'. Because the exchange rate is simply the international dollar price of Australian currency, changes in export levels may be sufficiently large to change the currency price. The exchange rate is then not exogenous, as assumed in section 12.1, but endogenous. In this section, a simple model of endogenous exchange rates is used to examine changes in the relationship between agriculture and the rest of the economy.

Bewley and White (1990, pp. 24–7) utilised a simple model of a trading economy in which exports and imports were functions of the exchange rate (figure 12.4, upper-left quadrant). On the vertical axis, exchange rate (rx) appreciates as rx increases (a given amount of domestic currency buys more overseas currency or commodities) and depreciates as rx falls (domestic currency buys less overseas currency or commodities). By comparison with figure 12.1, for example, as the exchange rate depreciates, the level of exports rises *ceteris paribus* because the excess demand curve denominated in Australian dollars shifts outwards. Thus there is a negative relationship between exports and the exchange rate in figure 12.4 (upper-left quadrant, curve X). Conversely, if the domestic country was a net commodity importer, it would have an excess demand curve denominated in its own currency, and the world's excess supply schedule would be converted into domestic currency via the exchange rate. A declining exchange rate increases the domestic currency price of imports and thus contracts the excess supply schedule, thus decreasing the equilibrium level of imports. An appreciating currency decreases the domestic currency price of imports and shifts the excess supply schedule outwards, thus increasing equilibrium imports. Thus there is a positive relationship between imports and the exchange rate *ceteris paribus* (figure 12.4, upper-left quadrant, curve M).

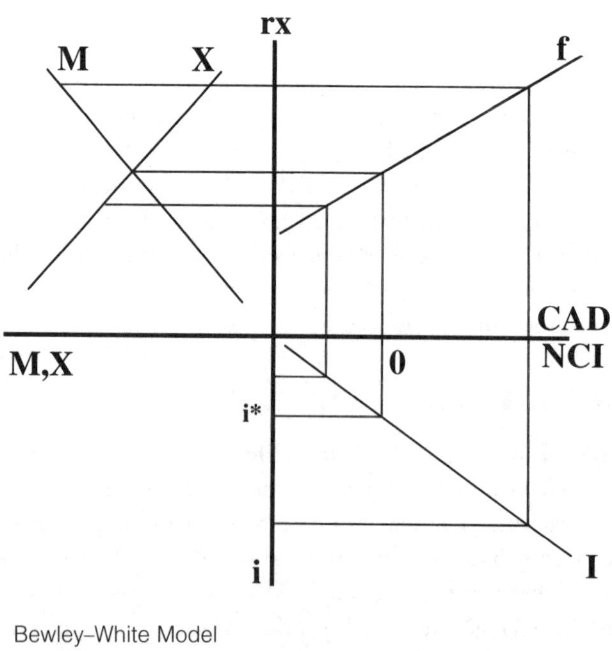

Figure 12.4 Bewley–White Model

At any given exchange rate, the difference between imports and exports—the horizontal distance between curves M and X in figure 12.4—is the current account deficit. At the intersection of M and X, the current account deficit (CAD) is zero (note: the zero point for CAD is *not* at the origin of the rx/CAD set of axes). Above the intersection of M and X, the current account deficit is positive (imports exceed exports); below the M/X intersection, CAD is negative (exports exceed imports). There is a positive relationship between the exchange rate rx and the level of the current account deficit (curve f, upper-right quadrant of figure 12.4).

If a current account deficit can only be financed by net capital inflow (NCI in figure 12.4), the real trading sector (upper-left quadrant) is linked to financial markets (bottom-right quadrant). A zero CAD will correspond to an equilibrium in financial markets for which there is a given interest rate (say i*). An interest rate higher than i* would be required for positive net capital inflow. To finance a positive CAD (imports exceeding exports) via net capital inflow, an interest rate higher than i* must be available to overseas investors; the curve I in the bottom-right quadrant shows the relationship between NCI and the interest rate (i).[5]

In this model there is no reason why a zero CAD should be optimal. If there are good investment opportunities within Australia for overseas investors, net capital inflow may be required to finance imports exceeding exports to enable that investment to proceed without other domestic effects (e.g. inflation).

This model can be used to investigate the aggregate consequences of two government policies concentrating on particular economic variables. First, if government intervened to force interest rates higher (e.g. for macroeconomic stability reasons), net capital inflow would also be higher than without intervention. The consequence of maintaining a higher NCI is that the current account deficit would also be higher *ceteris paribus*, i.e. higher than would be required to finance real capital investment. But a higher CAD not required to finance real investment necessitates a surplus of non-capital imports over exports, and forces up the exchange rate. Had the alternative policy been an interest rate compatible with a zero CAD, the elevated interest rate would have resulted in an appreciated exchange rate, a higher level of imports and a lower level of exports. The policy consequence of a lower level of exports is a lower level of profitability for the exporting sector than would otherwise occur. Conversely, an artificially lower interest rate would reduce NCI and corresponding CAD, reduce the exchange rate, and raise exports and reduce imports.

Second, if government intervened to maintain an over-valued currency other than through interest rates, imports would be higher than otherwise, and exports lower; this pattern would lead to a higher than otherwise interest rate in order to maintain the NCI to finance the CAD. Conversely, an artificially depreciated currency would favour exports over imports, leading to a relative depreciation of the exchange rate, and a lower than otherwise CAD and NCI (cf. Bewley & White 1990, pp. 26–7).

However, government policy may not operate simply on specific variables like interest rates or exchange rates. Government policy might attempt to shift the export schedule X (e.g. export enhancement programs), the import schedule M (e.g. import replacement programs), or the lending schedule I (e.g. policies to enhance Australia's reputation as a safe haven for liquid investments) (cf. Bewley & White 1990, pp. 26–7). The effects on agriculture of export expansion in non-farm industries are initially discussed using this model (section 12.4). Discussion of the effects of tariff barriers in the Australian non-farm sector, and the long-running debate about tariff reform and its potential effects on agriculture, also initially uses this model (Chapter 13).

12.3 GENERAL EQUILIBRIUM

Sections 12.1 and 12.2 present highly simplified models of the economy. Even though these models are relatively aggregated or have a macroeconomic orientation, they are partial equilibrium models because the rest of the economy is assumed to be held constant (*ceteris paribus*). In the models of figures 12.1–12.3, for example, the exchange rate is an exogenous parameter which, if it changes, simply shifts the excess demand schedule denominated in domestic currency. Similarly, in figure 12.4, the domestic production and demand conditions that give rise to the excess demand and supply schedules are similarly exogenous, as are the savings and investment decisions that give rise to the domestic interest rate with a zero current account deficit.

The effect of general economic policy on agriculture could be modelled by nesting an agricultural sector model within a macroeconomic model. The macroeconomic model could generate aggregate economic outcomes—e.g. aggregate output and the sectoral composition of output, consumption, exports, imports and investment, savings and interest rates, inflation rates and exchange rates, the level of employment and unemployment rates—and these settings would then be used as parameters within an agricultural sector model. The particular output levels in each of the agricultural industries would have to be constrained within the parameter settings of the macroeconomic model, or the outcomes be sufficiently similar to the aggregate model, so that any deviations could be assumed to have no macroeconomic effects.

The impact of general economic policy on agriculture, where policy affects agriculture through general economic activity and agriculture affects the aggregate economy, would preferably be assessed in a general equilibrium setting. This setting is similar to the general equilibrium model mentioned briefly in a closed economy setting in Chapter 1 to establish the conditions for economic efficiency. The setting is also similar to that utilised in Chapter 9 to examine the effects of general economic growth on the agricultural sector. The agricultural sector interacts with the rest of the economy, including the international trade sector, so that agricultural output is determined simultaneously with all other economic variables. A highly disaggregated general economic framework is desirable so that at least some of the components of the agricultural sector can be examined individually in case there are differential effects of aggregate economic change, or aggregate economic policy, on particular industries. A computable model is preferred to a theoretical model so that the magnitude of economic effects can be investigated.

A useful Australian general equilibrium model is ORANI (Dixon et al. 1982).[6] The ORANI family of models disaggregated the Australian economy into 23 sectors comprising about 110 industries. Sectors relevant to the present context are agriculture, forestry and logging, fishing and hunting, input services provided to agriculture (e.g. shearing, aerial seeding and crop dusting), agricultural machinery, and primary product processing industries such as meat products, milk products, fruit and vegetable products, margarine, oils and fats, flour and cereal products, alcoholic beverages, tobacco products, cotton ginning, wool scouring and topmaking, textiles, clothing and footwear, and wood and paper products (Higgs 1986, table 5.2). The agricultural sector in ORANI comprises eight industries, four of which are zones based on ABARE survey collections (pastoral, wheat–sheep, high rainfall and northern beef) and four 'composite product' industries (table 12.1). Three of the four composite product industries produce only that composite product (sugar, fruit and nuts; vegetables, cotton, oilseeds and tobacco; poultry) while the fourth, 'milk cattle and pigs' produces both its own commodity and also meat cattle. The northern beef 'survey' zone produces only one commodity, meat cattle, while the other four 'survey' zones each produce all commodities except poultry.[7]

Table 12.1 Agricultural Industries and Composite Commodities in ORANI

Commodity	*Industry#*							
	PZ	WS	HR	NB	MC&P	SF&N	VCOT	P
Wheat	*	*	*					
Barley	*	*	*					
Other grains	*	*	*					
Wool	*	*	*					
Sheep	*	*	*					
Meat cattle	*	*	*	*	*			
Milk cattle and Pigs	*	*	*		*			
Sugar, Fruit & Nuts	*	*	*			*		
Vegetables, Cotton, Oilseeds & Tobacco	*	*	*				*	
Poultry								*

Source: Higgs (1986, tables 1.1 and 2.1)

Notes: # PZ-Pastoral Zone; WS-Wheat-Sheep Zone; HR-High Rainfall Zone; NB-Northern Beef; MC&P-Milk Cattle and Pigs; SF&N-Sugar, Fruit and Nuts; VCOT-Vegetables, Cotton, Oilseeds and Tobacco; P-Poultry

The general structure of ORANI is depicted in figure 12.5. ORANI industries represent firms in the primary (agriculture, forestry, fishing, mining), secondary (manufacturing, including primary product processing) and tertiary (services, margins) industries. Firms are assumed to be price-takers in both input and output markets, and to maximise profit in a two-stage process. The optimal combination of inputs is selected in the first stage, and the optimal level of output is selected in the second. For given relative input prices, firms select the optimal composition of an input bundle (Higgs 1986, figure 2.1); the optimal number of input bundles is then determined by relative output–input prices. Households are assumed to be price-takers and to maximise utility subject to a budget constraint. Exports of major commodities are endogenous to the model, and there are investment and government sectors (cf. Higgs 1986, chapter 2; Dixon et al. 1982).

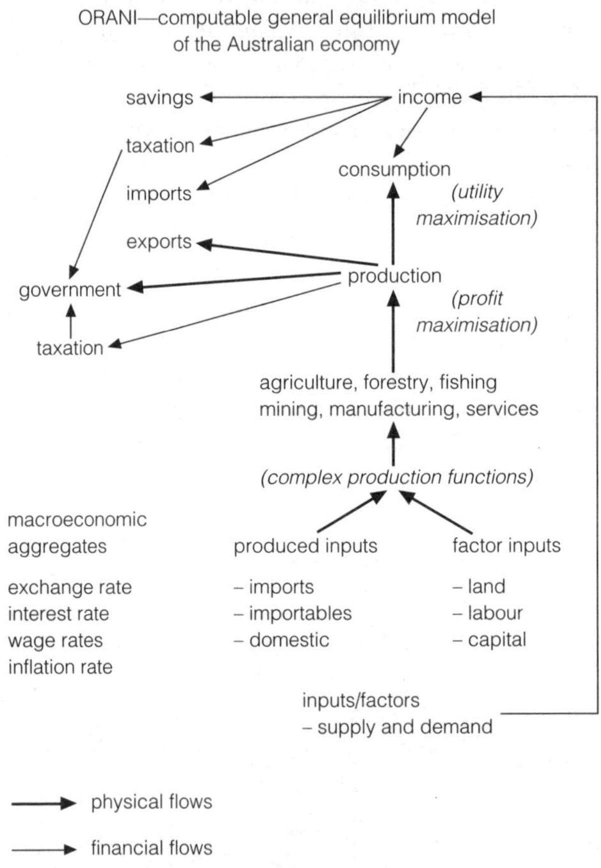

Figure 12.5 Schematic Representation of ORANI

ORANI is used by establishing an economic context in which to run it, since the model has more endogenous variables than equations. A 'closure' is determined for the model, and the model run to determine its behaviour under this closure. The model achieves equilibrium in a short run of approximately eight quarters. This initial run corresponds to a control treatment in a standard scientific experiment. The level of an individual variable of the ORANI model such as GDP may be tracked in the control run until short-run equilibrium is attained (e.g. curve 'control' in figure 12.6). The model is then changed—an assumed 'shock'—to reflect an economic event whose effects are being examined (e.g. a change in government policy). The model is again run for eight quarters under the revised scenario, and variables in the control run and the shock are compared to investigate the consequences of the shock. Continuing the example illustrated in figure 12.6, depending on the nature of the policy shock, GDP might be higher (curve 'shock1') or lower (curve 'shock2') once equilibrium is attained under the alternative scenario.

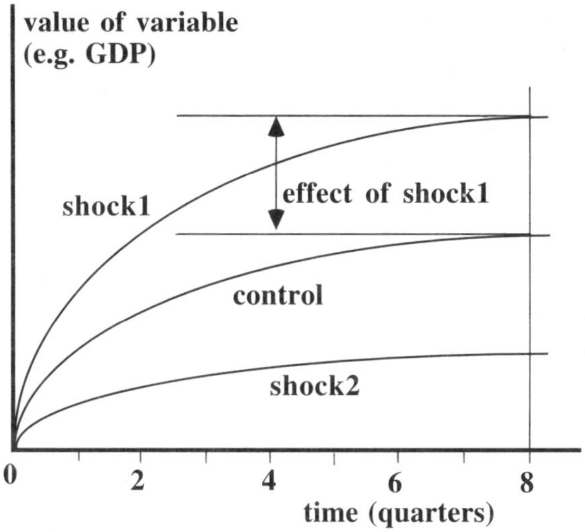

Figure 12.6 Schematic Representation of an ORANI 'Shock' Experiment

12.4 STRUCTURAL CHANGE IN AN AGGREGATE CONTEXT

In the 1950s, agricultural economists argued that in a closed economy growth in agricultural supply would proceed faster than demand growth for agricultural commodities, and agricultural product prices would fall (Chapter 9). This fall in output price would induce changes in the structure of agriculture. In Australia's case, the export orientation of its major agricultural industries created a price floor to mitigate continual reductions in price forced by supply growth relative to domestic demand growth. This price floor did not provide absolute protection against either short or long-run price reductions, but it did prevent agricultural prices from falling as fast or as far as they would have fallen without export opportunities. The price floor was supported by world economic growth, especially in the former Axis powers and those previously poor countries which grew rapidly in the period after the Second World War. In the latter, income elasticity of demand for food was relatively high and there was a continuing growth in aggregate demand for the kind of products produced by Australian agriculture. Thus Australian agriculture continued more profitably than if export opportunities had not been available.

There were, however, some constraints on this growth—e.g. quotas on meat exports to the USA and the slow liberalisation of Japanese agricultural markets retarded growth possibilities in potential Australian markets. Further, the intense trade war between the USA and the European Communities in the 1980s and early 1990s arising out of US determination to confront the EC's Common Agricultural Policy caused collateral damage through intensified competition in markets which Australia supplied. Australia invested heavily in international trade reform by playing a major role in organising the 'Cairns Group' of agricultural exporting nations, who lobbied strongly for freer international agricultural markets during the Uruguay round of negotiations for the reform of the General Agreement on Tariffs and Trade (GATT). Australia also played a leading role in forming the Asia–Pacific Economic Cooperation forum (APEC). From Australia's perspective, these initiatives may be understood as an attempt to shift outwards the excess demands for individual commodities (ED in figures 12.1–12.3) with a commensurate increase in the equilibrium level of Australian exports. Alternatively, these international trade initiatives could be seen as an attempt to shift the export demand schedule (X in figure 12.4) thus leading to a higher level of aggregate exports, and permitting a higher level of imports *ceteris paribus*.

However, there were domestic constraints on the potential growth of Australian agricultural exports. Beginning in the late 1950s, and accelerating through the 1960s and 1970s, there was a second wave of development of Australia's mineral resources, particularly iron ore, coal, bauxite, alumina and aluminium, copper, nickel, gold, beach sands and diamonds. These developments had several effects, including an initial effect on capital inflow and capital investment, and subsequent effects on the repatriation of profits, since much of the investment in minerals development was based on foreign equity or foreign borrowings. Apart from the capital inflow and profit repatriation effects, however, there was a longer term impact on the trade sector of the Australian economy, which affected the agricultural industries.

The Australian excess supply schedule (X) in figure 12.7 comprises traditional agricultural exports (XA), and mineral exports (ignoring other components). Suppose, for simplicity, that the import demand schedule is M, and trade was initially at equilibrium at exchange rate rx0 with exports X0 equal to imports and a zero current account deficit at interest rate i*. At the initial equilibrium, the level of agricultural exports was Xa, and the minerals component of exports was (X0–Xa). Suppose minerals exports growth shifted the *minerals* component of the export schedule outwards (to X1) so that, at some new equilibrium X*, the agricultural export schedule remained unchanged at XA. If the import demand schedule remained unchanged at M, the new equilibrium would be characterised by a higher level of aggregate exports X* and an appreciated exchange rate rx1. But at the appreciated exchange rate, agricultural exports are only Xa1, lower than the initial level Xa. Not only has the mineral sector expanded via capital investment but, by the induced appreciation of the Australian dollar because of the unchanged import demand schedule, the appreciated currency chokes off otherwise-profitable exports for the agricultural sector. This argument was first developed in a partial equilibrium context by Gregory (1976) (cf. Higgs 1986, figure 7.1).

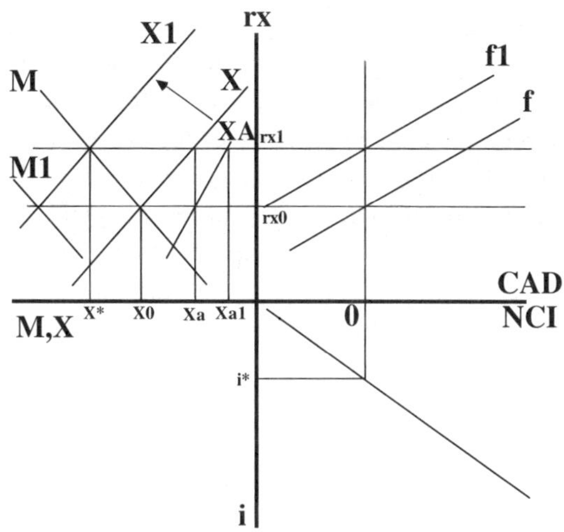

Figure 12.7 Effect of Mineral Exports Growth

The simplified economy in figure 12.7 captures only part of the change to the Australian economy since the rapid growth of the minerals sector. For example, if the import demand schedule M was infinitely elastic at the initial exchange rate rx0, there would there be no exchange rate effect and thus no reduction in agricultural exports. Effects that could occur simultaneously with the growth of mineral exports include:

- import demand schedule shifted outwards to M1—if import demand expanded to M1 due to higher domestic incomes, a reduction in the international price of imported goods, or reduction in the domestic production of import competing goods while export supply curve shifted to X1, the equilibrium exchange rate would remain at rx0 and there would be no reduction in agricultural exports;
- agricultural production costs declined—if the agricultural supply schedule shifted outwards (e.g. through adoption of new technologies) at the same time as mineral exports expanded, at least some of the exchange rate effect induced by rising mineral exports could be offset. Similarly, if international agricultural prices rose secularly—e.g. through rising world population—or the composition of agricultural exports changed so that the index of agricultural prices received by Australian exporters rose, the agricultural export schedule would also shift outwards and again at least partially offset the exchange rate effect. Of course, in both these cases, the

shift in the agricultural exports curve would contribute to the shift in the aggregate exports schedule and thus intensify the appreciation of the exchange rate.

Higgs (1986, chapter 7) reported an ORANI simulation of the effects of an increase in foreign exchange earnings of 5 per cent of GNP resulting from an expansion of the minerals resources sector. The results of this simulation are summarised in table 12.2. A selection of macroeconomic impacts indicates that the economy expands because of the growth of the mining sector. The selection of sectoral effects in table 12.2 shows that the outputs of primary-related sectors are generally diminished by expansion in minerals; this diminution also characterises the *existing* minerals sector which, like agriculture, also suffers from the exchange rate appreciation. These relatively large percentage sectoral output effects relate, however, to relatively small individual sectors. Although there is only a relatively small percentage expansion of the margins and services sectors, because these sectors account for a large proportion of GDP, they absorb much of the beneficial impact of the export expansion.[8] The output of all the agricultural industries and sectors is detrimentally affected by the export expansion. The effect for export industries comes from the direct exchange rate effect; however, even domestically oriented industries like poultry face negative effects arising from increased costs as the expanded minerals sector bids resources away from other industries, thereby increasing their costs. Where commodities—e.g. meat—undergo a high degree of domestic processing prior to export, the exchange rate effect on output is severely exacerbated by the resource competition effect (table 12.2).

Table 12.2 Impacts of Additional Foreign Exchange Earnings (equivalent to 5 per cent of GNP) resulting from Expansion of the Minerals Resources Sector

Variable	*Short-run effect (2 years)[a]*	*Long-run effect (10 years)[a]*
Macroeconomic impacts		
Real gross domestic product	1.29	0.32
Aggregate employment	1.36	exog.
Real household consumption	6.75	5.78
Consumer price index	15.80	2.76
Aggregate imports (foreign currency value)	13.31	5.27
Sectoral output effects		
Agriculture	–10.23	–4.56
Forestry and logging	10.75	–3.38
Fishing and hunting	–2.68	–0.077
Mining (traditional mining)	–11.27	–30.36
Meat products	–18.11	–7.17
Cotton ginning, wool scouring & top making	–16.72	–4.79
Agricultural machinery	–56.28	–5.01
Margins	2.22	0.72
Services	5.01	3.77
Employment—rural workers	–14.34	–5.36

Table 12.2 Cont'd

Variable	*Short-run effect (2 years)[a]*	*Long-run effect (10 years)[a]*
Agricultural industry output		
Pastoral zone	−10.22	−4.34
Wheat–sheep zone	−8.46	−5.44
High rainfall zone	−12.14	−3.34
Poultry	−9.47	−3.74
Agricultural commodity output		
Wool	−9.12	−4.01
Sheep	−9.96	−3.96
Wheat	−9.13	−6.22
Meat cattle	−17.57	−6.99
Milk cattle and pigs	−5.81	−2.21
Sugar, fruit and nuts	−18.26	−7.95
Vegetables, cotton, oilseeds and tobacco	−3.18	−0.38
Real net farm earnings		
Pastoral zone	−28.16	−3.28
Wheat–sheep zone	−30.12	−6.04
High rainfall zone	−32.91	−2.35
Northern beef	−59.48	−16.48
Milk cattle and pigs	−14.75	−0.19
Sugar, fruit and nuts	−35.64	−6.86
Vegetables, cotton, oilseeds and tobacco	−5.28	3.12
Poultry	−22.21	−1.69
All agriculture	−27.64	−3.94

Source: Higgs (1986, Tables 7.1–7.2, 7.5, 7.9)

Notes: a: percentage deviation from control

Output reductions occur simultaneously with reductions in the farmgate prices of export-dominated agricultural products, which face highly elastic international demand schedules, and where the cost of the 'margins' industries rises with minerals expansion (e.g. wheat, barley, other cereal grains, meat cattle; the latter is exacerbated by the above-noted effect in the meat products industry). The reduction in wool output leads to a modest increase in the farmgate price of wool (wool has a less than perfectly elastic demand). The largely (but not complete) domestic orientation of the other agricultural industries means that their demand curves are not perfectly elastic and so output reductions following from the minerals expansion leads to increases in their farmgate prices (Higgs 1986, p. 119). The combination of the consistently negative output effects, and price effects, means that real net farm returns decline for all agricultural industries as a consequence of minerals export expansion (table 12.2).

Preceding analyses demonstrate why a successful government response to the effect on agriculture of mineral sector growth could not take the form of mitigating it through subsidies. The exchange rate effect of mineral sector growth was likely to exist at least for decades, and was most likely to intensify over time. A subsidy-oriented response would, therefore, have had to persist over time, and the level of subsidy required to achieve a given level of protection against the exchange rate effect would have needed to intensify over time. Australian governments had previously experienced subsidy demands for assistance against currency revaluation. When the British pound was devalued in 1967, the long-standing relativity between Australian and British currencies was finally broken. Australian agricultural industries received 'devaluation' assistance, which was actually assistance for *appreciation* of the Australian currency relative to the pound sterling (Harris et al. 1974, para. 2.56).

The likely impacts of mineral export growth on the agricultural sector may partly explain government interest in 'value adding' as a possibly beneficial path for agricultural industries to pursue. Rather than entertain the prospect of continuing subsidies, government may have decided that a once-for-all injection of assistance via support for export enhancement based on greater domestic processing was appropriate. At the level of the individual firm, or even industry, this may have been so as long as the value of the additional processing exceeded the cost of the additional resources used (cf. section 8.1). However, if all agricultural industries were encouraged to take this path, the aggregate effects of all industries seeking additional processing resources would raise the per unit cost of processing by increasing the demand for resources used in these activities. These effects would be similar to those of mineral export expansion, and would particularly affect industries like meat and meat processing which rely heavily on post-farmgate processing. Thus, what appears appropriate from the perspective of a single firm or industry may not be possible for all industries in aggregate. Alternative opportunities, such as the search for alternative enterprises requiring limited post-farmgate services, may be superior alternatives—again, as long as the total benefits of the alternative enterprise exceed its total costs.

The Australian Government has recently strongly supported restructuring of the Australian manufacturing sector, and encouraged it to be more focused on export performance. If this reorientation is successful, a more export-oriented manufacturing sector will reinforce the trend of an appreciating Australian currency *ceteris paribus*. Possible offsetting factors to this trend would be reductions in manufacturing output and an outward shift in the demand for imports, which might result from reduced assistance to manufacturing industry arising from completion of the Uruguay Round of GATT. Conversely, the effects of trade liberalisation from the latest GATT agreements may intensify import competition for some Australian agricultural industries. For some agricultural industries at least, relief may be possible via quarantine requirements for imported products that could jeopardise the current disease status of agricultural production, general Australian health, or the absence of pests, pathogens and undesirable plants in the Australian ecology.

12.5 CONCLUSIONS

The agricultural and resource sectors interact with the rest of the economy through the direct effects of macroeconomic variables (e.g. interest rates and wage rates), the costs of farm inputs, the effects of post-farmgate services on the derived demand for agricultural commodities, and the effect of macroeconomic variables such as the exchange rate on the domestically denominated excess demand for exports. Unless all these influences are adequately accounted for in a general equilibrium framework, and government policy is correctly evaluated within this wider framework, the actual effects of general economic change and government policy may be seriously misunderstood.

ISSUES FOR DISCUSSION

1 In late March 1996, the Governor of the Reserve Bank warned employers and trade unions that, 'if wages growth continues to pick up, rather than come back a notch, there will be little option but to raise interest rates'. Assuming the macroeconomics behind this warning are sound, evaluate whether or not this warning makes sense outside a 'corporatist' state.

2 One of the Howard Coalition Government's first actions in 1996 was to establish a 'National Commission of Audit', apparently similar to that pioneered by the Greiner Coalition Government in 1988 (headed by a designer of a tax avoidance scheme in the 1970s) and those established by incoming state Coalition governments in Victoria, South Australia, Western Australia and Tasmania in the 1990s. One media commentator was moved to observe that the results of the National Commission of Audit would be to provide the incoming Federal Government with a 'mandate' for major tax reform, possibly including a goods and services tax. Discuss the possible economic and political objectives of establishing such an inquiry.

Endnotes

1 G. Blight, interviewed by Romona Koval on Radio National's 'Drive' program at approximately 5:30 p.m. on 11 May 1994 (transcript unavailable).

2 The approximate percentage volume of exports in major extensive industries' production is: wool 90 per cent (1988–89); beef and veal 65 per cent (1992); wheat 60 per cent (1992–93); sugar 70 per cent (1992–93); cotton 90 per cent (1991–92). The value of dairy exports as a percentage of the value of dairy production was approximately 60 per cent. The gross value of farm production was $20 966 million in 1991–92 while the value of farm exports in same year was $15 985 million; the latter figure includes a significant portion of post-farm services value added (ABARE 1993).

3 Ignoring curve shifters, a conventional price–quantity diagram depicts the solution of two simultaneous equations in two unknowns—i.e. price and quantity. Figure 12.1 is simply a graphical representation of a system of two simultaneous equations in two unknowns: excess demand: P (in $US) = a; excess supply: P (in $A) = c + d.X; and an exchange rate identity: $A = e.$US.

4 As before, figure 12.2 is simply a graphical representation of a system of two simultaneous equations in two unknowns together with an exchange rate identity: \$A = e.\$US. In this case, however, the excess demand is: P (in \$US) = a – b.X.
5 Like the models of figures 12.1–12.3, Figure 12.4 represents a model of three endogenous equations in three unknowns—for X, M and I in variables rx, CAD = M – X, and I—together with an identity CAD = NCI.
6 And its variants and later manifestations in the form of the Monash series of models.
7 Some of the agricultural industries in ORANI are odd composites which reflect the model's origins in the late 1960s when some enterprises were of minor importance in Australian agriculture. For example, the 'Vegetables, cotton, oilseeds and tobacco' industry reflects a period when oilseeds and cotton were minor crops. Similarly, 'Sugar, fruit and nuts' originally mainly aggregated across the tropical sugar industry and mainly temperate fruit industry, and now aggregates across sugar, a diminished temperate fruits industry and a much expanded tropical fruits and nut industry. The composite 'Milk cattle and pigs' reflects the structure of the old manufacturing dairy industry when milk was separated on farm to produce cream for sale, and the waste skim was fed to farm pigs; the Australian pig industry is now almost exclusively based on grains and protein meals, and the market milk sector is a major component of the dairy industry.
8 The margins sector comprises wholesale and retail trade; road, water, air, and rail and other transport; investments and services; and restaurants and hotels. The services sector comprises electricity, gas, water sewerage and drainage, residential building, other construction, mechanical repairs, other repairs, communication, banking, non-bank finance, insurance, other business services, ownership of dwellings, public administration, defence, health, education and libraries, welfare services, entertainment and leisure, personal services, non-competing imports.

13 Differential sectoral rates of protection

Australian governments provided extensive assistance to industry after Federation, following the earlier lead of protectionist colonies like Victoria. Although it has been substantially reduced since the 1970s, industry assistance—and possible combative policies—is worth reviewing. Demands for assistance have not disappeared, although they may take new guises and may be resurgent in the future. A knowledge of the recent policy past is a essential foundation for understanding the present, and avoiding an action replay of the past. Since the principal assistance measure has been the tariff, this chapter begins with analysis of its economic effects.

13.1 PARTIAL EQUILIBRIUM FRAMEWORK

The effect of a tariff may be illustrated in a partial equilibrium context (figure 13.1). Supply and demand conditions in Australia for some commodity Q are represented by curves S and D respectively (an atomistically competitive market is assumed). If world price exceeds the closed economy equilibrium price (P*), Australian firms export Q. If world price is less than P*, Australian firms import Q. At price Pw, Australian consumers demand Qt, but at this price Australian producers are only prepared to supply Qd, and hence (Qt – Qd) is imported. Suppose a fixed tariff of t per unit of commodity were applied to all imports. The price of the marginal unit imported therefore becomes Pd = Pw + t; at this price, domestic consumption falls to Qc and domestic producers are prepared to supply Qp. Total imports fall to (Qc – Qp). Efficiency and distributional effects of the tariff are illustrated in figure 13.1. Increased domestic production at a higher cost than is available without the tariff indicates a production efficiency loss (triangle abc). Reduced consumption because of the higher domestic price indicates a consumption efficiency loss (triangle def). Because consumers pay a higher price for the reduced level of imports, there is a transfer from consumers to taxpayers (rectangle bedc). Finally, because domestic producers of Q are paid a higher price for all their production, there is a transfer from consumers to producers (rectangle PdbcPw;

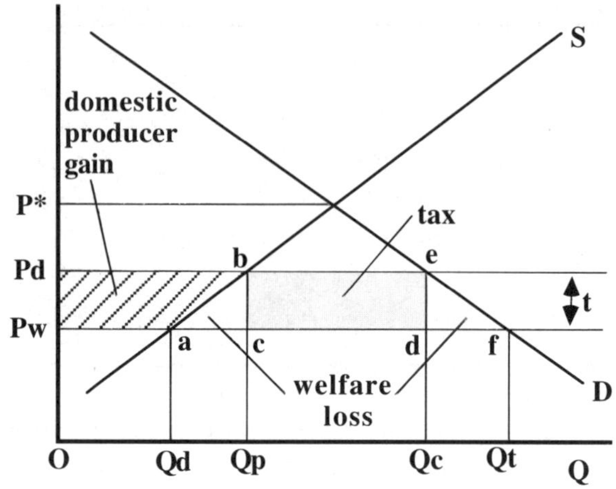

Figure 13.1 Partial Equilibrium Effects of a Tariff

the net producers' benefit is trapezium PdbaPw). If a non-tariff-protected industry uses the imported commodity in production, production costs in the non-protected industry will rise because of the tariff and its supply curve will contract.

Demand curves for an input may be derived from final product demand curves. Suppose the commodity illustrated in figure 13.1 is a consumer good and is produced by domestic industry m as well as being imported. At the zero-tariff equilibrium price Pw, derived demand for input Y used in producing Q is Dm (figure 13.2). The demand for the same input used in some other domestic industry is Da, and Dt is the aggregate demand curve for the input in industries m and a. Given the supply curve S, the input's equilibrium price is Pt. With a tariff on the importable commodity Q, the greater profitability of production shifts the input's derived demand curve to Dm′. The aggregate demand for the input shifts to Dt′ and, unless the supply curve for the input is infinitely elastic, the equilibrium price of the input will rise (Pt′ in figure 13.2). In the non-tariff-protected industry, use of the input will fall as the price of the input rises. Even if the tariff-protected commodity is not used as an input in the non-protected industry, costs in the unprotected industry will rise because of the tariff, and supply curves of these industries will thus contract. As these supply curves contract, prices for their products may change depending on the elasticity of demand for their output. Industries that are price-takers on international markets will have their prices unaffected by supply contraction, but their profitability will be reduced because of increased production costs. Industries selling on domestic markets, or exporters such as wool, which have non-infinite demand elasticities, may experience price increases for their products. These increases may partially counteract the cost increases caused by tariff-induced supply contraction. In this case, a general equilibrium solution is required to

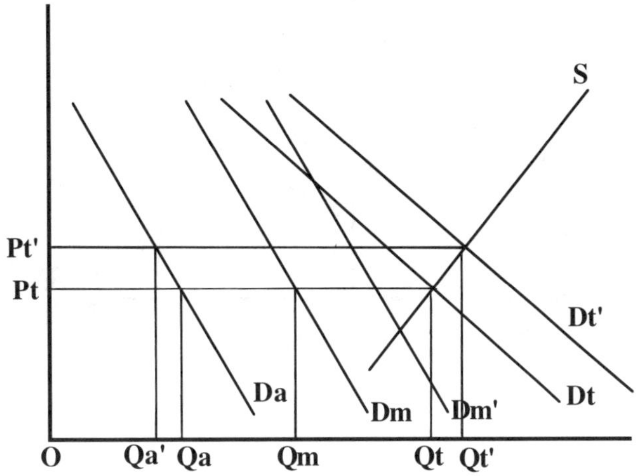

Figure 13.2 Input Demands with Tariff Protection

determine the equilibrium effect of the tariff on supply costs, and output prices and quantities (cf. section 13.5).

Industries that use tariff-protected inputs in production are thus doubly affected by tariffs: their input costs rise directly because of the tariff impost, and their input costs also rise because of increased input demands from tariff-protected industries.

The effect of tariffs can be examined using the Bewley–White model (cf. section 12.2). The effect of tariffs is to reduce the level of imports at a given exchange rate (figure 13.1). As the exchange rate appreciates, however, the level of imports will still rise *ceteris paribus*. Thus, in figure 13.3, the effect of tariffs is to create an effective import demand curve M1, which is closer to the vertical axis than an imports curve with zero tariffs (e.g. M). For any given exchange rate, imports are lower with tariffs than without. Given export supply, say X, and a specified current account deficit (for simplicity, let CAD = 0), the equilibrium level of imports and exports is lower than without tariffs, and the exchange rate is higher than otherwise (rx1 cf. rx0). This solution represents the direct effects of tariffs, but ignores the effect of high tariffs on increasing the costs of non-tariff-protected industries. These latter effects would contract the exports schedule towards the origin, further reducing the equilibrium level of exports.

Suppose tariffs had been maintained simultaneously with export sector expansion from the development of new mineral industries—e.g. from X to X1 in figure 13.3. With tariffs retained, exports and imports increase at the equilibrium solution as the export schedule expands, but so too does the exchange rate (to rx2) *ceteris paribus*. An alternative solution can be found with substantially higher imports and exports, and a substantially lower exchange rate, if industry protection is permitted to fall as the mineral export industry expands.

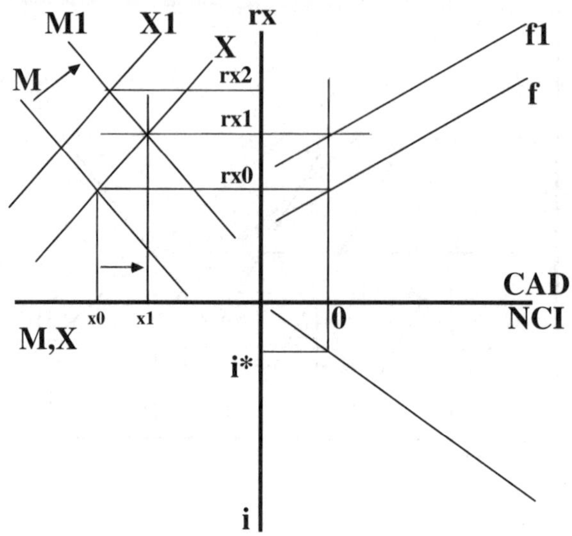

Figure 13.3 Tariffs in a Bewley–White Model

13.2 EFFECTIVE RATES OF ASSISTANCE

Assistance to industry provided by government policy may be measured in a variety of ways. The assistance measures summarised in table 13.1 are all partial equilibrium measures because they do not take into account producers' and consumers' responses to the changed relative prices that might be induced by the assistance (Industries Assistance Commission 1987, pp. 18–19; cf. section 13.3 below).

While the major form of assistance to manufacturing was the tariff, other government interventions included import quotas, industry plans such as local content schemes for motor vehicles, and regulation of interstate trade (Industries Assistance Commission 1987, pp. 58–60). Three sectors of manufacturing industry had high levels of protection—textiles; clothing and footwear; and transport equipment (cf. table 13.2). Most other industry sectors had rates of assistance at or below the industry average—i.e. in the range 10–25 per cent. Two industry sectors had low rates of assistance—food, beverages and tobacco; and non-metallic mineral products. Industry sectors differed in the variability of the effective rate of assistance within each sector, as well as their average effective rate of assistance. Thus, for example, the highly protected clothing and footwear sector had a low coefficient of variation of its effective rate of assistance indicating that, within this sector, all 4-digit ASIC[1] sub-sectors also had high average effective rates of assistance. The fabricated metal products and miscellaneous manufacturing sectors also had low coefficients of variation, suggesting that their respective 4-digit ASIC sub-sectors had similar, moderate average effective rates of assistance. By contrast, the food, beverages and tobacco sector had a very high coefficient of variation, suggesting that their 4-digit ASIC sub-sectors had highly disparate average effective rates of protection.

Table 13.1 Classification of Assistance Measures

Assistance measure	*Definition & examples*
Nominal rate of assistance on outputs	percentage change in gross return per unit of output relative to the (hypothetical) situation of no assistance
	tariffs, import quotas, domestic pricing arrangements; production bounties
Nominal rate of assistance on materials	percentage change in the prices paid for materials used in the production process
	tariffs and excise taxes on intermediate inputs raise prices; consumption subsidies reduce prices
Effective rate of assistance	percentage change in returns to value-adding factors due to assistance; net effect of output and input assistance, and penalties from tariffs and taxes on inputs
Gross subsidy equivalent	change in producers' gross returns from assistance; estimate of direct subsidy which would provide assistance equivalent to nominal rate of assistance on output
Tax on materials	net change in costs to user industries due to effects of policy on altering prices paid for inputs
Net subsidy equivalent	change in returns to value added resulting from assistance; estimate of direct subsidy which would provide assistance equivalent to effective rate of assistance
Consumer tax equivalent	value of the transfer from final consumers paying higher prices due to assistance

Source: Industries Assistance Commission (1987, pp. 17–18)

Table 13.2 Average Effective Rates of Assistance and Corresponding Coefficients of Variation for Sectors of Manufacturing, 1984–85 (per cent)

Manufacturing sector	*Average effective rate*	*Coefficient of variation*
Food, beverages and tobacco	6	3.33
Textiles	74	0.89
Clothing and Footwear	>250	0.36
Wood, wood products and furniture	17	0.76
Paper, paper products, printing and publishing	16	0.94
Chemicals, petroleum and coal products	12	1.42
Non-metallic mineral products	4	1.50
Basic metal products	10	1.50
Fabricated metal products	22	0.27
Transport equipment	66	1.56
Other machinery and equipment	23	0.52
Miscellaneous manufacturing	24	0.29
Total manufacturing	22	2.18

Source: Industries Assistance Commission (1987, tables 3.2–3.3); coefficients of variation derived from IAC data.

Aggregate estimates of assistance to Australian agriculture are summarised in table 13.3. The first two columns furnish estimates of Commonwealth assistance to agriculture. Most of the Commonwealth's positive assistance was provided as domestic pricing arrangements; the major beneficiaries of these arrangements in 1984–85 were manufacturing milk producers ($123m), market milk producers ($220m), egg producers ($25m), sugar ($68m) and wheat ($43m) (Industries Assistance Commission 1987, tables 2.11–2.12, 2.14). About 70 per cent of the value of domestic pricing arrangements was offset by the negative impact of tariff protection in non-agricultural industries (table 13.3), although some agricultural industries—citrus, wine grapes and potatoes (Industries Assistance Commission 1987, Table 2.10)—also had beneficial tariff arrangements.

Table 13.3 Assistance to Agriculture ($ million)

Form of assistance	*Commonwealth*		*State*
	1983–84	*1984–85*	*1984–85*
Assistance to outputs			
Domestic pricing arrangements	555	515	
Export incentives	8	3	
Export inspection services	55	29	
Local content schemes	26	17	
Marketing support	22	27	16
Price stabilisation and underwriting	2	13	
Tariffs	62	56	
Freight concessions			27
Other			1
Assistance to value-adding factors			
Adjustment assistance	29	33	
Agricultural extension services	–	–	111
Agricultural research	99	95	149
Concessional credit	5	7	27
Income tax concessions	96	126	
Natural disaster relief	102	43	26
Farm water supply and development			8
Soil conservation			23
Other			9
Assistance to inputs[(a)]			
Disease control etc.	11	18	55
Fertiliser subsidies	49	52	
Tariffs on materials	–158	–185	
Tariffs on plant & machinery	–204	–208	

Sources: Industries Assistance Commission (1987, Table 2.1; 1988, Table 4.1)

Notes: a: No estimates are available concerning any subsidy element associated with the price paid by farmers for irrigation water (cf. Industries Assistance Commission 1988, p. 3)

Nearly 60 per cent of state government assistance to agriculture was as research and extension services (table 13.3). A major form of state government assistance not accounted for in the IAC's estimates was provision of irrigation water since 'there are major conceptual problems in establishing appropriate benchmark prices to compare with the actual prices paid for water' (Industries Assistance Commission 1988, p. 3; cf. chapter 4).

Within the agricultural sector, and based only on Commonwealth Government assistance, effective rates of assistance were moderate to high in citrus, dried vine fruits, wine grapes, deciduous canning fruits, and sugar; and high to very high in tobacco, manufacturing milk, and market milk (cf. table 13.4).

Effective rates of assistance have fallen in both agriculture and manufacturing industry (figure 13.4). By the early 1990s, average effective rates of assistance were similar in both sectors.

Table 13.4 Nominal and Effective Rates of Assistance to Agricultural Industries, 1984/85 (per cent)

Industry	*Nominal rate*	*Effective rate*	*Industry*	*Nominal rate*	*Effective rate*
Apples and Pears	3.3	4.7	Wheat	1.7	1.7
Citrus	24.3	46.5	Barley	0.1	−1.0
Dried vine fruits	20.8	40.9	Oats	0.1	−0.1
Wine grapes	16.0	31.2	Maize	–	−3.9
Deciduous canning fruits	16.0	40.8	Sorghum	0.1	−2.5
Bananas	0.7	0.5	Oilseeds	–	−1.6
Tobacco	36.8	61.0	Sugar	15.3	38.7
Potatoes	8.0	16.1	Cotton	2.2	3.1
Onions	1.4	−0.3	Rice	7.7	13.8
Tomatoes	1.4	−1.4			
Other vegetables	1.4	−0.9			
Sheepmeat	1.0	1.8	Eggs	11.9	14.6
Wool	1.2	2.8	Poultry	–	−7.7
Beef	1.0	6.6	Pigs	–	−17.6
			Manufacturing milk	33.9	90.9
			Market milk	75.0	>250.0
			Honey	1.6	1.6

Source: Industries Assistance Commission (various dates) Annual report.

Notes: Nominal rate of assistance = assistance to outputs/(value of output - assistance to outputs)
Effective rate of assistance = (assisted value added – unassisted value added)/ unassisted value added = (assistance to value-adding factors + assistance to outputs + assistance to inputs)/unassisted value added
effective rates of assistance above 250% are highly sensitive to small changes in components and their calculated value is not reported.

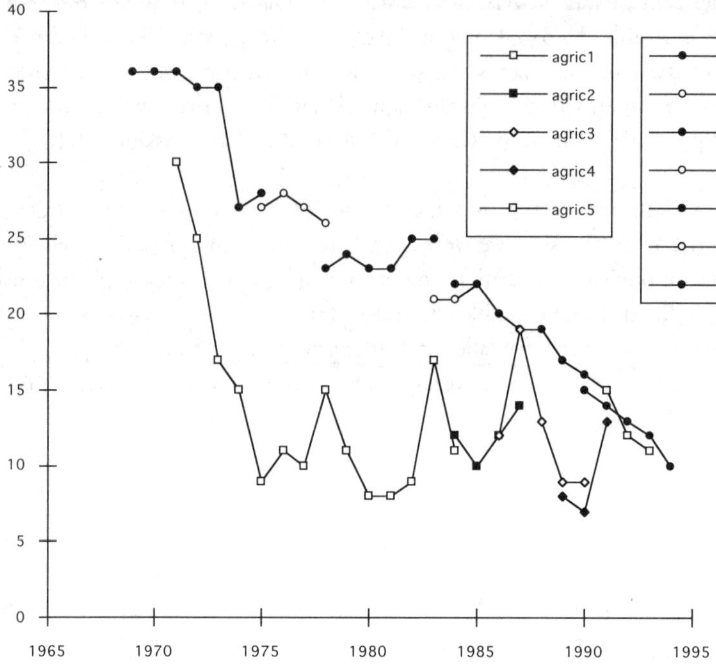

Source: Industries Assistance Commission (various dates), *Annual Report*

Figure 13.4 Estimated Effective Rates of Assistance, Australian Agriculture and Manufacturing

13.3 GENERAL EQUILIBRIUM ANALYSIS

Following Warr (1978), the general equilibrium effect of tariff assistance can be illustrated by figure 13.5. Consider a simple two-sector economy producing wheat and machines which has a production possibility frontier w0–m0. For convenience, welfare is expressed directly over wheat and machines by the social welfare indifference curves W0 and W1, rather than indirectly over these commodities using individuals' utility functions defined over wheat and machines. In a closed economy, welfare maximisation occurs at point A, where the social indifference curve W0 is just tangential to the production possibilities frontier w0–m0, and w1 wheat and m1 machines are produced. The common slope of frontier and indifference curves at A represents the relative price of wheat and machines.

Suppose international trade becomes possible, and the economy illustrated in figure 13.5 is a sufficiently small country that it is a price-taker in the world market at a relative wheat–machine price given by the slope of line pw. The maximum welfare point is then Y, achieved by the economy producing where the international price line pw is just tangential to the production possibility frontier w0 – m0 (point Z in figure 13.5).

The economy produces w2 wheat and m2 machines, and exports (w2 – w2′) wheat in exchange for imports of (m2′ – m2) machines, enabling consumption to occur at Y, where w2′ wheat and m2′ machines are consumed, providing utility W1.

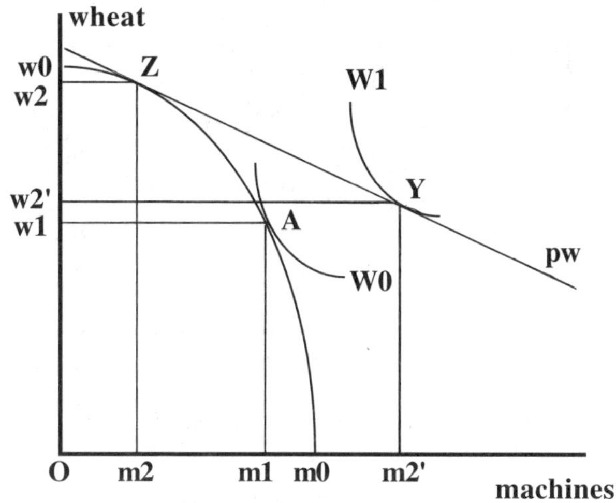

Figure 13.5 Trade in Simple General Equilibrium Model

The effect of a tariff on machine import is illustrated in figure 13.6. The tariff raises the relative price of machines from pw to pt—the quantity of wheat required for purchasing one machine is higher on the price line pt than on the international relative price line pw. The international price line remains the relevant trade line—i.e. it represents the rate at which wheat can be exchanged for machines in international markets. But the relevant domestic relative price line is pt. With the tariff, the economy maximises welfare by producing at point B where the price line pt is just tangential to the production possibility frontier. The economy then trades wheat for machines along the international price line pw′ (parallel to pw, i.e. unchanged relative price for wheat and machines in international trade). The efficient point for the economy with a tariff on machine imports is now a second-best optimum. This occurs where relative production and consumption prices are equal—i.e. a point of tangency on a social welfare indifference curve with the given internal relative prices pt—which is achievable via international trade along the price line pw′ starting from the point B. Point D on social indifference curve W3 is such a point. Point K on social indifference curve W2 is not feasible since the relative prices perceived by domestic consumers (given by the international price line pw′ at point K) are different from the relative domestic prices perceived by producers (given by pt). At K, producers and consumers would attempt to readjust their consumption and production to equalise domestic consumer and producer prices at the margin. Only point D on W3 is consistent with equal prices in consumption and production at the margin.

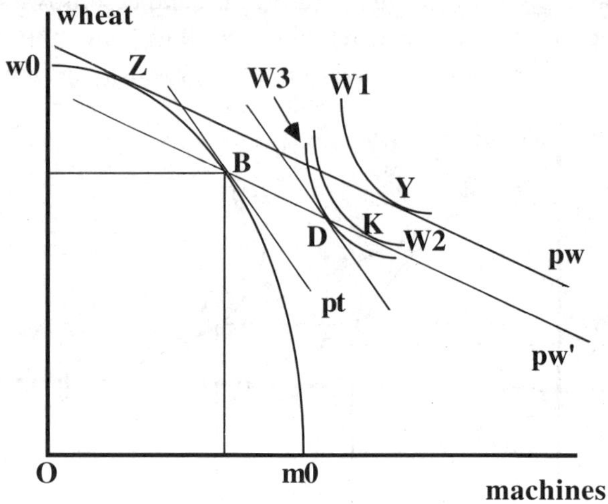

Figure 13.6 Effect of Tariff in Simple General Equilibrium Model

In this simple economy, tariff protection has reduced economic welfare in the economy in which the tariff was instituted. Why would a society consciously adopt policies that reduce aggregate welfare? While aggregate welfare might have been reduced by the tariff, the welfare of every individual and group has not necessarily been detrimentally affected. In a partial equilibrium setting, producers protected by the tariff are better off because of both higher prices and greater levels of output (cf. figure 13.1). Suppliers of inputs to the tariff-protected industry may also benefit because of the higher value of their input (figure 13.2). Those who suffer under the tariff in this model are consumers, who pay more for the imported commodity (and any domestically produced counterpart), and industries which either use the tariff-protected commodity in production or whose costs rise because of the greater competition for inputs used in the tariff-protected industry.

Where a country can influence its terms of trade—i.e. where the 'small country' assumption does not hold—a variety of trade interventions, including tariffs, might be used to improve its welfare. In Australia's case, it is inconceivable that trade policy action taken by this country alone could influence relative world prices for any imported commodity. Australia could theoretically influence the export prices of some commodities such as wool but, because no effective supply control is practised in this industry of atomistic producers, such price control has not been effectively undertaken. In the case of wool, even attempts to smooth supply and prices eventually collapsed (cf. Chapter 8). Similarly, the Australian Wheat Board, and CSR Ltd acting as the sugar export agent for the Queensland Government—each affecting approximately 10 per cent of world trade in their commodity—are unlikely to be sufficiently dominant to have a beneficial impact on Australia's price for those commodities, let alone Australia's terms of trade. Indeed, it is possible that quite the reverse has happened. Much of Australia's trade is controlled by foreign-owned firms who may use transfer pricing to manipulate import and export prices to the advantage of an individual firm, irrespective of the effects of such actions on Australia's economic welfare. One *prima facie* case of transfer pricing involving exports was dismissed in the High Court in 1980 (cf. Murphy J. in Blackshield et al. 1986, pp. 224–30).

If social welfare is reduced from W1 to W3 because of the tariff (cf. figure 13.6), those detrimentally affected by the policy may seek an amelioration strategy. What strategies are available to reduce the costs of tariff protection? The most obvious is to reduce the tariff itself. Ignoring the costs of achieving tariff reduction, social welfare would be maximised in the economy depicted in figure 13.6 if the tariff were eliminated, and domestic prices were made identical to the international trade prices pw. Alternatively, the cost of the tariff would be reduced if the rate of the tariff were reduced—equivalent to reducing the slope of the domestic price line pt so that its slope tended towards the slope of pw. An alternative solution is to consider leaving the tariff in place, and to find ways of mitigating its undesirable effects (cf. section 13.4).

Why might it be thought desirable to search for alternatives to tariff reduction? First, not all the welfare benefits of having a tariff might be captured by the analysis of figures 13.4 and 13.5. If there were social benefits from tariff protection independent of economic welfare effects, tariffs might be desirable despite their effects in depressing economic welfare. For example, one effect of extensive protection for the textile, clothing and footwear industries was their regional location, particularly in Victoria. If decentralisation was sufficiently highly valued by society, then the intangible benefits of decentralisation might offset reductions in economic welfare caused by tariffs. The conventional economic counter to this argument is two-fold:

- i without estimates of the social welfare benefits of decentralisation, it is difficult to determine whether reductions in economic welfare stemming from the tariff are greater or less than the increased welfare from decentralisation; and
- ii if decentralisation is desirable, it is preferable to achieve it directly by subsidising decentralised industries. The cost of decentralisation is then explicitly known and voters can intuitively compare it with the intangible benefits. Also, the intervention does not interfere directly with relative prices, and thus economic efficiency, as does tariff protection.

Interest groups benefiting directly and indirectly from tariff protection are, however, unlikely to prefer direct government payments in place of tariff protection because such payments are more obvious to taxpayers and more vulnerable to policy change. Politicians are also likely to prefer tariff protection as a strategy because it can be provided without the annual scrutiny demanded of budget expenditures. However, implementing other policies which provide the same indirect benefits as tariff protection is likely to be desirable because politicians must take account of the opportunity cost of indirect benefits such as decentralisation—this political opportunity cost takes the form of policies or programs which cannot be undertaken because of expenditure on such policies as decentralisation.

Second, the costs of achieving tariff reduction may simply be too high. The preceding analysis is comparative static and ignores the dynamic costs of reducing tariffs. Dynamic costs include the costs of retraining and relocating workers who would become unemployed as previously tariff-protected industries contracted. Significant transfer costs might also be involved—e.g. costs of permanent unemployment benefits if redundant workers were permanently unemployable outside tariff-protected industries. The implementation of tariff policy may also be seen as a contract between the government and tariff-protected industries. A change in tariff support might be argued to be a change in the terms of this contract having significant wealth effects and therefore requiring compensation (cf. Forbes 1995; Strong et al. 1990 concerning compensation for the abolition of NSW hen quotas). Compensation for the attenuation of 'property rights' deriving from the tariff may be justified on the general efficiency ground that contracts in the form of economic policy between government and investors need to be credible to attract the desired industry response. So rather than simply abandoning policies such as tariff protection when they are no longer required, government should compensate those detrimentally affected to maintain government's reputation with regard to the implementation of future policies. Retaining tariffs may be justified on efficiency grounds if the dynamic costs of their abolition or reduction exceed the benefits of abolition or reduction, or if the costs of uncompensated abolition/reduction, in the form of diminished trust in government, exceed the benefits of the reform. From government's private perspective, the benefits it obtains from tariff abolition/reduction may not be sufficiently great to outweigh the accompanying economic pain of compensating those affected by the policy change (with associated opportunity costs of reductions in other potential government programs), or the political pain of undertaking tariff reform without compensation. The Australian Government has undertaken significant expenditure in the form of assistance for industry restructuring and employee retraining in the face of tariff reform.

Both these arguments about whether or not tariff reduction should be undertaken depend on the economic costs of maintaining tariffs. General equilibrium estimates of the costs of the tariff are reviewed in section 13.5.

In addition to efficiency arguments over the desirability of tariff reform are the distributive issues raised by the tariff. Industries and workers benefiting from the tariff, and dependent local communities, are likely to support tariff retention even if the aggregate economic effect is negative. The potential problems associated with reducing tariffs were well summarised by Ingram (1966, p. 32, original emphases):

> In their zeal to make out a case for free trade, economists sometimes fall into a trap, and attempt to persuade (for example) a cotton-textile manufacturer that *his* firm will be better off without the tariff. Many words have been wasted in this vain (and mistaken) attempt!

> . . . Individuals, hurt by a specific tariff cut, will fight fiercely to protect their positions and, in a democratic country, they may join with other groups who are threatened with tariff cuts to make a common cause for the retention of tariffs. The very fact that injury is concentrated upon a relatively small group tends to make that group strong, vociferous, and politically effective . . . The argument that a tariff cut will harm a particular group within the economy is difficult to combat precisely because the *argument is valid and correct*!

13.4 TARIFF COMPENSATION

Faced with possible barriers to tariff reform in Australian manufacturing industry, 'tariff compensation' was considered as an alternative strategy. The general argument was that if substantial reductions to or abolition of tariffs were unlikely, other government policies could reduce the impact of the tariff. Consider figure 13.7, a simplified version of figure 13.6. Suppose the tariff were left in place so that domestic prices remained as represented by pt, and international prices remained as pw. However, instead of producing at point B, government intervened with other policies to increase access to resources by the non-protected wheat industry. Production would not necessarily have to occur on the production possibility frontier for social welfare to be increased. Consider the interior point V: the wheat–machines production combination corresponding to V would permit wheat export and machine import, enabling a higher social welfare indifference curve than W3 to be achieved (e.g W4). The 'closer' a point like V is to the social optimum production point Z, the closer the final solution will be to the maximum level of social welfare, W1.

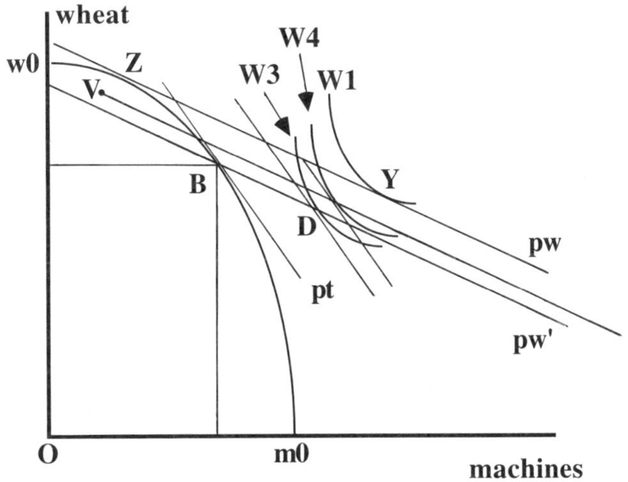

Source: Warr (1978)

Figure 13.7 Tariff Compensation

Three comments may be made on this solution. First, the tariff compensation solution can never exactly replicate the zero-tariff solution—if the tariff is maintained, the zero-tariff solution on W1 is infeasible because of the constraint that relative domestic prices are given by pt. This distortion would only be eliminated if the tariff were removed, and domestic prices equalled international prices. Second, even if the tariff compensation solution closely approximated the zero-tariff solution, resources would still be required to administer the tariff and the tariff compensation measures. Once the resources to administer both policies are considered, tariff compensation might be considered a less attractive alternative than tariff reduction. The more closely the tariff compensation policy mimics the zero-tariff solution, the more unattractive tariff compensation becomes because of the double dose of administrative costs. Third, for tariff compensation to result in a better solution, the corresponding policy would need to result in a production point above the international price line passing through B. It would not necessarily be easy, however, to ensure that a policy would achieve this outcome, particularly if the adopted policy was available to all industries and not just those that enjoyed—or endured—low levels of protection.

These issues were thrashed out by economists in a series of policy debates commencing with the 1974 'Green Paper'. Harris et al. (1974, paras. 3.57–3.81) argued that, while the 'first best' solution to the economic problems created by the tariff in manufacturing industries lay with its removal, some arguments suggested that achieving this goal might be difficult or even impossible. Difficulties included the costs of structural adjustment in highly protected industries and the fact that high levels of tariff protection were not confined to manufacturing industries—there were also some highly tariff-protected agricultural industries. Further, because tariff compensation measures were most likely to be provided from general government revenue, competitive demands on this revenue would mean that the level of compensation implemented might be less than desirable. The authors also noted that tariff compensation was generally not made available to non-exporting industries, which were also adversely affected by manufacturing tariffs. However, the Green Paper's authors concluded that 'There is a case on economic grounds, a second-best course of action, for providing some compensating protection, i.e. assistance, to the export sector' (Harris et al. 1974, para. 3.64).

Lloyd (1975) argued that, while second-best considerations for tariff compensation justified selected assistance to some primary producers, such grounds also justified targeted assistance to any producers receiving low levels of assistance, whether in primary or secondary industry. Indeed, since tariffs also affected costs in the non-traded goods sector, a case could also be made for targeted assistance to low-assistance industries in the tertiary sector. Harris (1975) noted that tariff compensation was offsetting assistance for any form of protection, not simply for tariff protection. Lloyd (1975) also argued that intervention to offset existing distortions may itself create new distortions. In particular, a serious potential disadvantage of tariff compensation is that—if domestic activity is successfully reoriented towards lowly assisted industries—beneficiaries of pre-existing assistance would notice the effects of tariff compensation in the form of increased costs and reduced profits. Since high assistance industries had been sufficiently influential to initially obtain such assistance—and sufficiently powerful to preclude tariff reduction as a first-best policy option—it could be expected that successful tariff compensation would lead to demands for higher tariffs or other forms of assistance to offset the effects of tariff compensation. The existence of both tariffs and tariff compensation would maximise opportunities for rent-seeking behaviour, and maximises resource misallocation into non-productive compensation activities (cf. Warr 1978).

Other arguments favouring tariff reduction rather than tariff compensation are that gains from tariff reduction are relatively more certain than those of tariff compensation (Warr 1978). Further, the gains from tariff reduction are large (cf. section 13.5), whereas tariff compensation measures, which rely on direct government expenditure, were only likely to be implemented at a level that would have little overall impact on resource allocation (Warr 1978).

Irrespective of the success of the intellectual case against tariff compensation, it appeared to have died a natural death in the 1980s. This death may have partly stemmed from a sustained commitment by the Commonwealth Government—and eventually the states—to reduce overall intervention in the economy. Such commitment by a Labor Government—with support from, and indeed generally demands by, the Federal Coalition Opposition for faster and deeper reductions in assistance—swamped the tariff compensation alternative in a political spree of deregulation.

13.5 GENERAL EQUILIBRIUM ESTIMATES OF TARIFF EFFECTS

Estimates using ORANI of the general equilibrium impacts of an across-the-board 25 per cent tariff cut on early 1980s levels are presented in table 13.5. Higgs (1986, chapter 6) also analysed the individual impact of 25 per cent tariff cuts in the three most highly protected industries (i.e. textiles, clothing and footwear; motor vehicles and parts; and metal trades). Because of the linearity of the ORANI model, the reported impacts may be converted into those of any specified tariff cut. For example, dividing all impacts in table 13.5 by 25 indicates the impact of a 1 per cent across-the-board tariff cut; or multiplying all results by –1 indicates the impact of a 25 per cent *increase* in tariffs. The principal macroeconomic impacts of the simulated tariff cut comprise increased imports (and exports, not shown), and reduced increases in the general price level. Both these effects could be anticipated from an action that would both reduce prices of imports and reduce prices in import-competing industries.

Table 13.5 Impact of a 25 per cent Across-the-board Tariff Cut

Variable	*Short-run effect (2 years)*[a]
Macroeconomic impacts	
Real gross domestic product	0.17
Aggregate employment	0.28
Consumer price index	–2.33
Aggregate imports (foreign currency value)	1.54
Sectoral output effects	
Agriculture	1.60
Forestry and logging	–1.11
Fishing and hunting	0.61
Mining	1.79
Meat products	2.82
Cotton ginning, wool scouring & top making	1.72
Agricultural machinery	9.99
Margins	0.30
Services	0.03
Employment—rural workers	
Agricultural industry output	
Pastoral zone	1.50
Wheat–sheep zone	1.29
High rainfall zone	1.85
Poultry	1.49
Agricultural commodity output	
Wool	1.35
Sheep	1.57
Wheat	1.37
Meat cattle	2.74
Milk cattle and pigs	0.94
Sugar, fruit and nuts	2.94
Vegetables, cotton, oilseeds and tobacco	0.49
Real net farm earnings	
Pastoral zone	4.14
Wheat–sheep zone	4.60
High rainfall zone	5.02
Northern beef	9.31
Milk cattle and pigs	2.40
Sugar, fruit and nuts	5.75
Vegetables, cotton, oilseeds and tobacco	0.82
Poultry	3.51
All agriculture	4.29

Source: Higgs (1986, Tables 6.1, 6.3, 6.7, 6.10)

Notes: a: percentage deviation from control

At the sectoral level, export-related industries benefit from the tariff cut, whereas import-competing industries—such as forestry and logging; textiles, clothing and footwear; and transport equipment (not shown, but –3.49 per cent and –5.60 per cent respectively)—suffer output declines. All agricultural industries and commodities enjoy output increases, and exports of all agricultural commodities except other cereal grains also increase (Higgs 1986, table 6.6). The prices of export-oriented agricultural commodities, except for wool, also increase; prices fall for milk, cattle and pigs; sugar, fruit and nuts; vegetables, cotton, oilseeds and tobacco; and poultry (Higgs 1986, table 6.5). The net farm returns of ORANI industries all increase as a result of the tariff cut, especially for those industries where both output and prices increase (table 13.5).

The simulated results confirm the complaints of farmers and their lobby groups over an extended period of time that tariffs severely reduced their output and incomes.

13.6 FUTURE ASSISTANCE ISSUES

While the tariff compensation debate is largely dead, and industry assistance has been substantially reduced (figure 13.4), industry assistance is a continuing issue. Future contentious issues include trade policy, non-tariff trade barriers, and arbitrary exemption from microeconomic reform.

Internationally, a key issue for Australian agriculture is to defend gains already made towards freer agricultural trade and, if possible, to maintain the momentum for trade liberalisation. Overseas agricultural interests, however, may not necessarily share Australia's enthusiasm for greater liberalisation of agricultural trade. For example, in October 1995 major nations in the Asia–Pacific Economic Cooperation (APEC) forum were signalling a desire not to proceed with previously decided trade liberalisation on agricultural products. As this conflict intensified, 'public choice' solutions were desperately sought, as indicated by the following news report:

> However, the [Japanese] vice-minister for international trade and industry (MITI), Mr Norifumo Kato, has hinted that there was still room for a compromise in the weeks leading up to the APEC summit in Osaka on November 19.
>
> 'There are subtle differences of nuance between English and Japanese,' he said. 'We may have to capitalise on the differences and think up wording that can be interpreted as acceptable by both sides' (Hills 1995b).

Since international trade negotiation is clearly a public good, there is a good reason for the Australian Government to continue arguing for trade liberalisation. Other areas of trade policy, or taking advantage of trade opportunities, where public goods may be involved include the reputation of some Australian agricultural industries as 'clean and green' (e.g. benefiting from the 'mad cow disease' turmoil in the UK); continuing rigorous analysis of the value of extra value-adding for agricultural and resource products; and continuing assessment of the net benefits of export monopolies in major products (e.g. wheat, sugar).

There is also a domestic aspect of assistance policy. The national government has wound back many of the assistance measures previously available to agricultural industries. However, the northeastern Australian drought of the first half of the 1990s exposed the difficulties of deciding the boundary between 'normal' and 'abnormal' seasonal variability. Managing 'normal' seasonal variability could reasonably be expected to be the responsibility of individual businesses in farming-related industries—and result in the efficient management of risk. Since European farming in Australia has a short history and there appear to be low frequency but imperfectly understood seasonal cycles (cf. Spark 1995, figures 3–5), it may be unreasonable to expect managers of farm and related businesses to bear the risk of seasonal conditions well outside 'normal' experience. However, judging the appropriate boundaries will never be easy, and the appropriate level of justifiable assistance not easy to determine. A related, and increasingly popular, form of assistance to agricultural and resource industries (and also urban areas) has been Landcare. Where Landcare involves market failure, implementation of government policy may be appropriate. However, where Landcare is primarily devoted to rehabilitation of privately owned natural resources, or preventing their degradation, Landcare simply becomes an assistance measure. Even where there are significant externalities in association with natural resource degradation—e.g. rising water tables and associated salinity—government funding of mitigation measures is not the only rehabilitation strategy available, nor is it necessarily the most efficient response. For example, with irrigation-induced salinity, raising the price of irrigation water would reduce the subsidy for irrigation activities and encourage allocatively efficient enterprise selection in irrigation areas; it would also, by encouraging economies in the use of water, assist in the reduction of water tables. Thus the most efficient 'landcare' policy measure might not be for the government to expend funds but actually to create efficient input prices.

ISSUES FOR DISCUSSION

1 'High levels of protection for Australian manufacturing and some rural industries raise domestic profits and wages, and this benefits agriculture by increasing the demand for Australian rural products.' Discuss.
2 How does the subsidisation of agricultural and manufacturing industries impose costs on the agricultural sector? Compare the relative merits of tariff reductions and tariff compensation policies.

Endnote

1 Australian Standard Industry Classification.

14 Microeconomic reform

14.1 INTRODUCTION

Throughout the 1980s and early 1990s, political parties and interest groups increasingly became promoters—or reluctant acolytes—of 'microeconomic reform'. But what is this microeconomic reform which is now *de rigeur*? Brennan (1990, p. 35) defined it as covering:

> . . . a wide range of policies and programs designed to make public sector activities and markets more efficient. Its main focus is to provide opportunities for growth and to remove impediments preventing resources from being employed most productively or limiting the flexibility of workers and enterprises. Many of these activities result from government regulations and so require action by governments. Others can be addressed directly by business and unions.

Measures can be as varied as reducing tariffs and assistance to industries, strengthening competition in markets, restructuring taxation, reforming union–employer relationships and public sector operations and ownership (cf. Forsyth 1992).

Microeconomic reform is thus all about improving the efficiency of the economy and increasing economic well-being. By improving economic outcomes, microeconomic reform has the capacity to deliver improved per capita incomes. Several preliminary comments must, however, be made:

1 Just because economic activity increases, and per capita incomes improve, not all individuals will necessarily be made better off through microeconomic reform. There may be (possibly substantial) negative effects, which explains why particular groups may bitterly oppose its implementation.[1] A Pareto-safe implementation of microeconomic reform would require adequate compensation of its victims.
2 Brennan's argument comes close to committing the GDP fallacy—that is, *only* measured economic welfare is relevant to social welfare. However, society may validly choose to exchange the level of economic well-being for distributional equity or economic stability. Or society may validly choose to trade off economic wel-

fare for less tangible dimensions of welfare, such as maintaining environmental assets, additional to that appropriate for rectifying market failures. Or society may recognise that some activities—e.g. education and health services—are poorly handled in national economic accounting,[2] and so greater levels of these activities may be warranted than would appear to be 'economic'. Or economic 'inefficiency' may be the least cost way of delivering policies implemented for distributional purposes.

3 'Non-economic' activities are important to individuals as well as in aggregate. The optimal allocation of an individual's resources may include not only income, consumption expenditure and consumption of publicly provided goods, but also leisure, workplace safety and income variability and security. Observed workplace 'bargains' which may initially appear to be 'uneconomic' may appear quite rational when all participants' objectives are accounted for. Some cases apparently requiring microeconomic reform may optimally allocate resources from the participants' perspectives.

Where there are rational reasons for (apparent) economic inefficiency, microeconomic reform may not be appropriate policy. Genuine microeconomic reform requires either unintended efficiency effects—e.g. changes over time have invalidated previously optimal resource allocations—or cases where eliminating economic or political bargains provides potential Pareto improvements. Apparent inefficiencies stemming from trade-offs between efficiency and other objectives are unlikely to be candidates for microeconomic reform unless the non-efficiency objectives can be achieved in some lower cost way.

This is not to argue that there is no room for microeconomic reform. At an aggregate level, or the microeconomic level, decision-makers—whether governments, firms or individuals—may have made inappropriate trade-offs between efficiency and non-efficiency objectives, or changing circumstances may have outmoded previous optima. Or interest groups may cleverly conceal adverse distributional objectives under the cloak of distributional or non-marketed goods (especially environmental) objectives.

14.2 AGRICULTURE AND NON-AGRICULTURAL MICROECONOMIC REFORM

Agricultural and resource policy interest in microeconomic reform occurring outside this sector arises because of the potential effects on it from such reform. The simplest illustration of the potential benefits of such reform can be seen in figure 14.1. Microeconomic reform may have the following effects for agriculture:

- Shift the final product demand curve. If economic inefficiencies have resulted in constraints on final product demand—e.g. arising from inappropriate tariff policy, or restrictions on the consumption of agricultural commodities within Australia (perhaps arising from restraints on interstate trade)—microeconomic reform may result in an expansion of final product demand. This expansion would, *ceteris paribus*, expand the derived demand for farm output. Conversely, microeconomic reform would depress any agricultural industries that have benefited from inappropriate policies. For example, eliminating Commonwealth and state policy which has discouraged the sale of Victorian fresh milk products in New South Wales would expand final product demand for Victorian milk while depressing farmgate demand for milk in New South Wales.

- Shift the marketing services supply curve. If past policy has made the supply of marketing services more expensive than it might otherwise be, microeconomic reform would lower the marketing services supply curve, reducing the equilibrium price of marketing services and possibly increasing the equilibrium level of marketing services added to the final product. The pre-1988 wheat marketing arrangements in Australia, which protected domestic bilateral monopolies in wheat storage and transport, contributed to higher costs of storing and handling grain. Deregulating these bilateral monopolies, and subsequent reductions in the labour force in grain handling, has apparently delivered substantial reductions in grain handling costs.
- Shifts in the demand curve for marketing services. Regulated industries sometimes restrict the degree of product innovation. For example, the growth of product-differentiated market milks based on fat level (full- versus low-fat milks),[3] the origin of the fat component (animal versus vegetable fats), modification to the sugar component (e.g. lactose modified milks), addition of components such as minerals, or storage method (e.g. refrigeration versus UHT treatment) took several decades to

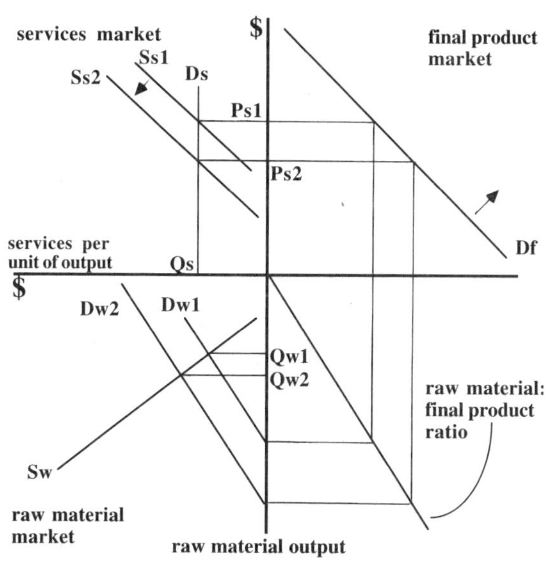

Figure 14.1 Effect of Increasing Efficiency in Post-farmgate Industries

develop in Australia, and was retarded at various stages by a reluctance to change the legislation governing milk processing. Microeconomic reform which eliminated such restrictions would result in the outwards movement of the demand curve for marketing services.

Microeconomic reform might be illustrated via the Bewley–White model. The effects of tariffs were previously analysed as contracting the import demand schedule, raising exchange rates and reducing trade (cf. figure 13.3). Tariff reduction is one form of microeconomic reform, whose effects reverse the effect of the tariff—reducing the exchange rate and increasing trade *ceteris paribus*. Microeconomic reform thus has potential macroeconomic effects, for example on the exchange rate, which may affect exporting or import competing industries such as agriculture. Conversely, if microeconomic reform increased efficiency and reduced supply costs in the import competing sector then, without tariff reform, the imports schedule might contract from M to M1, appreciating the exchange rate (from rx0 to rx1) and reducing trade (from x0 to x1) *ceteris paribus* (Figure 13.3).[4] A combination of tariff reform and supply cost reductions in the import competing sector would have an ambiguous impact on imports/exports and the exchange rate; quantitative analysis—e.g. in a computable general equilibrium model like ORANI—would be required to determine the net effect of both changes taken together.

Microeconomic reform which improved agricultural sector efficiency might expand the exports schedule (from X to X1 in figure 13.3), increasing trade and appreciating the exchange rate *ceteris paribus*. Conversely, microeconomic reform that eliminated subsidies in the agricultural sector and reduced the profitability of exports would contract the exports schedule, and reduce trade and depreciate the exchange rate.

Deregulation might deliver other benefits affecting the allocation of resources to farm production, or the relative profitability of farm and non-farm production, leading to resource shifts between agriculture and other sectors. Such changes included deregulation of the financial services sector of the Australian economy during the 1980s (section 14.3), and possible tax reform (section 14.4).

14.3 FINANCIAL SECTOR DEREGULATION

Agriculture largely comprises small businesses, and what applies to small businesses is generally true of agriculture. Johns et al. (1989, pp. 149, 143 respectively) commented that:

> Since the deregulation of Australian financial markets in the early 1980s complaints about the availability of bank credit to small business have been more muted. There has been a rapid growth in business lending by the banks and other financial institutions . . .
>
> As a result [of financial deregulation and other factors], small firms are typically finding credit more readily available, but at a higher cost. Real interest rates are significantly higher than they were at the beginning of the decade.

This experience was mirrored in the experience of Australian agriculture: real farm debt increased steadily throughout the period 1977–92 and, while more volatile, real interest rates increased substantially at the beginning of the 1980s and continued on an upward trend until the early 1990s (figure 14.2). Real farm debt increased approximately 56 per cent from 1977 to 1992, while real interest rates increased 10 percentage points. These changes are substantially explained by deregulation of Australian financial markets from the early 1980s.

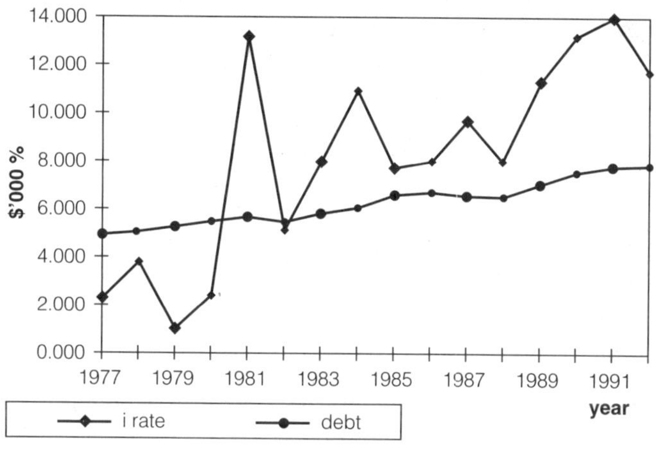

Source: Australian Bureau of Agricultural and Resource Economics (1994, tables 12, 89, 90)

Figure 14.2 Real Farm Debt and Real Interest Rates, Australia, 1977–91

There is some debate about the objectives of financial market deregulation. Hawtrey et al. (1991, p. 11) argued that:

> According to the Campbell and Martin Reports, deregulation of the financial system during the 1980s was aimed at improving the efficiency in the banking system via the introduction of competition (implicitly capping profits from banking by the oligopoly of the four major banks) and, correspondingly, to ensure that banks were as responsive as possible to customer needs (i.e. consumer sovereignty rather than producer sovereignty).

Conversely, Ackland and Harper (1992, p. 46) argued that the principal reason for government pursuing deregulation of financial markets was not to achieve gains in allocative efficiency (i.e. microeconomic reform) although such gains were available, but for macroeconomic management objectives, since 'it was obvious that regulations imposed on financial markets were impeding rather than facilitating the implementation of monetary policy'. Because financial sector regulation was implemented through the banks, and innovation in the financial sector led to the growth of non-bank institutions beyond the direct reach of monetary policy, even the banks 'began to question the balance of burden and privilege conferred by financial regulations' and argued forcefully for the abolition of the regulations (Ackland & Harper 1992, p. 47).

The process of financial market deregulation is recorded in Industry Commission (1991a, Appendix E) and the key events are summarised in Box 14.1. While deregulation aimed at reducing restrictions on financial institutions, new forms of restriction were occasionally introduced as limitations of the deregulated financial environment became apparent. Considerable product innovation accompanied financial market deregulation (Box 14.2). Financial deregulation was itself shaped by the 1980s worldwide economic boom and associated inflationary expectations, and the stockmarket crash of 1987. These events both fashioned and were fashioned by the demand for debt, and corporate behaviour. On the supply side, there was fierce competition for market share in finance markets between existing banks and near-bank institutions, and the new financial market entrants including foreign banks.

Box 14.1 Key Events in Financial Sector Deregulation

Year	Month	Event
1980	May	Removal of interest rate ceilings on deposits with trading and savings banks
1981	July	Removal of restrictions on overseas investment by Australians
	December	Final report of the Campbell Committee tabled in Parliament
1983	October	Reserve Bank withdrew from the Australian forward currency market and relaxed restrictions on foreign currency holdings and the setting of exchange rates by trading banks
	December	Float of the Australian currency. Trading banks no longer required to settle foreign exchange position each trading day. All substantive exchange rate controls removed.
1984	April	Deregulation of Australian Stock Exchanges
	June	Forty non-bank institutions permitted to trade in foreign exchange
	August	Savings banks permitted to offer cheque accounts. Removal of limits on size of fixed deposits with savings banks.
	August	Trading banks permitted to borrow funds for terms exceeding four years.
	September	Abolition of tax concessions for life and superannuation funds holding at least 30 per cent of their assets as government or semi-government securities.
1985	January	Relaxation of limits on foreign investment in Australian securities

	February	Sixteen banking licences offered to foreign banks
	April	All bank lending ceilings abolished, except for housing loans less than $100 000
	September	Capital Gains Tax introduced
1986	April	Removal of interest rate ceilings on new bank housing loans less than $100 000; subsidy to savings banks to maintain rates on existing loans at 13.5 per cent
1987	April	Removal of limits on ownership of stockbroking firms by Australian financial intermediaries
1988	May	Introduction of 15 per cent tax on income of life and superannuation funds
1989	July	Introduction of restrictions on superannuation fund assets and liabilities
	November	Reserve Bank given explicit powers of prudential supervision of banks. Distinctions removed between trading and savings banks.
1990	May	Establishment of Australian Banking Ombudsman
1991	January	Establishment of Australian Securities Commission

Source: Industry Commission (1991a, Appendix E)

Box 14.2 Some Banking Product Innovations

Year	Innovation
1980	Automatic teller machines
1981	Monthly income term deposits; card (not passbook) savings accounts
1982	Variable repayment home loans; Visa card for domestic and international use; reduced terms and minimum balances for term deposit; PIN for credit and debit card access
1983	MasterCard; compounding term deposits
1984	Daily interest cheque account
1985	EFTPOS; MasterCard ATM linkage; specialised agribusiness and rural budgeting centres; telephone banking; cash management accounts
1986	Housing bonds; equity mortgage loans
1987	Life insurance; fixed rate mortgage lending; home banking
1988	Bankcard debit/credit card; enhanced retirement services;
1989	Payroll system; business credit cards
1990	Mortgage offset account

Source: Harper (1991, table 1)

Explaining the effects of financial deregulation on agriculture requires a description of the earlier highly regulated agricultural and general financial markets. Prior to deregulation, the financial sector was a predominantly domestic market; there were interest rate constraints leading to the rationing of bank lending, with considerable ingenuity being expended in modifying or concealing the consequences of interest rate capping on deposit rates (cf. Standen 1982 on lending for farmers). In a simple model of the aggregate supply of and demand for borrowings, a free market in a closed economy would result in an equilibrium interest rate i* with savings equal to borrowings (figure 14.3). But if the interest rate on deposits is capped below i* (say i#) the demand for savings exceeds its supply (B#>S#), and savings must be rationed. Further, at the level of savings available at the capped interest rate i#, the borrower's rate i** could exceed the lender's rate (figure 14.3), hence Ackland and Harper's (1992, pp. 52–3) comment on the banks' high interest margins prior to deregulation. In the agricultural sector, savings rationing and restrictions on bank lending to farmers led to the establishment of institutions such as the Commonwealth Development Bank with access to Commonwealth Government funds to expand lending to farmers (Standen 1982). Powell and Milham (1990, p. 229) commented that regulated financial markets were characterised by a 'focus on institutional and regulatory measures to facilitate the financing of farmers'. These measures, while principally aimed at farmers, had spin-off effects for rural and regional Australia; Harper (1991, pp. 80–1) commented on cross-subsidies that were rife in the regulated system:

> . . . banks were expected to maintain a presence in isolated rural communities irrespective of profitability . . . urban/rural cross subsidies were feasible only because regulations on banks, including, especially, market closure, enabled them to extract economic rent from particular groups of customers. This surplus could then be used to subsidise services provided to others at prices below marginal cost. It is no surprise that deregulation has seen the closure of many isolated rural bank branches. It is not that banks have lost their sense of social obligation but rather that the customers who bore the cost of the cross-subsidy will do so no longer.

The implications for the farming sector of financial sector deregulation might be explained using a simplified agricultural credit market (figure 14.4). The demand for savings in farming is Dsa. If the equilibrium level of borrowers' interest rates in the interest-rate-capped financial sector is i**, then lending in the agricultural credit market could be set at Qsr, and credit would therefore be made available to those farmers who valued it most. Even if credit were rationed in this way, however, the actual level of interest rates in the agricultural credit market could vary between the equilibrium borrowers' rate (i**) and the lenders' rate (i#). In a government-preferred credit market like agriculture, banks might (be subsidised to) shave their margins so that interest rates for farmers lay between i** and i#.

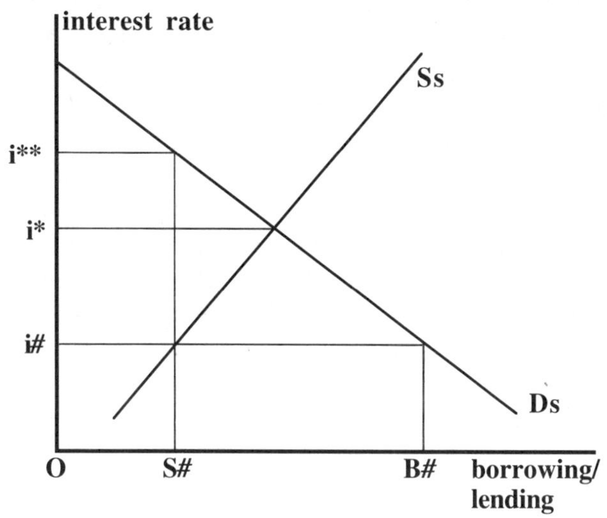

Figure 14.3 Aggregate Financial Market

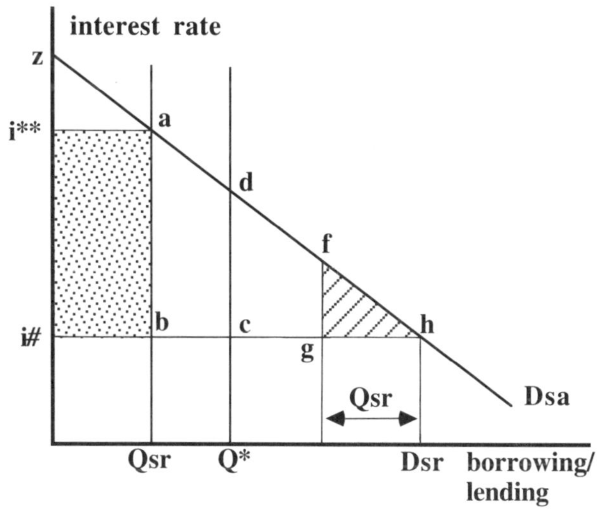

Figure 14.4 Agricultural Credit Market

Suppose, for simplicity, that farmers' borrowing rates were i#. With credit supply at Qsr, agricultural credit demand at interest rate i# is Dsr>Qsr, i.e. a borrowing gap exists. This excess demand requires that available savings Qsr are rationed in the agricultural credit market. In the best-case scenario, a rationing mechanism is devised in which available credit is lent to those who value it most—i.e. those prepared to borrow at or above the interest rate i**. If credit is actually supplied at interest rate i#, these borrowers receive an interest rate subsidy equivalent to the shaded area i**abi# (figure 14.4) which corresponds to an income loss to lenders (e.g. as reduced bank profits). If Q* were the unregulated level of agricultural credit, the social loss from the interest rate cap just in the agricultural credit market would be abcd. In the worst-case scenario, available savings are lent to those who value it least: borrowers gain by the amount of the triangle fgh and, given the restrictions on available lending, society loses by an amount (zabi# – fgh). In figure 14.4, triangles zai** and fgh are congruent, and hence the net social loss from lending to borrowers who value it least is equal to the income transfer i**abi#, and lenders continue to lose income equivalent to i**abi#. Thus, the extent of potential losses in a regulated financial market depends on the ability of lending institutions to mimic the operations of a market and lend to those clients who value their borrowings the highest.

Conversely, the effect of deregulating financial markets in general, and the agricultural credit market in particular, depends on how well the rationing mechanism in the regulated market mimicked the market. If rationing mimicked the market well—i.e. the best-case scenario in a rationed credit market—then the gains from deregulation would be least. Suppose, for simplicity, deregulation led to an increase in available agricultural credit so that Dsr was the supply of borrowings in a deregulated environment. The previous regulated interest rate i# would then be the equilibrium interest rate in the deregulated market (figure 14.4). In this previous 'best case' scenario, the gains from deregulation are abh. In the previous worst case—where lending was confined to borrowers who valued it least—then the gains from deregulation are greatest, equivalent to area zfgi# = i**ahi#.

The effects of deregulation on the Australian agricultural and resource sector thus depended on a number of related factors:

- the aggregate increase in available funds in the Australian financial sector and the corresponding net increase in agricultural credit (i.e. increased commercial lending less any change in previous special agricultural credit schemes);
- the extent to which interest rates rose above the previous regulated rate (cf. the 'for simplicity' assumption above that post-deregulation rates were the same as the regulated rates); and
- the degree to which the credit rationing system of the regulated market mimicked the allocation function of private markets—the better this mimicry, the lower the benefits of deregulation.

The extent of these changes is difficult to discern from figure 14.2 because of other substantial changes occurring simultaneously with financial deregulation. Some sections of the agricultural industry have apparently yet to realise that the financial sector has been substantially deregulated—cf. Box 14.3, outlining a call made from corporate agriculture for a financial instrument through which the Commonwealth Government (i.e. taxpayers generally) would bear the cost of the riskiness of Australian agricultural production.

Box 14.3 New Call for Government Financial Sector Intervention to Benefit Agriculture

Australian farmers are placed at an international disadvantage because of the comparatively short-term borrowing facilities available to them, according to the managing director of the country's biggest cotton producer.

Mr David Farley of Colly Farms Cotton, which also operates farms in the US, is an advocate of a US-style aggie bond—a 20- to 30-year loan for farmers that has a fixed interest rate and tax advantages to the investors.

He believes such a bond would be attractive to superannuation and corporate investors seeking long-term investments, and would facilitate rationalisation within the rural sector.

'We have limits in Australia, limits in horizons, that make financing long-term cyclical businesses difficult,' he said.

'Most other developed nations have addressed the importance of agriculture and [have] structured agricultural finance to suit the nature of the business. If you look at the US farm program, it finances agriculture with 30-year time horizons, with fixed interest rates for that period of time.'

Similar schemes exist in Japan, Europe and some African nations.

'If our Government is keen to measure us on international standards and on international practices, we must develop a strategy and then, ultimately, the tools that can keep us competing on a level playing field.

. . .

'Australia currently has not got a long-term debt market for agriculture or other industries.'

Mr Farley said fund managers seeking long-term investments were investing money outside Australia, some in US farm bonds.

'What we need is to be able to develop a market that can use that capital here. We should transfer the risk of managing adversity away from the Government and put it in people's hands who can manage a portfolio of risk.

'The risk is taken with the bond holder. Most governments have addressed the risk profile involved and have given tax relief on the earnings from those bonds.'

Source: Wahlquist (1994)

14.4 TAX REFORM

Petroleum taxes

Excises on petroleum products were originally solely directed towards ('hypothecated' for) road construction and maintenance. While a component of these taxes continues to be hypothecated, the bulk of the funds raised is directed to the Commonwealth's Consolidated Revenue. The effect of petroleum excises is to raise the price of petroleum-based fuels above the level which they would otherwise be.

In a simple partial equilibrium model of a single industry or sector, the effect of an excise (ignoring substitution effects away from the now higher priced fuel) is to increase production costs, shifting individual producers' marginal cost curves upwards to the left, and contracting the industry supply curve. However, in the case of agricultural production (and for other primary sector 'off-road' uses), the principal petroleum product used is diesel and the effect of the excises is reduced by producers being entitled to a rebate (introduced in 1982) for the excise paid on the fuel they use. The justification for the rebate was presumably not to penalise exporting industries which had no influence over the prices paid for their products on international markets. However, the same argument applies to import competing industries that are not tariff protected, which similarly have no control over the prices at which their products are sold, but are not eligible for diesel fuel excise rebates. Excise is still payable on petroleum used in manufacturing inputs for primary industries (e.g. fertiliser and chemicals) and for any services added to primary products after they leave the farmgate (e.g. transport and processing). In a partial equilibrium model with a services sector separated out from agricultural production, the effect of excises on the sector processing agricultural products is to increase its costs, contracting its supply curve towards the vertical axis (cf. figure 14.1, a contraction from Ss2 to Ss1), thus increasing the cost of adding services and contracting the derived demand for agricultural products (from Dw2 to Dw1). Diesel rebates leave the raw material supply curve (Sw) unaltered.

The aggregate effect of fuel excises in a model of endogenous exchange rates is ambiguous (cf. figure 13.3). Where fuel excises increase the costs of producing or adding services to exportable primary products, the export demand schedule will contract towards the origin (e.g. X1 contracting to X in figure 13.3). Conversely, increased costs of domestic production of importables will increase the demand for imports where petroleum-intensive activities have been undertaken overseas (e.g. M1 expanding to M in figure 13.3). The direction of change for the exchange rate is likely to be towards depreciation, but the net effect of fuel excises on trade or the precise effect on the exchange rate cannot be determined analytically in this simple model.

The effect of fuel excises must therefore be analysed in a model of the entire economy, taking into account the full range of possible effects. Mannion et al. (1988) used the ORANI computable general equilibrium model of the Australian economy to examine the effect of removing petroleum excises. There are two key assumptions underlying their analysis: (i) the level and composition of real consumption, investment and government expenditure ('real absorption') is assumed exogenous; and (ii) there are no labour market constraints but there is real wage rigidity (money wages were assumed to be fully indexed to changes in the consumer price index, reflecting the Wages Accords of the 1980s). Because their analysis was critically dependent on these assumptions, the results cannot be generalised to other policy frameworks—e.g. where there is substantially less transmission of inflation effects into wages. While these difficulties considerably reduce the contemporary relevance of their results, the results are reviewed here because they illustrate the use of computable general equilibrium results in policy analysis where policy effects clearly have general equilibrium dimensions.

Table 14.1 Impact of Removal of Petroleum Taxes with Revenue Neutrality, or Abolition of Diesel Rebate, short-run effect (2 years)[a]

Variable	*Removing of all petroleum product taxes (1)*	*Reduced government expenditure with (1)*	*Revenue-neutral removal of petroleum taxes[b]*	*Removing diesel fuel rebate[c]*
Macroeconomic impacts				
Real gross domestic product	1.4	0.6	–2.3	–0.1
Aggregate employment	1.9	0.5	–3.1	–0.2
Consumer price index	–4.6	–5.7	9.8	0.0
Imports	–1.4	–2.4	2.5	0.0
Exports	6.5	8.1	–9.1	–0.7
Real government revenue	–3.0	3.3	0.0	0.5
Sectoral output effects				
Agriculture	3.0	3.7	–4.7	–0.6
Forestry and logging	1.2	0.5	–2.0	–.03
Fishing and hunting	3.1	3.7	–4.3	–0.5
Mining export	4.2	5.1	–6.8	–0.6
Food export	7.5	9.1	–9.8	–0.6
Real net value of output				
Agriculture	9.1	11.3	–12.7	–2.3
Forestry and logging	1.6	2.5	–1.8	–0.3
Fishing and hunting	3.8	4.3	–5.5	–0.6
Mining export	14.6	19.2	–21.0	–2.1
Food export	24.5	28.7	–3.1	not reported

Source: Mannion et al. (1988, tables 3 and 4)

Notes a: percentage deviation from control
b: achieved via removal of petroleum excises together with a revenue-neutral value-added tax on consumption
c: not revenue neutral

Mannion et al.'s (1988) analysis proceeded in two phases. First, the effect of simply removing the petroleum excises was estimated (table 14.1, column 1). In this phase, the economy expands—particularly the export sectors—but this expansion is achieved at the expense of a 3 per cent real decline in government revenue. However, there is no corresponding reduction in the government's expenditure (Mannion et al 1988, p. 126). Second, such a reduction in government revenue could be accommodated either by an offsetting decrease in real government expenditure (column 2)—further benefiting exporting industries—or by offsetting increased taxation. The effect of a uniform consumption tax to offset the loss in petroleum excise taxation is reported in column 3. The combination of these two measures significantly depresses the export sector and the economy overall and substantially raises the consumer price index. The issue of a uniform consumption tax (or value added or goods and services tax) is examined below.

The effects of removing the diesel fuel rebate are shown in the final column of table 14.1. Because the rebate forms a small proportion of Commonwealth Government expenditure (or a small reduction to its revenue), abolishing the rebate has a small macroeconomic effect. Removing the diesel rebate also has relatively small percentage effects on the output of the industries such as agriculture which suffer from its removal—agricultural output, for example, declines by 0.6 per cent. However, because the rebate also increases costs at the same time as output falls, the real net value of production declines by substantially more than the output declines; in the case of agriculture, real net value of output declines by 2.3 per cent.

Both petroleum excises, and the manner of their possible removal, have substantial potential impacts for the exporting primary industries in general, and agriculture in particular. Simply because of its much smaller aggregate magnitude, elimination of the diesel rebate has a smaller aggregate impact, but still a potentially important one for industries like agriculture.

Value added tax

Approximately 10 per cent of Commonwealth Government revenue is raised from wholesale sales taxes on commodities. Sales tax is levied at three standard rates, varying depending on the commodity in question. The effect of differential wholesale sales taxes is to change the relativity of prices perceived by consumers, and change the relative prices consumers face compared to those faced by producers. A value added tax (VAT, or consumption tax, or goods and services tax GST) extends the taxation net to all consumer goods and services, and preserves relative prices by using a common tax rate.[5] By maintaining the same relative prices in production and consumption, the conditions for achieving a Pareto-efficient allocation of resources are preserved. Compared to the Australian situation, where consumption taxes are predominantly applied to commodities whereas services are exempt, a consumption tax does not favour consumption of services or commodities. Additionally, while creative accounting may be used to avoid income taxation, it is less easy to avoid consumption taxes, since the latter can only be avoided by not consuming.[6]

The effect of consumption taxes on agriculture and other primary industries is difficult to determine in a partial equilibrium setting, since the combined effects on primary production, processing and consumption of primary products, or export of primary products are unclear. The effects of a consumption tax, and its particular effects on the primary industries, can only be properly assessed in a general equilibrium setting. Higgs (1986, chapter 8) attempted such an analysis using a version of the ORANI computable general equilibrium model of the Australian economy. There are three key aspects of such an analysis: (i) a careful definition of how a consumption tax would be implemented in the model; (ii) whether the resulting increase in taxation would increase government gross revenues, or whether other taxes would be reduced to leave the government's revenues unchanged (and, if so, which taxes would be reduced); and (iii) the nature of the macroeconomic setting in which the simulation would proceed. Higgs (1986, p. 135) chose the following scenarios:

- An additional sales tax of 5 per cent of initial purchasers' prices, imposed on household purchases of all goods and services, represented the effect of a consumption tax.
- An offsetting tax cut which, when combined with the increased indirect taxes, left unchanged 'real private absorption' (i.e. the sum of CPI-deflated consumption, and price deflated investment).
- Three wage tax packages were analysed separately: pre-tax wages were fully indexed to the consumer price index; take-home pay was fully indexed to the CPI; or direct tax cuts restricted to labour incomes only.
- The labour supply curve was assumed to be infinitely elastic at the going real wage rate.

The results of these simulations are summarised in table 14.2. Under full pre-tax wage indexation, in an environment where all prices have increased as the result of the imposed consumption tax, the CPI rises. This is as a result of the direct effects of the consumption tax and the indirect effects of the initial increase in the CPI being fed through production costs and supply curve shifts in markets where demand is less than perfectly elastic (table 14.2, column 1). As wage rates increase more than the CPI (cf. Higgs 1986, table 8.1), employment falls, as does real GDP. Exporting industries are particularly badly hit because they are assumed to sell in price-takers' markets. It is less costly to maintain post-tax real income when the CPI rises because of consumption tax, so the macroeconomic effects of consumption taxes in the second scenario are less than with full pre-tax indexation (cf. table 14.2, column 2). Finally, where only labour is compensated for the price effects of the consumption tax, the macroeconomic effects are less severe and real GDP and exports actually expand because the CPI rises less than the tax-induced price increases.

Table 14.2 Impact of a 5 per cent consumption tax, short-run effect (2 years)[a]

Variable	*Full pre-tax wage indexation*	*Full post-tax wage indexation*[b]	*Full post-tax wage indexation*[c]
Macroeconomic impacts			
Real gross domestic product	–3.06	–0.21	0.67
Aggregate employment	–4.48	–0.30	0.99
Consumer price index	13.01	5.49	3.17
Aggregate imports (foreign currency value)	4.26	0.26	–0.97
Aggregate exports (foreign currency value)	–12.23	–0.84	2.66
Sectoral output effects			
Agriculture	–8.60	–0.56	1.91
Forestry and logging	3.06	0.14	–0.75
Fishing and hunting	–3.38	–0.23	0.74
Mining	–9.11	–0.71	1.87
Meat products	–14.47	–0.96	3.19
Cotton ginning, wool scouring & top making	–14.26	–1.00	3.07
Agricultural machinery	–44.38	–2.96	9.78
Margins	–2.60	–0.18	0.56
Services	–1.01	–0.06	0.23
Employment—rural workers	–13.07	–0.86	2.90
Agricultural industry output			
Pastoral zone	–8.29	–0.60	1.76
Wheat–sheep zone	–6.74	0.43	1.51
High rainfall zone	–9.79	–0.66	2.14
Poultry	–7.70	–0.51	1.70
Agricultural commodity output			
Wool	–7.34	–0.55	1.54
Sheep	–8.08	–0.54	1.78
Wheat	–7.10	–0.34	1.74
Meat cattle	–14.09	–0.93	3.11
Milk cattle and pigs	–5.02	–0.33	1.11
Sugar, fruit and nuts	–15.05	–0.97	3.36
Vegetables, cotton, oilseeds and tobacco	–4.90	–0.32	1.09
Real net farm earnings (pre-tax)			
Pastoral zone	–22.84	–6.38	–1.32
Wheat–sheep zone	–24.02	–6.25	–0.78
High rainfall zone	–26.55	–6.52	–0.36
Northern beef	–47.42	–7.52	4.75
Milk cattle and pigs	–12.67	–5.55	–3.36
Sugar, fruit and nuts	–29.38	–6.61	0.38
Vegetables, cotton, oilseeds and tobacco	–8.35	–5.26	–4.32
Poultry	–18.07	–5.92	–2.18
All agriculture	–22.81	–6.21	–1.10

Source: Higgs (1986, tables 8.1, 8.3.)

Notes: a: percentage deviation from control
b: equal percentage cuts in effective capital and labour income tax rates
c: cuts in labour income tax rates only

The sectoral effects of these three consumption tax scenarios reflect the macroeconomic results. In the first scenario, where exports fall substantially, export-oriented primary industries suffer substantial reductions in output (cf. table 14.2, column 1). Under the second scenario, the effects of the consumption tax are substantially reduced (cf. table 14.2, column 2) and under the third scenario, macroeconomic expansion leads to small to modest increases in primary sector activity (except for forestry and logging) (cf. table 14.2, column 3). The output effects on agricultural industries and commodities mirror the output effects for agriculture as a whole. Under the first consumption tax scenario, real net farm incomes decline severely because the output reductions are accompanied by substantial cost increases (cf. table 14.2, column 1); these effects are less severe in the second scenario. However, even in the third consumption tax scenario, where there were modest output increases for agriculture a whole and for all ORANI agricultural industries and commodities, real net farm incomes fell for all industries except northern beef and sugar, fruit and nuts.

The efficiency effects of introducing a consumption tax depend not only on the direct and indirect effects of the tax itself, but also on the manner in which offsetting tax reductions—if any—are implemented. Not only do these offsets affect the macroeconomic outcome, but there are also potentially significant effects for individual industries such as agriculture and other primary industries.

There are also probable distributional effects of introducing a consumption tax. A consumption tax of x per cent imposed on all consumption expenditure requires an individual's consumption expenditure to be increased by x per cent *ceteris paribus* for welfare neutrality—hence the offsetting tax policies in the preceding ORANI simulations. However, offsetting tax policies in a sector of largely self-employed operators like agriculture requires that firms pay at least as much income tax as the increased tax on gross consumption expenditure incurred via the consumption tax. Low income consumers, paying less income tax than a consumption tax would raise on their consumer expenditure, would have their tax burden increased by a consumption tax, and their welfare would be correspondingly reduced. A policy response that focused only on reducing tax to offset the welfare-reducing effects of a consumption tax would not prevent welfare reductions for low income taxpayers. In a highly variable enterprise like agriculture, low or negative income years are expected, and tax offsets would not be available to counteract the cost of increased consumption implemented by the consumption tax. The existing income averaging approach used to manage income tax is not readily available for consumption taxes. In agriculture at least—and also for other low income groups—implementing a consumption tax would require direct social security payments to offset the effect of increased taxation on consumer expenditure.

14.5 NATIONAL COMPETITION POLICY

Following agreement between the Commonwealth, states and territories, the Commonwealth established the Committee of Review of the Application of the *Trade*

Practices Act 1974 (Hilmer Committee) to enquire into national competition policy and law. Agreed national competition principles underlay this review (Box 14.4). The Committee saw its role as being to articulate a framework within which effective competition would flourish, thus promoting efficiency and economic growth. It also sharply distinguished between competition policy and consumer protection, focusing exclusively on the former (Hilmer et al. 1993, p. xvi). The Committee identified six specific elements of competition policy, and contemporary progress in implementing these elements (Box 14.5). The imperatives for a national competition policy arose from Australia effectively being a single integrated market; the growing disparity between increased competitiveness in the internationally traded goods sectors and the sheltered competition in public utilities, professions and some domestically oriented (including agricultural) industries; and the limitations of sector-by-sector reform (Hilmer et al. 1993, pp. xvii–xviii). The Committee approached its task under three broad headings:

- generally applicable competition conduct rules, their sphere of application and the enforcement regime;
- specific policy proposals and mechanisms for principles and processes governing the reform of regulatory restrictions on competition, the structural reform of public monopolies, competitive neutrality between government and private businesses, a general access regime, and a focused prices oversight mechanism; and
- implementation issues such as institutional, legal, transitional and resource matters (Hilmer et al. 1993, p. xxi).

The Review Committee explicitly recognised the public choice dimension of its task (Hilmer et al. 1993, p. 18):

- It is possible that 'in some cases competitive market outcomes will not meet the national interest' because of undesirable effects on efficiency or 'some other valued social objective'—the Committee argued that the onus of proof should always be to demonstrate why competition should be inhibited to prevent such undesirable effects.
- Vested interests 'often have powerful incentives to resist reforms that deliver substantial but sometimes more dispersed benefits to the wider community'—the Committee argued for the establishment of processes that would increase the 'transparency' of policy making.
- The possibly competing interests of the Commonwealth, state and territory governments had to be accommodated (and also the interests of local government which the Review Committee appeared to ignore)—the Committee proposed a mechanism by which all nine Commonwealth, state and territory governments 'owned' the national competition policy.

Box 14.4 National Competition Policy—government-agreed principles

1 No participant in the market should be able to engage in anti-competitive conduct against the public interest.

2 As far as possible, universal and uniformly applied rules of market conduct should apply to all market participants regardless of the form of business ownership.

3 Conduct with anti-competitive potential said to be in the public interest should be assessed by an appropriate transparent assessment process, with provision for review, to demonstrate the nature and incidence of the public costs and benefits claimed.

4 Any changes to the coverage or nature of competition policy should be consistent with, and support, the general thrust of reforms:

- to develop an open, integrated domestic market for goods and services by removing unnecessary barriers to trade and competition; and
- in recognition of the increasingly national operation of markets, to reduce complexity and eliminate administrative duplication.

Source: Hilmer et al. (1993, p. 361 and repeated pp. xvii–xix and Box 1.2)

Box 14.5 Elements of competition policy

Policy element	Example	Current implementation
1 Limiting anti-competitive conduct of firms	Competitive conduct rules of Part IV of Trade Practices Act	Numerous exemptions
2 Reforming regulation which unjustifiably restricts competition	Deregulation: e.g. domestic aviation, egg marketing, telecommunications; also grain storage, handling and transport	Reviews by individual governments without systematic, national focus

3	Reforming the structure of public monopolies to facilitate competition	e.g. proposed restructuring of energy utilities in several states	Mostly case-by-case by individual governments; some intergovernmental work for rail and electricity
4	Providing third-party access to certain facilities that are essential for competition	Access arrangements to natural monopolies: e.g. telecommunications	Industry specific; no general mechanism for dealing economy-wide with these issues
5	Restraining monopoly pricing behaviour	Prices surveillance: Prices Surveillance Authority, NSW Government Pricing Tribunal	Declared firms only with important exemptions for PSA
6	Fostering competitive neutrality between government and private businesses when they compete	Tax-equivalent payments to government: e.g. dividend payments from electricity generation and metropolitan distribution, and metropolitan water distribution, in NSW	Largely addressed ad hoc by individual governments; corporatisation using disparate models

Source: based on Hilmer et al. (1993, Boxes 1 and 1.1), with additions

In August 1994, the Council of Australian Governments requested the Industry Commission to undertake 'an assessment of the benefits to economic growth and revenue from implementing Hilmer and related reforms' (Industry Commission 1995, pp. 1, 8–9). The 'related' reforms involved fostering competition in national infrastructure (e.g. electricity, gas, water and road transport) and promoting the free movement of goods and occupations among the states (including marketing of agricultural products). In its report, the Commission pointed out the difficulties of reducing the impact of the proposed changes to a single number arising from the difficulty of modelling the proposed changes. The Industry Commission (1995, pp. 14–16, 18–21) argued that the modelling needed to account for the scope of the reforms, and the nature of both direct and indirect effects of proposed changes, and that the indirect effects could be subtle and pervasive. To model the effects of the proposed reforms, the Industry Commission used a modified version of ORANI (Industry Commission 1995, pp. 32ff) and a variety of alternative assumptions, but results did not vary markedly across the scenarios whose effects are summarised in table 14.3. The macroeconomic effects of the 'Hilmer and related reforms' at the state level are generally three to five times greater than the effects of such reforms at the Commonwealth level (with some exceptions, especially investment and capital stock). The large 'state' effect arose from the substantial changes modelled in the delivery of infrastructure services. For individual agricultural industries, however, there are several cases where the effect of Commonwealth reforms substantially outweighs the effects of state-level reforms—especially for dairy-related industries—because of the impact of national agricultural statutory marketing arrangements. While the economic effect of Commonwealth-implemented 'Hilmer and related reforms' was generally small relative to the effect of state reforms, the Commonwealth was estimated to be a major beneficiary of increased tax revenues arising from an expanded economy ($5900 million for the Commonwealth compared to $3000 million for the states). The disparity arose largely because state-level reforms predominantly involved government business enterprises (e.g. electricity, gas and water) which some states had used as surrogates for indirect taxation (cf. section 3.2).

Table 14.3 Aggragate Effects of Hilmer and Related Reforms[a]

	State reforms	*Commonwealth reforms*	*Total*
Real GDP	4.5	0.9	5.5
Real consumption	2.6	0.8	3.4
Real investment	5.3	0.4	5.7
Real government spending	0.0	0.0	0.0
Export volume	12.9	2.5	15.4
Import volume	1.2	0.0	1.2
CPI	0.0	0.0	0.0
GDP deflator	–1.0	0.0	–1.0
Nominal exchange rate	2.9	1.4	4.3
Real post-tax wage	2.3	0.7	3.0
Aggregate employment	0.3	0.1	0.4
Aggregate capital stock	5.3	0.4	5.7
Sectoral output			
Agriculture	4.2	0.2	4.4
Mining	14.3	4.2	18.6
Manufacturing	4.4	1.0	5.4
Services	2.7	0.7	3.4
Output			
Pastoral	0.0	1.4	1.4
Wheat–sheep	7.3	1.4	8.7
High rainfall	–0.3	1.5	1.2
Northern beef	9.8	5.3	15.0
Milk cattle	0.5	–15.5	–15.0
Pigs	5.9	2.2	8.1
Sugar cane	14.7	6.6	21.3

Table 14.3 Cont'd

Other farming export	4.0	1.1	5.1
Potatoes	2.5	0.6	3.1
Other farming import	2.7	–1.1	1.6
Poultry	3.9	1.4	5.3
Services to agriculture	2.1	0.6	2.7
Forestry and logging	1.9	0.9	2.7
Fishing and hunting	2.4	0.7	3.1
Meat products	5.6	2.0	7.6
Pasteurised milk	0.7	–3.0	–2.3
Milk products	0.6	–20.0	–19.4
Raw sugar	13.8	6.2	19.9
Cotton ginning	5.8	2.3	8.0
Wool scouring	5.6	2.2	7.8
Cotton fabrics	5.6	2.1	7.6
Wool worsted fabrics	2.9	0.9	3.7
Sawmill products	4.4	1.4	5.8
Agricultural machinery	1.1	3.1	4.2

Source: Industry Commission (1995, Tables A4.1–A4.2)

Notes. a. assuming monetary accommodation to changes; percentage change from control

Implementation of the 'Hilmer and related reforms' began in 1995 following agreement at COAG. The Commonwealth enacted its *Competition Policy Reform Act 1995*, which revised conduct rules, established infrastructure access and pricing regimes, removed the shield of the Crown, and established a competition code. At the state level, for example, New South Wales enacted the *Competition Policy Reform (NSW) Act 1995*, which applied conduct rules of Part IV of the Trade Practices Act to previously exempt entities within the state. Through COAG, intergovernmental agreements included the Competition Principles Agreement, which encompassed a structural reform agreement, independent price monitoring, competitive neutrality for government business enterprises, a review framework for systematically justifying regulatory regimes, and access to essential facilities. A Related Reforms Agreement under the National Competition Policy established a process of payments to states contingent on reforms under the Competition Principles Agreement, and specific reforms in infrastructure (e.g. electricity, water, gas and road transport). A conduct code established procedures affecting cooperation between the Commonwealth and the states.

In late 1995, the Australian Competition and Consumer Commission (ACCC) was formed by merging the former Trade Practices Commission and the Prices Surveillance Authority. The ACCC was charged with fostering competition across the whole economy. The National Competition Council, an advisory, research and policy body, was created at the same time. The Council was to review areas covered by intergovernmental agreements such as anti-competitive regulation, structural reform of government monopolies, and competitive neutrality. It was to advise governments on applying third-party access regime and price surveillance arrangements; assess the progress of state and territory governments in implementing national competition policy and related reforms; and provide policy advice about the Competition Principles Agreement to Commonwealth, state and territory governments.

In the process of implementing national competition policy, the Industry Commission increasingly focused on efficiency in delivering human services, and the Bureau of Industry Economics on the delivery of infrastructure services by business (cf. its agglomeration of reports utilising partial productivity indicators on infrastructure industries).[7] During the undeclared election campaign of 1995–96 leading up to the federal election of 2 March 1996, the leader of the National Party argued it would be dangerous to fully implement national competition policy in areas such as newsagents, pharmacists and some rural marketing schemes (Seccombe 1996, p. 4).

ISSUES FOR DISCUSSION

1 Discuss the relationships between microeconomic reform of the economy and the tariff compensation debate.
2 'Agriculture has everything to gain and nothing to lose from comprehensive microeconomic reform in the Australian economy.' Discuss.
3 'Microeconomic reform is a major plank of the current government's broader economic strategy.' What have been the major components of this program and what implications might there be for the agricultural sector?
4 The Australian Government deregulated the financial sector during the 1980s. Discuss whether or not this deregulation was advantageous for the agricultural sector.
5 'The drive for microeconomic reform of the agricultural sector, and for microeconomic reform of sectors affecting agriculture, has been more a religious crusade than a rational debate about improving economic welfare.' Discuss.

Endnotes

1 cf. the comment by Ingram 1966 on tariff cuts quoted in section 13.3.
2 Publicly provided and non-marketed services are valued at cost, and so there is no estimate of value added for them in the national accounts.
3 In NSW, before processors could legally manipulate the fat content of milk, regulators were naturally ensuring moderate fat milk by blending lower fat milk from areas where Friesians predominated with milk from areas where Channel Island breeds (Jersey, Guernsey) predominated. Choice of breed was itself a consequence of milk pricing policy because market milk producers were paid for the volume of milk they delivered, with a minium fat composition requirement. The producer's optimal supply decision was to sell market milk with the lowest possible fat content, but dairy industry regulators feared this milk would be unacceptable to consumers and so milk supplies were blended.
4 Forsyth (1992, p. 7) commented that economists tend to discount the effect of microeconomic reform on productive efficiency.

5 Not all these objectives are necessarily achieved in practice; in some countries necessities such as food, clothing—and even books—may be exempted from VAT, or differential rates may apply.

6 A consumption tax effectively implements a single rate taxation system, rather than the current progressive income tax system. In the latter, however, rising marginal income tax rates add to the incentive to avoid income tax as incomes rise, and rising incomes increase the ability of taxpayers to seek tax minimising strategies.

7 But cf. the amalgamation of the Industry Commission, Bureau of Industry Economics, and Economic Planning and Advisory Council in 1996.

References

ABARE 1990, *Wool: outlook and options for the future*, Australian Bureau of Agricultural and Resource Economics, Canberra.

ABARE 1992a, *Determinants of wool imports by the Commonwealth of independent states*, Research Report 92.9, Australian Bureau of Agricultural and Resource Economics, Canberra.

ABARE 1992b, *World wool trade*, Research Report 92.12, Australian Bureau of Agricultural and Resource Economics, Canberra.

ABARE 1993, *Commodity Statistical Bulletin 1993*, Canberra.

ABARE 1994, *Commodity Statistical Bulletin 1994*, Canberra.

ACIL Australia Pty Ltd. 1989, *International Agribusiness Trends and their Implications for Australia*, Discussion paper prepared for the Primary and Allied Industries Council, AGPS, Canberra.

Ackland, R. & Harper, I. R. 1992, 'Financial Deregulation in Australia: boon or bane?', chapter 3 in P. Forsyth (ed.), *Microeconomic Reform in Australia*, Allen & Unwin, Sydney.

Aitkin, D. 1985, 'Countrymindedness'—the spread of an idea, *Australian Cultural History* 4, 34–41.

Anderson, K. 1992, 'International Dimensions of the Political Economy of Distortionary Price and Trade Policies', chapter 123 in I. Goldin & L. A. Winters (eds) *Open Economies: structural adjustment and agriculture*, Cambridge University Press. See also Anderson, K. & Hayami, Y. 1986, *The Political Economy of Agricultural Protection: East Asia in international perspective*, Allen & Unwin, Boston.

anon. 1992, *The High Court Decision on Native Title*, Discussion Paper, [Aboriginal Reconciliation Unit, Department of Prime Minister and Cabinet, Canberra], October.

anon. n.d., *Codex Alimentarius*, second edition, Joint World Health Organization/Food and Agriculture Organization Food Standards Programme, Rome.

Arndt, H. W. 1994/95, 'Review of Public Goods, Public Enterprise, Public Choice', in *Policy: a journal of public policy and ideas* 104, Summer, 54–5.

Asimov, I. 1953, *Foundation*, Panther, London.

Auslig. 1992, Public Land: Aboriginal and Torres Strait Islander Land, Preliminary Edition—1992.

Australian Bureau of Statistics 1992, *Australia's Environment: issues and facts*, Cat. No. 4140.0 [Canberra].

Australian Bureau of Statistics 1993, *Australia's Aboriginal and Torres Strait Islander Population. Census of Population and Housing, 6 August 1991*, Cat. No. 2740.0, Canberra.

Australian Bureau of Statistics 1994a, *National Income, Expenditure and Product*, Australian National Accounts, Cat. No. 5206.0, Canberra.

Australian Bureau of Statistics 1994b, *Year Book, Australia*, AGPS, Canberra.

Australian Bureau of Statistics 1995, *National Aboriginal and Torres Strait Islander Survey 1994: detailed findings*, Cat. No. 4190.0, Canberra.

Australian Chamber of Commerce and Industry 1993, *Mabo*: native title and the community, a paper prepared for Australian Chamber of Commerce and Industry, Australian Coal Association, Australian Mining Industry Council, Australian Petroleum Exploration Association, Australian Tourism Industry Association, National Association of Forest Industries and National Fishing Industry Council, 24 June.

Australian Nature Conservation Agency, Department of Primary Industries and Energy, Department of Environment, Sport and Territories 1993, Landcare Information: land, water and vegetation programs 1994–95, Commonwealth of Australia [Canberra] [document captured off ERIN, the Department of the Environment's Internet network].

Baker, I. (chair) 1990, *A Review of Agricultural Research in Victoria*, Government Printer, Melbourne.

Balderstone, J. S. et al. 1982, *Agricultural Policy: issues and options for the 1980s*, Working Group Report to the Minister for Primary Industry, September, AGPS, Canberra.

Bardsley, P. 1991, Wool on the brink: the public cost of underwriting the wool market, 35th Annual Conference, Australian Agricultural Economics Society, Armidale, 11–14 February, roneo.

Batterham, R. L. & MacAulay, T. G. 1994, 'Price-linked farm and spatial equilibrium models', *Australian Journal of Agricultural Economics* 38(2), 143–70.

Baxter, K. P. 1994, 'Failure, chaos and leadership: ingredients of democratic reform', *Occasional Paper 49*, Centre for Independent Studies, Sydney.

Beale, B. 1995, 'Rising water threatens irrigation areas', *Sydney Morning Herald*, 18 March, p. 4.

Bennett, S. 1986, 'Parties and Elections', chapter 11 in B. Galligan (ed.), *Australian State Politics*, Longman Cheshire, Melbourne.

Bewley, R. & White, G. 1990, 'Do high interest rates improve or worsen the current account?', *Economic Papers* 9(4), December, 19–33.

Binswanger, H. P. 1978, 'Measured Biases of Technical Change: the United States', chapter 7 in H. P. Binswanger, V. W. Ruttan et al., *Induced Innovation: technology, institutions, and development*, Johns Hopkins University Press, Baltimore.

Black, J. 1995, *The Country's Finest Hour: fifty years of rural broadcasting in Australia*, ABC Books, Sydney.

Blackshield, A. R., Brown, D., Coper, M. & Krever, R. 1986, *The Judgements of Justice Lionel Murphy*, Primavera Press, Sydney.

Blainey, G. 1976, *Triumph of the Nomads: a history of ancient Australia*, Sun Books, Melbourne.

Blue-green Algae Task Force. 1992, *Blue-green Algae: Final Report of the New South Wales Blue-green Algae Task Force*, Department of Water Resources, Parramatta.

Boulding, K. E. 1970, 'Economics as a Moral Science', chapter 6 in his *Economics as a Science*, McGraw-Hill, New York.

Braddock, R. 1994, 'The Process of Microeconomic Policy Formulation and Co-ordination', chapter 2 in Jones 1994.

Brandolini, G. V. 1994, 'Quality Systems in the Agro-Food Chain', in *Food Policies and the Food Chain: structures and inter-relationships*, 36th Seminar of the European Association of Agricultural Economists, University of Reading, 19–21 September.

Brennan, J. P. & Fox, P. N. 1995, Impact of CIMMYT Wheats in Australia: evidence of international research spillovers, Economics Research Report No. 1/95, NSW Agriculture, Wagga Wagga.

Brennan, P. 1990, 'Microeconomic reform', *Economic Papers* 9(4), December, 34–50.

Brooks, J. C. & Carter, C. A. 1994, *The Political Economy of US Agriculture*, ABARE Research Report 94.8, Australian Bureau of Agricultural and Resource Economics, Canberra.

Brownhill, D. 1994, 'Why we need another Snowy', Letters, *The Land* 23 June, p. 17.

Brunton, R. 1992, '*Mabo* and Oral Traditions', roneo; reprinted from P. Durack, R. Brunton & T. Rutherford 1992, *Mabo and After*, Institute of Public Affairs, Melbourne.

Campbell, K. O. & Fisher, B. S. 1982, *Agricultural Marketing and Prices*, 2nd edition, Longman Cheshire, Melbourne.

Campbell, K. O. 1973, 'The State marketing board—relic or prototype?', *Australian Journal of Agricultural Economics* 17(3), December, 179–88.

Cashin, P. 1988, 'Australian agriculture: is there any value in "high-value" commodities?', *Economic Papers* 7(2), June, 21–32.

Centre for Independent Studies. n.d., *The Centre for Independent Studies*, St Leonards.

Chamberlin, P. 1993, 'More WA–federal friction on Mabo', *Sydney Morning Herald* 5 November, p. 6.

Chisholm, T., Haszler, H., Edwards, G. & Hone, P. 1994, The Wool Debt, the Wool Stockpile and the National Interest: did Garnaut get it right?, 38th Annual Conference, Australian Agricultural Economics Society, Wellington, NZ, 8–10 February.

Choi, E. K. & Jensen, H. H. 1991, 'Modeling the Effect of Risk on Food Demand and the Implications for Regulation', chapter 2 in J. A. Caswell (ed.), *Economics of Food Safety*, Elsevier, New York.

Commission of Inquiry into Poverty (Henderson Inquiry) 1975, *Poverty in Australia*, AGPS, Canberra.

Commonwealth of Australia (Davis Committee) 1990, *Review of the Commonwealth Primary Industry Statutory Authorities*, Report to the Minister of Primary Industries and Energy, Review Committee, AGPS, Canberra.

Commonwealth of Australia 1986, *Economic and Rural Policy: a government policy statement*, AGPS, Canberra.

Commonwealth of Australia 1993a, *Mabo: the High Court decision on native title, Discussion Paper*, AGPS, Canberra, June. ('33 principles' in appendix)

Commonwealth of Australia 1993b, *Mabo: proposed legislation on native title*, Canberra, September, roneo.

Commonwealth of Australia, *Native Title Bill 1993.*

Coper, M. 1978, 'Constitutional obstacles to organised marketing in Australia', *Review of Marketing and Agricultural Economics* 46(2), August, 71–102.

Coper, M. 1985, 'In the beginning there was interstate trade . . .', *Review of Marketing and Agricultural Economics* 53(3), December, 103–12.

Coper, M. 1992, 'The Economic Framework of the Australian Federation: a question of balance', chapter 7 in G. Craven (ed) *Australian Federation: towards the second century*, Melbourne University Press.

Costar, B. & Woodward, D. (eds) 1985a, *Country to National: Australian rural politics and beyond*, George Allen & Unwin, Sydney.

Costar, B. & Woodward, D. (eds) 1985b, Introduction, chapter 1 in Costar & Woodward 1985a.

Council for Aboriginal Reconciliation 1993, *Making Things Right: reconciliation after the High Court's decision on native title*, Canberra.

Court, R. 1993, *Land (Titles and Traditional Usage) Bill*, Second Reading Speech, [Parliament of Western Australia, Perth], roneo.

Craven, G. 1992 (ed.) *Australian Federation: towards the second century*, Melbourne University Press.

Crean, S. 1993, Government Announces Sugar Industry Package, Media Release DPIE 93/7J, Canberra.

Cullen, R. 1988, 'Section 92: quo vadis?' pp. 1–29 in A.M.J. Arch & W. W. Fisher (eds) *Agricultural Marketing—Section 92 of the Commonwealth Constitution*, Proceedings of the symposium on Section 92 of the Constitution: economic and legal implications. Australian Agricultural Economics Society (Victorian Branch) Melbourne, 13 October.

Davidson, B. R. 1984, 'A preliminary benefit–cost analysis of the inland diversion of the coastal rivers of New South Wales', *Review of Marketing and Agricultural Economics* 52(1), April, 23–47.

Davidson, B. R., 1969, *Australia: wet or dry?*, Melbourne University Press.

Davies, A. 1993, 'Farmers "betrayed" on Mabo', *Sydney Morning Herald* 20 December, p. 1

Davis, J. S., Oram P. A., & Ryan, J. G. 1987, *Assessment of Agricultural Research Priorities: an international perspective*, Australian Centre for International Agricultural Research, Canberra (especially chapter 2).

de Janvry, A. 1978, 'Social Structure and Biased Technical Change in Argentine Agriculture', chapter 11 in H. P. Binswanger, V. W. Ruttan et al., *Induced Innovation: technology, institutions, and development*, Johns Hopkins University Press, Baltimore.

Department of Primary Industries and Energy 1994, *The Rural Book*, AGPS, Canberra.

Dixon, P. B., Parmenter, B. R., Sutton, J. & Vincent, D. P. 1982, *ORANI: a multi-sectoral model of the Australian economy*, North-Holland, Amsterdam.

Drought Policy Review Task Force. 1990, *Final Report*, volume 1: National Drought Policy; volume 2: Managing for Drought; volume 3: Annexes, AGPS, Canberra.

Drynan, R. G. 1995, Department of Agricultural Economics, University of Sydney, personal communication.

Duncan, C. J. & Epps, W. R. 1992, 'The demise of "countrymindedness"; new players or changing values in Australian rural politics', *Political Geography* 11(5), 430–48.

Edwards, G. 1993, Sweet or Sour? microeconomic reform in the sugar industry, 37th Annual conference, Australian Agricultural Economics Society, University of Sydney, 9–11 February, roneo.

Edwards, G. W. & Freebairn, J. W. 1981, *Measuring a Country's Gains from Research: theory and application to rural research in Australia*, A Report to the Commonwealth Council for Rural Research and Extension, AGPS, Canberra (especially chapter 4).

Edwards, G. W. & Watson, A. S. 1978, 'Agricultural Policy', chapter 4 in F. H. Gruen (ed.), *Surveys of Australian Economics*, George Allen & Unwin, Sydney.

Edwards, G. W. 1980, 'The IAC's approach to agricultural policy', *Australian Journal of Agricultural Economics* 24(2), August, 94–114.

Environment Protection Authority 1993, *New South Wales State of the Environment 1993*, Sydney.

Ewing, G. 1992, 'The likely impact of the *Mabo* case on Aboriginal land rights claims', *The Mining Review*, November, 8–12.

Fischer, T. 1993, 'Mabo and the Monarchy, and don't forget the economy!', speech to the NSW National Party State Conference, 20 June, Wagga Wagga, roneo.

Fisk, E. K. 1985, *The Aboriginal Economy in Town and Country*, George Allen & Unwin, Sydney.

Flugge, T. 1996, 'Future opportunities and strategies for Australian wheat—"investing in the future"', *Outlook 96: volume 2 Agriculture*, Proceedings of the National Agricultural and Resources Outlook Conference, Canberra, 6–8 February, pp. 229–305.

Forbes, J. 1995, 'Taking without paying: interpreting property rights in Australia's Constitution', *Agenda* 2(3), 313–20.

Forsyth, P. 1992, 'A Perspective on Microeconomic Reform', chapter 1 in P. Forsyth (ed.), *Microeconomic Reform in Australia*, Allen & Unwin, Sydney.

Freebairn, J. W., Davis, J. S. & Edwards, G. W. 1982, 'Distribution of research gains in multistage production systems', *American Journal of Agricultural Economics* 64, 39–46.

Galligan, B. & Walsh, C. 1992, 'Australian Federalism—Yes or No?', chapter 10 in G. Craven (ed.) *Australian Federation: towards the second century*, Melbourne University Press.

Galligan, B. 1986, 'The Political Economy of the States', chapter 13 in B. Galligan (ed), *Australian State Politics*, Longman Cheshire, Melbourne.

Galligan, B. 1995, *A Federal Republic: Australia's constitutional system of government*, Cambridge University Press.

Garnaut, R., Bennett, S. & Price, R. 1993, *Wool: structuring for global realities*, Report of the Wool Industry Review Committee, Commonwealth of Australia, Canberra, August.

Gifford, K. H. & Gifford, D. J. 1982, *How to Understand an Act of Parliament*, 5th ed. Law Book Co., Sydney.

Godden, D. 1991, Government, public servants, interest groups and the public, Annual Conference, Australian Agricultural Economics Society, Armidale, February.

Godden, D. 1993, Some Economic Aspects of the High Court's 1992 *Mabo* Decision on Native Title, 1993 conference of economists, Murdoch University, September.

Godden, D. 1994a, *Mabo* and the Rural Sector, Annual Conference, Australian Agricultural Economics Society, Wellington NZ, February, roneo.

Godden, D. 1994b, *Plant Breeder's Rights Bill 1994: Submissions*, Standing Committee on Rural and Regional Affairs, Senate, Canberra, pp. 120–86.

Godden, D. 1995, 'Beyond the conventional wisdom: a reply to Sher and Sher 1994', *Journal of Research in Rural Education* 111, Spring, 45–54.

Godden, D. P. 1987, 'Technological change embodied in plant varieties: a review of *ex post* studies', *Oxford Agrarian Studies* 16, 1–22.

Goesch, T., Podbury, T. & Nicholas, S. 1996, 'Outlook for wool', *Outlook 96: volume 2 Agriculture*, Proceedings of the National Agricultural and Resources Outlook Conference, Canberra, 6–8 February, pp. 177–81.

Government Pricing Tribunal 1993, *Water: an interim report*, Sydney.

Gregory, R. G. 1976, 'Some implications of growth in the mining sector', *Australian Journal of Agricultural Economics*, 20(2), 71–91.

Greiner, N. 1990, Speech to the 41st Federal Council Meeting of the Liberal Party of Australia, Brisbane, 24 October.

Harper, I. 1991, 'Bank Deregulation in Australia: choice and diversity, gainers and losers', pp. 63–86 in I. Macfarlane (ed), *The Deregulation of Financial Intermediaries*, Proceedings of a Conference, 20–21 June 1991, Research Department, Reserve Bank of Australia, Sydney.

Harris, S. et al. 1974, *The Principles of Rural Policy in Australia: a discussion paper*, AGPS, Canberra.

Harris, S. F. 1975, 'Tariff compensation: sufficient justification for assistance to Australian agriculture?', *Australian Journal of Agricultural Economics* 19(3), 131–45.

Haward, M. 1991, 'The Offshore', chapter 6 in B. Galligan, O. Hughes & C. Walsh (eds), *Intergovernmental Relations and Public Policy*, Allen & Unwin, Sydney.

Hawke, B. 1989, *Rural and Regional Australia*, A Statement on Federal Government policy for rural and regional communities, Commonwealth of Australia, Canberra, December.

Hawtrey, K., Favotto, I. & Gilchrist, D. 1991, 'The impact of bank deregulation on Australian manufacturers', *Economic Papers* 10(4), December, 30–43.

Hayami, Y. & Ruttan, V. W. 1985, *Agricultural Development: an international perspective*, revised edition, Johns Hopkins University Press, Baltimore.

Henningham, J. (ed.) 1991, *Institutions in Australian Society*, Department of Journalism, University of Queensland, Brisbane.

Henningham, J. 1993, 'The Press', pp. 59–71 in S. Cunningham & G. Turner (eds), *The Media in Australia*, Allen & Unwin, Sydney.

Henson, S. & Traill, B. 1993, 'The demand for food safety: market imperfections and the role of government', *Food Policy* 18(2), April, 152–62.

Hewson, J. 1993, Mabo Issues Paper, Media Release, Leader of the Opposition, 83/93, 28 June.

Higgs, P. J. 1986, *Adaptation and Survival in Australian Agriculture*, Oxford University Press, Melbourne.

Hill, B. 1995a, 'Last of the Lockheed guilty finally sentenced', *Sydney Morning Herald* 23 February, p. 11.

Hill, B. 1995b, 'Japan digs in against open food markets', *Sydney Morning Herald* 24 October, p. 9.

Hilmer, F. G., Rayner, M. R. & Taperell, G. Q. 1993, *National Competition Policy*, Report by the Independent Committee of Inquiry, AGPS, Canberra, August.

Hooper, G. N. 1995, 'The National Registration Scheme—a guided tour', *NRA News* (Newsletter of the National Registration Authority for Agricultural and Veterinary Chemicals), 21, 3–5.

House of Representatives Standing Committee on Long Term Strategies 1994, *Australia's Population 'Carrying Capacity': one nation—two ecologies*, AGPS, Canberra.

Howard, M. 1995, forthcoming, 'A growth industry?: use of consultants reported by Commonwealth departments, 1974–1994', *Canberra Bulletin of Public Administration*.

Hulme, S. E. K. 1993, 'Aspects of the High Court's handling of Mabo', paper delivered to the Samuel Griffith Society, in anon. (ed.) 1993, *The High Court of Australia in Mabo*, Association of Mining & Exploration Companies Inc., Leederville, WA.

Human Rights and Equal Opportunity Commission 1988, *Toomelah Report*, Canberra.

Human Rights and Equal Opportunity Commission 1990, *A Report by the Race Discrimination Commissioner on a visit to Baryulgil & Malabugilmah*, New South Wales, [Sydney], July.

Human Rights and Equal Opportunity Commission 1995, *Mornington Island Review Report*, A Report by the Race Discrimination Commissioner & the Aboriginal and Torres Strait Islander Social Justice Commissioner, Sydney, April.

Hutcheon, S. 1995, 'Taiwan MPs in $500m vote-buying scandal', *Sydney Morning Herald* 1 December, p. 11.

Industries Assistance Commission 1984, *Rural Adjustment*, AGPS, Canberra.

Industries Assistance Commission 1987, *Assistance to Agricultural and Manufacturing Industries*, AGPS, Canberra, June.

Industries Assistance Commission 1988a, *The Wheat Industry*, AGPS, Canberra.

Industries Assistance Commission 1988b, *State Government Assistance to Agriculture*, AGPS, Canberra, April.

Industries Assistance Commission 1989a, *Food Processing and Beverages Industries*, Report No. 424, Canberra, December.

Industries Assistance Commission 1989b, *Using Input–Output Analysis and Multipliers*, Working Paper No. 12, August.

Industry Commission 1991a, *Availability of Capital*, Draft Report, Canberra.

Industry Commission 1991b, *Statutory Marketing Arrangements for Primary Products*, Report No. 10, AGPS, Canberra.

Industry Commission 1992a, *The Australian Sugar Industry*, Report No. 19, AGPS, Canberra, 6 March.

Industry Commission 1992b, *Water Resources and Waste Water Disposal*, Draft Report, 26 March.

Industry Commission 1993, *Draft Report on Impediments to Regional Adjustment*, AGPS, Canberra.

Industry Commission 1994, *Research and Development*, Draft Report, in 3 volumes, Canberra, 8 December.

Industry Commission 1995, *The Growth and Revenue Implications of Hilmer and Related Reforms*, A Report by the Industry Commission to the Council of Australian Governments, Canberra, March.

Ingram, J. C. 1966, *International Economic Problems*, John Wiley & Sons, New York.

International Public Relations. n.d. *The Governments of Australia: the people and the departments*, Melbourne. ISBN 0 9598904 24.

Jaensch, D. 1995, *Election!: how and why Australia votes*, Allen & Unwin, Sydney.

James, D. E. & Throsby, C. D. 1973, *Introduction to Quantitative Methods in Economics*, John Wiley & Sons, Sydney.

James, P. G. 1971, *Agricultural Policy in Wealthy Countries*, Angus & Robertson, Sydney.

Jarrett, F. G. & Lindner, R. K. 1982, 'Rural Research in Australia', chapter 4 in Williams 1982.

Johns, B. L., Dunlop, W. C. & Sheehan, W. J. 1989, *Small Busine$$ in Australia: problems and prospects*, 3rd edition, Allen & Unwin, Sydney.

Johns, G. 1994, 'Review of Public Service Act', media release 23/94, Assistant Minister for Industrial relations, Parliament House, Canberra, 30 June.

Johnson, D.B. 1991, *Public Choice: an introduction to the new political economy*, Mayfield Publishing, Mountain View, Calif. ch. 2, 4–10, 12.

Jones, R. 1994a, *The Housing Need of Indigenous Australians*, 1991, Research Monograph No. 8, Centre for Aboriginal Economic Policy Research, Australian National University, Canberra.

Jones, R. (ed) 1994b, *Australian Microeconomic Policies*, 4th edition, Prentice Hall, New York.

Keating, P. J. 1992, Speech by the Prime Minister, Australian Launch of the International Year for the World's Indigenous People, [Redfern Park Speech], Redfern, 10 December (roneo).

Keating, P. J. & Collins, B. 1994, Additional Drought Relief, Joint Statement by the Prime Minster, the Hon PJ Keating MP & the Minister for Primary Industries and Energy, Senator the Hon Bob Collins, Canberra, 8 December.

Kerin, J. & Cook, P. 1988, *Primary Industries and Resources: Policies for Growth—a government policy statement*, AGPS, Canberra, May.

Kinsey, J. 1993, 'GATT and the economics of food safety', *Food Policy* 18(2), April, 163–76.

Kitney, G. 1993, 'Mabo myopia wasn't budgeted for', *Sydney Morning Herald* 3 September, p. 11.

Lawson, K. et al. 1993, *Maximising the Return: adding value to Australian wool*, Report of the Wool Processing Task Force, AGPS, Canberra (also *Technical Annex* to this report).

Lembit, M. & Kingma, O. 1995, Commonwealth drought policies, Coping with Drought, 1995 Drought Forum, Australian Institute of Agricultural Science, Sydney.

Lewis, J. N. 1967, 'Milking the Australian Economy', chapter 6 in C. D. Throsby (ed.), *Agricultural Policy*, Pelican Readings in Australian Economic Policy, Melbourne.

Lloyd, A. G. 1982, 'Agricultural Price Policy,' chapter 18 in D. B. Williams (ed.), *Agriculture in the Australian Economy*, 2nd edition, Sydney University Press.

Lloyd, A. G. 1986, *Rural Economics Study*, A report to the Minister for Agriculture and Rural Affairs, Government Printer, Melbourne, July.

Lloyd, A. G. 1988, 'The importance of agriculture: what hope for agriculture, and what needs doing?', *Review of Marketing and Agricultural Economics* 561, April, 129–34.

Lloyd, A. G. 1990, 'Appendix II—Value Added for Primary Products', pp. 37–42 in A. Lloyd, M. Harris & D. Tribe 1990, *Australian Agricultural Research: some policy issues*, Crawford Fund for International Agricultural Research, Melbourne.

Lloyd, P. J. 1975, 'Tariff compensation: an undesirable policy', *Australian Journal of Agricultural Economics* 19(3), 146–53.

Longmire, J. 1995, Changing structure and location of Australian agriculture, 39th Annual Conference, Australian Agricultural Economics Society, University of Western Australia, Perth, February.

Lopez, R. E. 1980, 'The structure of production and the derived demand for inputs in Canadian agriculture', *American Journal of Agricultural Economics* 62(1), February, 38–45.

Lucas, R. E. 1976), 'Econometric Policy Evaluation: a critique', pp.19–46 in K. Brunner & A. H. Meltzer (eds), *The Phillips Curve and Labor Markets*, North-Holland, Amsterdam.

Lucy, R. 1985, 'New South Wales', chapter 6 in Costar & Woodward 1985a.

Mabo and Others v. State of Queensland. 1992, *Australian Law Reports* 107(1), 1–170.

Males, W. P., Davidson, H., Knopke, P., Loncar, T. & Roarty, M. J. 1990, *Productivity Growth and Developments in Australia's Primary Industries*, Discussion Paper 90/06, EPAC, AGPS, Canberra, September.

Mannion, R., Tillack, R. & Vincent, D. 1988, 'Effects of petroleum product excises on agriculture', *Review of Marketing and Agricultural Economics* 56(1), April, 121–8.

Marglin, F. A. 1990, 'Smallpox in Two Systems of Knowledge', chapter 4 in F. A. Marglin & S. A. Marglin (eds), *Dominating Knowledge: development, culture, and resistance*, Clarendon Press, Oxford.

Martin, W. 1990, 'Public choice theory and Australian agricultural policy reform', *Australian Journal of Agricultural Economics* 34(3), December, 189–211.

Mauldon, R. 1991, 'Concluding Comments', pp. 99–104 in L. Copeland & F. Ahmadi (eds) *Products Processing Marketing*, Proceedings of the Grains Research Symposium 1991, Faculty of Agriculture, University of Sydney.

Mauldon, R. G. 1990, 'Price Policy', chapter 19 in D. B. Williams (ed.), *Agriculture in the Australian Economy*, 3rd edition, Sydney University Press.

McEwen, J. 1952, *Agricultural Production Aims and Policy*, A Statement Explaining the Agricultural Production Aims approved by the Australian Agricultural Council in April 1952, Minister for Commerce and Agriculture, AGPS, Canberra.

McGauran, P. 1993, The *Mabo* decision, Address to the RMIT Business Studies Course, Royal Melbourne Institute of Technology, [May?], roneo.

McGeough, P. 1995, 'One man's audacious plan: to slake the thirst of a parched continent', *Sydney Morning Herald*, 7 June, p. 7.

McLachlan, I. 1993, 'Rights—right or wrong?', Speech Notes, H R Nicholls Society, 14 May, Canberra.

McMaster, G. & McKay, J. 1991, 'Processed Products from Wheat Flour: an overview', pp. 31–7 in L. Copeland & F. Ahmadi (eds), *Products Processing Marketing*, Proceedings of the Grains Research Symposium 1991, Faculty of Agriculture, University of Sydney.

Meyer, W. S. 1992, Sustainability of Land and Water Resources used for Australian Irrigated Agriculture: a research strategy position paper, CSIRO Water Resources Series No. 8, CSIRO Division of Water Resources, Griffith.

Millington, P. 1994, 'Hunter water quality', News Release, NSW Department of Water Resources, Sydney, 23 June.

Morris, L. 1995, 'Public works "destructive"', *Sydney Morning Herald* 11 March, p. 3.

Mortensen, K. 1994, *Agricultural Legislation in New South Wales*, Agfact M3.6, NSW Agriculture, Orange, January.

Mueller, D. C. 1979, *Public Choice*, Cambridge Surveys of Economic Literature, Cambridge University Press.

Mueller, D. C. 1989, *Public Choice II*, Cambridge University Press.

Mules, T. J. 1983, 'Input-output analysis in Australia: an agricultural perspective', *Review of Marketing and Agricultural Economics* 51(1), 9–30.

Murray, L. 1994, 'The Family Farmers' Victory: for Salvatore Zofrea', *The Independent Monthly*, June, p. 81.

Murray–Darling Basin Ministerial Council 1989, *Murray–Darling Basin Natural Resources Management Strategy*, Background Papers No. 89/1, Canberra, June.

Musgrave, W. 1982, 'Rural Adjustment', chapter 15 in D. B. Williams (ed.), *Agriculture in the Australian Economy*, 2nd edition, Sydney University Press.

Musgrave, W. 1990, 'Rural Adjustment', chapter 16 in D. B. Williams (ed.), *Agriculture in the Australian Economy*, 3rd edition, Sydney University Press.

National Farmers' Federation 1981, *Farm Focus: the 80s*, Canberra.

National Farmers' Federation 1995 *Australian Agricultural Year Book 1985*, Strand Publishing, Brisbane.

National Food Authority 1992, *The 1992 Market Basket Survey: a total diet survey of pesticides and contaminants*, Canberra.

National Food Authority 1993, *Final Report of the Policy Review*, AGPS, Canberra.

National Food Authority 1994, *Safe Food Handling Australia: a discussion paper on the future direction of food hygiene regulation*, Canberra, October.

National Health and Medical Research Council 1985, *Model Food Legislation*, (Commonwealth Department of Health), AGPS, Canberra.

Norton, G. W. & Davis, J. S. 1981, 'Evaluating returns to agricultural research: a review', *American Journal of Agricultural Economics* 63(4), 685–99.

NSW 1987–88, *Capital Works Program 1987–88: List of Projects*, Budget Paper No. 5, New South Wales Treasury, Sydney.

NSW Agriculture 1995, *Country Guide*, Orange.

NSW Farmers 1993, 'Aboriginal Land Rights—the costs to Aboriginal and the community', *The Primary Report* 03/93.

NSW Government 1994, *The Management and Regulation of Water in NSW*, NSW Government White Paper, Sydney, May.

NSW Ombudsman 1992, *The Ombudsman's Report on Toomelah*, Report under Section 26 of the Ombudsman Act, Investigation into the conduct of the Office of Aboriginal Affairs following a complaint by the Toomelah Local Aboriginal Land Council & the Toomelah Aboriginal Cooperative, Final, Office of the NSW Ombudsman, Sydney, 9 November.

NSW Ombudsman 1995, *Freedom of Information: the way ahead*, Sydney, January.

Office of State Owned Enterprises 1993, *Reforming Victoria's Water Industry. A competitive future—water*, Department of the Treasury, [Melbourne], October.

O'Keefe, J. A. 1968, *Bell and O'Keefe's Sale of Food and Drugs*, 14th edition, Butterworths, London.

Pearce, D. W. & Turner, R. K. 1990, *Economics of Natural Resources and the Environment*, Harvester Wheatsheaf, New York.

Piggott, R. 1990, 'Agricultural Marketing', chapter 18 in D. B. Williams (ed.), *Agriculture in the Australian Economy*, 3rd edition, Sydney University Press.

Posner, R. A. 1974, 'Theories of economic regulation', *Bell Journal of Economics and Management Science* 5(2), Autumn, 335–58.

Powell, R. A. & Jensen, R. C. 1989, 'Regional Economic Structure: an input–output approach', chapter 9 in B. Higgins & K. Zagorski (eds) *Australian Regional Developments: readings in regional experiences, policies and prospects*, No. 10, Australian Regional Developments, AGPS, Canberra.

Powell, R. & Milham, N. 1990, 'Capital, Investment and Finance', chapter 15 in Williams (ed.) 1990, *Agriculture in the Australian Economy*, 3rd edition, Sydney University Press.

Pusey, M. 1991, *Economic Rationalism in Canberra: a nation-building state changes its mind*, Cambridge University Press, Sydney.

Randall, A. 1987, *Resource Economics: an economic approach to natural resource and environmental policy*, 2nd edition, John Wiley & Son, New York.

Rausser, G. C. 1982, 'Political economic markets: PERTs and PESTs in food and agriculture', *American Journal of Agricultural Economics* 64(5), 821–32.

Reynolds, H. 1992, *The Law of the Land*, 2nd edition, Penguin, Ringwood.

Reynolds, P. 1991, 'State Government', chapter 2 in Henningham 1991.

Richardson, B. 1995, 'The wool outlook and future directions in raw wool marketing' *Outlook 95: volume 2 Agriculture*, Proceedings of the National Agricultural and Resources Outlook Conference, Canberra, 7–9 February, pp. 78–82.

Richardson, G. 1994, *Whatever it Takes*, Bantam, Sydney.

Roberts, T. & Foegeding, P. M. 1991, 'Risk Assessment for Estimating the Economic Costs of Foodborne Disease Caused by Microorganisms', chapter 6 in J. A. Caswell (ed.), *Economics of Food Safety*, Elsevier, New York.

Roberts, T. & Pinner, R. 1990, 'Economic Impact of Disease Caused by *Listeria monocytogenes*', chapter 22 in A. J. Miller, J. L. Smith & G. A. Somkuti (eds) *Foodborne Listeriosis*, Society for Industrial Microbiology, Elsevier, Amsterdam.

Rose, R. 1989, *Politics in England: change and persistence*, 5th edition, Scott, Foresman & Company, Glenview, Illinois.

Rowley, C. K. & Peacock, A. T. 1975, *Welfare Economics: a liberal restatement*, Martin Robertson, London.

Royal Commission into Aboriginal Deaths in Custody 1991, *National Report, Volume 2*, AGPS, Canberra.

Royal Commission into Grain Storage, Handling and Transport 1988a, *Executive Summary and Recommendations*, Commonwealth of Australia, Canberra.

Royal Commission into Grain Storage, Handling and Transport 1988b, *Report*, AGPS, Canberra.

Rutledge, D. 1996, 'The changing face of the Australian sugar industry', *Outlook 96: volume 2 Agriculture*, Proceedings of the National Agricultural and Resources Outlook Conference, Canberra, 6–8 February, pp. 426–33.

Ruttan, V. W. 1978, 'Induced Institutional Change', chapter 12 in H. P. Binswanger, V. W. Ruttan et al., *Induced Innovation: technology, institutions, and development*, Johns Hopkins University Press, Baltimore.

Ryan, T. J. 1993, Marketing Australia's Wheat Crop: the way ahead, 37th Annual Conference, Australian Agricultural Economics Society, University of Sydney, 9–11 February.

SBS 1994, *The SBS World Guide: 1994 Edition*, Text Publishing, Melbourne.

Seccombe, M. 1991, 'Resource plan may sink Labor', *Sydney Morning Herald* 8 March, p. 1.

Seccombe, M. 1996, 'Hilmer "no help to small business"', *Sydney Morning Herald* 22 January, p. 4.

Senate Economics References Committee 1994, *CSIRO: The Case for Revitalisation: administration and funding of rural research*, Parliament of the Commonwealth of Australia, Canberra, December.

Senate Rural and Regional Affairs and Transport References Committee 1994, *Rural Adjustment, Rural Debt and Rural Reconstruction: Report*, Parliament of the Commonwealth of Australia, Canberra, December.

Shamsullah, A. 1990, 'The Australian Democrats', chapter 12 in Summers et al. 1990.

Sharman, C. 1991, 'Executive Federalism', chapter 2 in B. Galligan, O. Hughes & C. Walsh (eds), *Intergovernmental Relations and Public Policy*, Allen & Unwin, Sydney.

Sharp, M. 1995, 'Water bans, drought prevail', *Sydney Morning Herald* 30 September, p. 5.

Shaw, A. G. L. 1990, 'Colonial Settlement 1788–1945', chapter 1 in D. B. Williams (ed), *Agriculture in the Australian Economy*, 3rd edition, Sydney University Press.

Sher, J. P. & Sher, K. R. 1994, 'Beyond the Conventional Wisdom: rural development as if Australia's rural people and communities really mattered', *Journal of Research in Rural Education* 10(1), Spring, 2–43.

Sieper, E. 1982, *Rationalising Rustic Regulation*, Centre for Independent Studies, St Leonards.

Simms, M. & Singleton, G. 1993, Trade Unions, chapter 18 in Smith 1993a.

Simons, M. 1995, 'Pick a card, any card', *Eureka Street* 5(6), August, 16–22.

Smith, R. 1993a, *Politics in Australia*, 2nd edition, Allen & Unwin, Sydney.

Smith, R. 1993b, 'The Party System', chapter 9 in Smith 1993a.

Sorensen, T. & Epps, R. (eds) 1993, *Prospects and Policies for Rural Australia*, Longman Cheshire, Melbourne.

Sorensen, T. 1993, 'Approaches to Policy', chapter 13 in T. Sorensen & R. Epps (eds), *Prospects and Policies for Rural Australia*, Longman Cheshire, Melbourne.

Sorensen, T. 1994, 'The folly of regional policy', *Agenda* 1(1), 33–44.

Spark, E. 1995, 'The NSW Climate Record', *Program and Abstracts*, New South Wales Climate Symposium, University of New South Wales/Bureau of Meteorology/ University of California, Los Angeles, Sydney, September.

Standen, B. J. 1972, An Assessment of Adjustment Problems in North Coast Dairy Farming, unpublished Master of Agricultural Economics thesis, University of New England, Armidale.

Standen, B. J. 1982, 'Credit and Agriculture', chapter 13 in Williams (ed.), *Agriculture in the Australian Economy*, 2nd edition, Sydney University Press.

Steahr, T. E. & Roberts, T. 1993, Microbial Foodborne Disease: hospitalizations, medical costs and potential demand for safer food, WP-32, Working Paper Series, Private Strategies, Public Policies & Food System Performance Food Marketing Policy Centre, University of Connecticut, January.

Steering Committee on National Agricultural Research Strategy for Australia 1995, *A National Strategy for Agricultural Research in Australia*, Communications Unit, NSW Agriculture, Orange, April.

Steering Group for the Development of Competency Standards for Professional Agriculturists 1993, *Standards for Professionals in Agriculture and Agri-industry*, Australian Institute of Agricultural Science, Melbourne, October.

Stigler, G. J. 1971, 'The theory of economic regulation', *Bell Journal of Economics and Management Science* 2(1), Spring, 3–21.

Stiglitz, J. E. 1988, *Economics of the Public Sector*, 2nd edition, W. W. Norton, New York.

Stoeckel, A., Borrell, B. & Quirke, D. 1990, *Wool into the 21st Century: implications for marketing and profitability*, Centre for International Economics, Canberra.

Strong, A., Johnston, J., Green, W. & Marshall, G. 1990, Estimating the Financial Impacts on Producers of Deregulation in the NSW Egg Industry, 34th Annual Conference, Australian Agricultural Economics Society, University of Queensland, 12–15 February, roneo.

Stuckey, E. B. 1994, *Bluett Local Government Handbook (New South Wales)*, 13th edition, Law Book Co., Sydney.

Sugar Industry Task Force 1993, Sugar Industry Task Force Report, [Minister for Primary Industries and Energy, Canberra], roneo.

Summers, J. Woodward, D. & Parkin, A. 1990, *Government, Politics and Power in Australia: an introductory reader*, Longman Cheshire, Melbourne.

Süssmuth, R. 1993, 'No final line to draw', Speech given at the German Bundestag, 9 November, reported in *Das Parliament* 42, No.46–7, p. 7, Bonn, 12–19 November (slightly edited from a translation kindly provided by Professor D. Manegold, Institut fur Landwirtschaftliche Marktforschung der Forschungsanstalt fur Landwirtschaft, Braunschweig-Völkenrode).

Swinbank, A. 1993, 'The economics of food safety', *Food Policy* 18(2), April, 85–94.

Taylor, J. 1989, *Quality Control Systems*, McGraw-Hill, New York.

Taylor, J. 1995, Parliament and the Auditor-General, Australian Senate Occasional Lecture Series, Canberra, 19 May.

Taylor, L. 1993, 'Mabo win for PM at last', *The Australian* 22 December, p. 1.

Thirtle, C. & Bottomley, P. 1988, 'Is publicly funded research excessive?, *Journal of Agricultural Economics* 39(1), January, 99–111.

Tietenberg, T. H. 1992, *Environmental and Natural Resource Economics*, 3rd edition, Harper Collins, New York.

Tisdell, C. A. 1982, *Microeconomics of Markets*, John Wiley & Sons, Brisbane.

Tully, K. 1995, *Every woman's guide to getting into politics*, Office for the Status of Women, Commonwealth of Australia, Canberra, March.

Turner, K. 1993, 'Parliament', chapter 6 in Smith 1993.

Turnovsky, S. J. 1978 'The Distribution of Welfare Gains from Price: a survey of some theoretical issues', chapter 6 in F. G Adams & S. A. Klein (eds), *Stabilizing World Commodity Markets: analysis, practice, policy*, Lexington Books, Lexington Mass.

Varian, H. R. 1990, *Intermediate Microeconomics: a modern approach*, 2nd edition, W. W. Norton & Co., New York.

Vernon, J. et al. 1965, *Report of the Committee of Economic Enquiry*, vols. I and II, Commonwealth of Australia, Canberra.

Verrall, D., Ward, I., & Hay, P. 1985, 'Community, Country, Party: roots of rural conservatism', chapter 1 in Costar & Woodward 1985a.

Vines, W., Davis, J. G. & Millar, F. W. 1991, *The Australian Wool Industry: recommendations for the future*, Report to the Minister for Primary Industries & Energy by the Committee of Review into the Wool Industry, Commonwealth of Australia, Canberra, March.

von Weizsäcker, R. 1986, 'The 8th of May 1945—Forty Years After', chapter 3 in *A Voice from Germany: speeches by Richard von Weizsäcker*, translated by Karin von Abrams, Weidenfeld & Nicolson, New York.

Wahlquist, A. 1994, 'Colly chief calls for "aggie bond"', *Sydney Morning Herald* 27 December, p. 29.

Walker, R. & Roberts, D. 1988 *From Scarcity to Surfeit: a history of food and nutrition in New South Wales*, New South Wales University Press, Kensington.

Ward, I. 1991, 'Federal Government', chapter 1 in Henningham 1991.

Warhurst, J. 1990, 'The Politics of Rural Australia: rural industries and rural areas', chapter 8 in D. B. Williams (ed), *Agriculture in the Australian Economy*, 3rd edition, Sydney University Press.

Warr, P. G. 1978, 'The case against tariff compensation', *Australian Journal of Agricultural Economics* 22(2), 85–98.

Watson, A., Hely, R., O'Keefe, M., Cary, J., Clark, N., & Allen, J, 1992, *Review of Field-based Services in the Victorian Department of Food and Agriculture*, Agmedia, [Melbourne], August.

Watson, A. S. & Parish, R. M. 1982, 'Marketing Agricultural Products', chapter 17 in D. B. Williams (ed) *Agriculture in the Australian Economy*, 2nd edition, Sydney University Press.

Weller, P. (ed.) 1994, *Royal Commissions and the Making of Public Policy*, Centre for Australian Public Sector Management/Macmillan, Brisbane/Melbourne.

Whitwell, G. & Sydenham, D. 1991, *A Shared Harvest: the Australian wheat industry*, 1939–1989, Macmillan, Melbourne.

Williams, M. 1995, 'Re Dingjan; ex parte Wagner', *Australian Journal of Corporate Law* 5, 121–3.

Winterton, G. 1994, 'A Constitution for an Australian Republic', chapter 1 in G. Winterton (ed.), *We, the People: Australian republican government*, Allen & Unwin, Sydney.

Wonder, B. 1995, Australia's Approach to Agricultural Reform, Department of Primary Industries and Energy, Canberra (paper delivered in Washington D.C. 13 March).

Woodward, D. 1985, 'The Federal National Party of Australia', chapter 4 in Costar & Woodward 1985a.

Woodward, D.1990, 'The Australian Labor Party', chapter 10 in Summers et al. 1990.

Wootten, H. 1993a, 'A cheer for the Mabo nudgers', *Sydney Morning Herald* 11 May, p. 17.

Wootten, H. 1993b, 'Mabo means simple justice, and no cause for hysteria', *Sydney Morning Herald* 29 July, p. 11.

Index